Black Holes: A Laboratory for Testing Strong Gravity

Cosimo Bambi

Black Holes: A Laboratory for Testing Strong Gravity

 Springer

Cosimo Bambi
Department of Physics
Fudan University
Shanghai
China

and

Theoretical Astrophysics
Eberhard-Karls Universität Tübingen
Tübingen
Germany

ISBN 978-981-13-5158-7   ISBN 978-981-10-4524-0   (eBook)
DOI 10.1007/978-981-10-4524-0

Printed on acid-free paper

This Springer imprint is published by Springer Nature
The registered company is Springer Nature Singapore Pte Ltd.
The registered company address is: 152 Beach Road, #21-01/04 Gateway East, Singapore 189721, Singapore

It doesn't matter how beautiful your theory is, it doesn't matter how smart you are. If it doesn't agree with experiment, it's wrong.

—Richard Feynman

Preface

Black holes are thought to be the final product of the complete gravitational collapse of massive bodies. In the framework of standard physics, the spacetime metric around astrophysical black holes should be well approximated by the Kerr solution. However, macroscopic deviations from standard predictions may be expected from a number of scenarios beyond Einstein's theory of general relativity. Astrophysical black holes are thus an ideal laboratory for testing strong gravity.

The main aim of this book is to discuss the electromagnetic techniques to study the strong gravity region around astrophysical black holes. For completeness, gravitational wave methods will be also reviewed, but only very briefly and without the necessary details to start working on the corresponding line of research. This book has not the ambition to be a complete manual on this research field. Hopefully, it may be a good starting point. The reader should be already familiar with the theory of general relativity (at a more advanced level than that one can learn in an introductory course in an undergraduate program), while it is not required a background in astronomy/astrophysics.

Chapters 1–5 provide a general introduction on some basic concepts. Some topics are not necessary to understand the rest of the book (in particular some topics in Chap. 2), and in such a case they are briefly reviewed without many details, but they may be useful to get a complete overview of current knowledge on black holes. Chapters 6–11 are the core of the book. Chapters 6–10 discuss the main techniques to test astrophysical black holes with electromagnetic radiation. Chapter 11 briefly reviews the approaches available with gravitational waves. Chapters 12–14 summarize the state of the art of tests of the Kerr black hole hypothesis. Appendices A–F provide more details on some particular calculations, briefly discuss some related topics not covered in Chaps. 1–14, or summarize some useful formulas. The Glossary at the end of the book can be useful to learn/recall some quite common terms.

I am particularly grateful to Alejandro Cardenas-Avendano for reading a preliminary version of the manuscript and providing useful feedback. I thank Matteo Guainazzi for useful comments and suggestions about the part devoted to X-ray missions and X-ray data analysis, and Kostas Kokkotas about the part on gravitational waves. I would like also to thank Javier Garcia-Martinez, Jiachen Jiang, Daniele Malafarina, Yueying Ni, James Steiner, and Jingyi Wang for valuable comments to improve the manuscript and for some figures. This work was supported by the NSFC (grants 11305038 and U1531117), the Thousand Young Talents Program, and the Alexander von Humboldt Foundation.

Shanghai, China   Cosimo Bambi
December 2016

Contents

Conventions

Throughout the paper, I employ units in which $G_N = c = 1$, unless stated otherwise, and the convention of a metric with signature $(-+++)$. Greek letters $(\mu, \nu, \rho,...)$ are used for spacetime indices and can assume the values 0, 1, 2, and 3. Latin letters $(i, j, k,...)$ are used for space indices and can assume the values 1, 2, and 3. The time coordinate can be indicated either as t or as x^0.

The Riemann tensor is defined as

$$R^{\mu}{}_{\nu\rho\sigma} = \frac{\partial\Gamma^{\mu}_{\nu\sigma}}{\partial x^{\rho}} - \frac{\partial\Gamma^{\mu}_{\nu\rho}}{\partial x^{\sigma}} + \Gamma^{\mu}_{\lambda\rho}\Gamma^{\lambda}_{\nu\sigma} - \Gamma^{\mu}_{\lambda\sigma}\Gamma^{\lambda}_{\nu\rho},$$

where $\Gamma^{\mu}_{\nu\rho}$s are the Christoffel symbols of the Levi-Civita connection

$$\Gamma^{\mu}_{\nu\rho} = \frac{1}{2}g^{\mu\lambda}\left(\frac{\partial g_{\lambda\rho}}{\partial x^{\nu}} + \frac{\partial g_{\nu\lambda}}{\partial x^{\rho}} - \frac{\partial g_{\nu\rho}}{\partial x^{\lambda}}\right).$$

The Ricci tensor is defined as $R_{\mu\nu} = R^{\lambda}{}_{\mu\lambda\nu}$. The Einstein equations read

$$G_{\mu\nu} = R_{\mu\nu} - \frac{1}{2}g_{\mu\nu}R = 8\pi T_{\mu\nu}.$$

Part I
Basic Concepts

Chapter 1
Introduction

A *black hole*, roughly speaking, is a region of the spacetime in which gravity is so strong that nothing can exit or even communicate with the exterior region. The *event horizon* is the boundary of the black hole. In the Universe, there are compact objects that can be naturally interpreted as black holes and they could be something else only in the presence of new physics. However, it is fundamentally impossible to confirm the existence of a black hole by observations. If someone inside a certain region is unable to communicate with an external observer at a certain time, we cannot conclude that such a region is a black hole. The communication between the interior and the exterior may be possible at a later time. The identification of a black hole and of its event horizon require the knowledge of the spacetime at any time, including also the faraway future. On the other hand, any human observation can only last for a finite time. For this reason it is fundamentally impossible to prove the existence of a black hole.

There are several conventions among the scientific community. Some authors like using the term *black hole candidate* to indicate any astrophysical black hole, just because it is impossible to prove a similar object meets the mathematical definition of black hole. Other authors adopt a less conservative perspective and call black hole any astrophysical compact object that is interpreted as a black hole when there is a dynamical measurement of its mass (without testing the possible existence of some kind of horizon), bestowing the term black hole candidate to compact objects for which there is no dynamical measurement of their mass but share common features observed in sources with a black hole.

The idea of the possibility of the existence of extremely compact objects such that even light cannot escape from their strong gravitational field can be dated back to the dark stars of Michell and Laplace at the end of the 18th century. General relativity was proposed by Albert Einstein at the end of 1915, and the simplest black hole solution was discovered immediately after, by Karl Schwarzschild, in 1916. However, the actual properties of black holes were understood only much later. In 1958,

© Springer Nature Singapore Pte Ltd. 2017
C. Bambi, *Black Holes: A Laboratory for Testing Strong Gravity*,
DOI 10.1007/978-981-10-4524-0_1

David Finkelstein was the first to realize the true physical meaning of the event horizon. Even the astrophysical implications of such solutions were initially not taken very seriously. Influent scientists, like Arthur Eddington, were more inclined to believe that "some unknown mechanism" could prevent the complete collapse of a massive body and the formation of a black hole in the Universe.

In 1964, Yakov Zel'dovich and, independently, Edwin Salpeter proposed that quasars were powered by a central supermassive black hole. In the early 1970s, Thomas Bolton and, independently, Louise Webster and Paul Murdin identified the X-ray source Cygnus X-1 as the first stellar-mass black hole candidate. Since then, an increasing number of astronomical observations have pointed out the existence of stellar-mass black holes in some X-ray binaries and of supermassive black holes at the center of many galaxies. Thanks to technological progresses and new observational facilities, in the past 10–15 years it was possible to start studying the physical properties of these objects and of their environments. Tests of general relativity in the strong field region of black holes are now at the beginning. The detection of gravitational waves from the coalescence of two black holes by the LIGO experiment in September 2015 has opened a new window to study these intriguing objects.

It is curious that the term black hole is relatively recent. While it is not clear who used the term first, it appeared for the first time in a publication in the January 18, 1964 issue of Science News Letter. It was on a report on a meeting of the American Association for the Advancement of Science by journalist Ann Ewing. The term became quickly very popular after it was used by John Wheeler at a lecture in New York in 1967.

1.1 Dark Stars in Newtonian Gravity

John Michell in 1783 and, independently, Pierre-Simon Laplace in 1796 were the first to consider the possibility of the existence in the Universe of extremely compact objects, so compact to be black because even light could not escape from their strong gravitational field. Their idea was discussed in the framework of Newtonian mechanics and these dark compact objects are usually referred to as *dark starts*.

The energy E of a test-particle of mass m in the gravitational field of a spherically symmetric body of mass M is simply (in this section about Newtonian gravity, we reintroduce G_N and c)

$$E = \frac{1}{2}mv^2 - \frac{G_N M m}{r} , \tag{1.1}$$

where v is the velocity of the test-particle and r is the distance between the test-particle and the center of the massive body. If the test-particle has energy $E < 0$, it is trapped in the gravitational field of the massive body. If $E > 0$, the particle can escape to infinity with a finite velocity. The escape velocity is the minimum velocity that a test-particle on the surface of the massive body needs to escape to infinity. If r_* is the radius of the massive body, the escape velocity is

$$v_{\text{esc}} = \sqrt{\frac{2G_N M}{r_*}}.$$ (1.2)

In the corpuscular theory of light developed in the 17th century, light would be made of small particles traveling in a straight line and with a finite velocity $v = c$. If the escape velocity in Eq. (1.2) is higher than c, the surface of the body cannot emit radiation and therefore the body looks black. This is the concept of dark star discussed by Michell and Laplace. The massive body is thus a dark star if its radius is smaller than the critical value

$$r_{\text{crit}} = \frac{2G_N M}{c^2}.$$ (1.3)

It is just a coincidence that the critical radius r_{crit} for the existence of a Newtonian dark star is exactly equal to the radius of the event horizon of a Schwarzschild black hole in Schwarzschild coordinates.

1.2 Black Holes in Theoretical Physics

1.2.1 Black Hole Solutions and No-Hair Theorem

Albert Einstein proposed the theory of general relativity at the end of 1915 [17]. In 1916, Karl Schwarzschild derived the spherically symmetric vacuum solution of the Einstein equations [42], now called the *Schwarzschild solution*, which describes the spacetime of a non-rotating black hole. The *Reissner–Nordström solution*, describing the spacetime of a non-rotating black hole with a possible non-vanishing electrically charge, was found immediately after: in 1916, Hans Reissner solved the Einstein equations for a point-like charged mass [38], and, in 1918, Gunnar Nordström found the metric for a spherically symmetric charged mass [34].

The solution found by Schwarzschild appeared singular at the Schwarzschild radius, but at that time it was not understood that such a surface was the black hole event horizon. In 1924, Arthur Eddington showed that this singular surface could be removed by a coordinate transformation. In 1933, Georges Lemaitre pointed out that the singularity at the Schwarzschild radius was an artifact of the choice of the coordinates and that the spacetime is regular there.

In 1958, David Finkelstein realized the actual nature of the singular surface at the Schwarzschild radius [19]. He described the event horizon as a one-way membrane and understood that, if something crosses the horizon, it cannot influence the exterior region any more. Finkelstein found an analytic extension of the Schwarzschild solution and, in 1960, Martin Kruskal found the maximal extension of the Schwarz-schild solution [31]. It is worth noting that the physical implications of the Kruskal solution is questionable, to say the least, because astrophysical black holes form from gravitational collapse, while the Kruskal solution makes sense only in the case of a static spacetime.

In 1963, Roy Kerr found the solution for a rotating black hole in general relativity [29], now called the *Kerr solution*. This was an important step, because astrophysical bodies have naturally a non-vanishing angular momentum. In 1965, Ezra Newman and collaborators found the solution for a rotating and electrically charged black hole [33], today known as the *Kerr–Newman solution*.

Starting from the work by Werner Israel [27], Brandon Carter [12], and David Robinson [40], it was understood that, under certain assumptions, the black holes of general relativity are relatively simple objects, in the sense that they can be fully characterized by a small number of parameters, namely the black hole mass M, the black hole spin angular momentum J, and the black hole electric charge Q. This result is known under the name of *no-hair theorem*, to indicate that black holes have only a small number of features (hairs). In the context of tests of astrophysical black holes, it is also relevant the *uniqueness theorem*, according to which black holes are only described by the Kerr–Newman solution. However, violations of the no-hair theorem or of the uniqueness theorem are possible if we relax some assumptions or we consider theories beyond general relativity.

At the end of the 1960s, Roger Penrose and Stephen Hawking proved, under quite general assumptions, the inevitability of the creation of singularities in a gravitational collapse in general relativity [25, 36]. This led Penrose to propose the *cosmic censor-ship conjecture*, according to which singularities produced by gravitational collapse must be hidden behind an event horizon and the final product of a collapse must be a black hole [37]. While today we know some physically reasonable counterexamples in which a naked singularity can be created for an infinitesimal time from regular initial data, see e.g. the review article [28], it is commonly thought that some form of the cosmic censorship conjecture holds and the final product of gravitational collapse in classical general relativity is indeed a black hole.

1.2.2 Beyond the Purely Classical Picture

In the 1970s, a number of studies pointed out a strong analogy between black holes and thermodynamic systems [6–8]. In particular, it was possible to formulate the *four laws of black hole mechanics*, in close analogy with the four laws of thermodynamics, by relating the black hole mass to energy, the area of the event horizon to entropy, and the surface gravity to temperature.

Hawking showed that black holes are not really "black" but radiate like a black-body[1] with a temperature proportional to the surface gravity of the black hole [22]. Such a radiation is due to quantum effects and it is today called the *Hawking radiation*. A black hole is expected to slowly evaporate. If this semiclassical treatment is correct, the evaporation of a black hole may not be a unitary process, violating the laws of quantum mechanics. This is the so-called information loss paradox [23].

Black hole thermodynamics is presumably irrelevant for astrophysical black holes. The temperature of a black hole is $T_{BH} = \kappa/(2\pi)$, where κ is the surface gravity of the event horizon. For a Schwarzschild black hole, we have

$$T_{BH} = \frac{1}{8\pi M} = 5.32 \cdot 10^{-12} \left(\frac{M_\odot}{M}\right) \text{ eV}. \tag{1.4}$$

A black hole should radiate any particle with a mass lower than the black hole temperature, while the emission of heavier particles should be exponentially suppressed, so the total emission rate depends on the black hole mass and the particle content of the theory. Neglecting these complications, the luminosity of a Schwarzschild black hole due to Hawking radiation is

$$L_{BH} \sim 10^{-21} \left(\frac{M_\odot}{M}\right)^2 \text{ erg/s}. \tag{1.5}$$

For stellar-mass black holes of 5–20 $M_\odot$ and supermassive black holes of 10^5–$10^{10} M_\odot$, this radiation is completely negligible and impossible to detect, even in a foreseeable future. For much smaller black holes, possibly created in the early Universe, the effect may be important, but there are today strong constraints on the possible cosmological abundance of these objects.

The spacetime singularity at the center of black holes is an inevitably consequence of the gravitational collapse and its existence is guaranteed by the singularity theorems pioneered by Penrose and Hawking. However, such a spacetime singularity may be a pathological prediction of classical general relativity and may be removed by yet unknown quantum gravity effects. While we do not have today any reliable and predictable theory of quantum gravity, there are some speculations on the possible fate of spacetime singularities and of classical black holes.

A number of different approaches suggest that black holes, strictly speaking, may never form. The gravitational collapse would never create an event horizon, but only an apparent horizon for a finite time [3–5, 20, 24, 26]. However, the timescale of this process would likely be so long that our observations may not be able to distinguish these objects from classical black holes.

It has been also argued that quantum gravity effects may not show up at the Planck scale, as it can be expected in the scattering of two particles, but at the gravitational

---

[1]Deviations from a perfect blackbody spectrum arise due to the finite size of the black hole, the mass and the spin of the emitted particles, etc.

radius of the system, and therefore the black holes predicted in classical general relativity may actually be intrinsically quantum objects [16, 32]. In such frameworks, the metric description simply breaks down inside the object, and the evaporation process slightly deviates from Hawking's predictions, making the process unitary and solving the information loss paradox.

1.3   Black Holes in the Universe

1.3.1   Discovery of Astrophysical Black Holes

Since normal stars have a radius much larger than their Schwarzschild radius, it was initially thought that the existence of the surface singularity in the Schwarzschild solution had no physical implications, because the solution holds in the vacuum only, while a different solution is necessary for the interior of the star.

In the 1920s, it was known that white dwarfs were dead stars without nuclear fusion and it was suggested that they were supported by the degenerate pressure of electrons. In 1931, Subrahmanyan Chandrasekhar pointed out the existence of a maximum mass, now called the *Chandrasekhar limit*, above which the degenerate electron pressure could not balance the gravitational force and the body had to collapse to a point [13]. This conclusion was however criticized by many physicists, like Arthur Eddington, arguing the existence of a yet unknown mechanism capable of stopping the collapse.

It was later realized that a dead star with a mass exceeding the Chandrasekhar limit could be supported by the degenerate pressure of neutrons and become a neutron star. In 1939, Robert Oppenheimer and George Volkoff found that even neutron stars have a maximum mass, and that the degenerate pressure of neutrons cannot support a star exceeding this limit [35]. Once again, it was advocated the existence of new physics to prevent the complete collapse of the body.

Quasars are a class of very luminous active galactic nuclei. They were discovered in the 1950s and called "quasi-stellar radio sources", later shortened to "quasars". Their nature was initially unknown. In 1964, Yakov Zel'dovich and, independently, Edwin Salpeter were the first to propose that quasars could be powered by the accretion disk around a black hole [41, 45]. Their idea was initially not taken very seriously, and other possibilities, like that in which quasars would have been supermassive stars, were considered more promising.

The first object identified as a black hole was Cygnus X-1, which is one of the strongest X-ray sources in the sky. It was discovered in 1964 [10]. In 1971, it was found that Cygnus X-1 had a massive stellar companion and, by studying the orbital motion of the companion star, it was possible to infer the mass of the compact object [9, 43]. Since the latter exceeded the maximum mass for a neutron star, the most natural interpretation was that it was a stellar-mass black hole. Such a discovery is a milestone in black hole astrophysics and helped convincing the astronomy

Fig. 1.1 An artist's illustration of Cygnus X-1. The stellar-mass *black hole* pulls material from a massive, *blue* companion star toward it. This material forms an accretion disk (in *red, orange*, and *yellow* in the picture) around the *black hole*. We also see a jet originating from the region close to the *black hole*. Credit: NASA

community about the existence of black holes in the Universe. Figure 1.1 shows an artist's illustration of Cygnus X-1.

1.3.2   Recent Studies and Future Prospectives

Today we have strong evidence for the existence of at least two classes of astrophysical black holes. *Stellar-mass black holes* have a mass $M \approx 5\text{--}100\ M_\odot$. Most of them are in X-ray binaries and, from the study of the orbital motion of the stellar companion, it is possible to see that the compact object exceeds the maximum mass for a neutron star [39]. Stellar-mass black holes can also be in binary systems black hole-black hole or black hole-neutron star, and in this case we can detect the gravitational waves in the last stage of the coalescence [1]. There are also attempts to find isolated stellar-mass black holes, for instance with microlensing techniques [21, 44]. *Supermassive black holes* have a mass $M \sim 10^5\text{--}10^{10}\ M_\odot$ and reside in galactic nuclei [30]. There may also exist a third class of objects, *intermediate mass black holes*, with a mass filling the gap between the stellar-mass and the supermassive ones, but their nature is more uncertain, because there are not dynamical measurements of their masses [14].

There are currently two popular techniques to probe the spacetime geometry around astrophysical black holes with electromagnetic radiation. They are the continuum-fitting method [46] and the analysis of the X-ray reflection spectrum (often called the iron line method) [11].

The continuum-fitting method is the analysis of the thermal spectrum of geometrically thin and optically thick accretion disks. This technique is normally applied to stellar-mass black holes only, because the temperature of the disk depends on the mass of the compact object. The spectrum of thin disks around stellar-mass black holes is in the soft X-ray band. In the case of supermassive black holes, the spectrum is in the UV/optical bands, where extinction and dust absorption limit the ability to make an accurate measurement. Under the assumption that astrophysical black holes are Kerr black holes, the continuum-fitting method can be used to measure the black hole spin parameter $a_* = J/M^2$. Relaxing the Kerr black hole hypothesis, this technique can constrain possible deviations from the Kerr solution.

The reflection spectrum is produced by illumination of the inner part of the accretion disk. Its features, in particular the shape of the iron $K\alpha$ line, are strongly determined by the relativistic effects occurring in the vicinity of the compact object (Doppler boosting, gravitational redshift, light bending). An accurate measurement of the reflection spectrum can potentially probe the metric around black holes.

In the future, other approaches could be available to test astrophysical black holes. For instance, the quasi-periodic oscillations (QPOs) observed in the X-ray power spectrum of black holes are a promising tool. Their frequency can be measured with high precision and this could permit one to get strong constraints on the properties of the compact object. However, today we do not know the exact mechanism responsible for these phenomena, and different models provide different measurements, which means that this technique cannot yet be used to test fundamental physics.

Today it is not possible to image the accretion flow around a black hole with a resolution necessary to detect the black hole "shadow", but this should not be out of reach in the next few years [15, 18]. The best candidate for this kind of observations is SgrA*, the supermassive object at the center of the Galaxy, because it should be the black hole with the largest angular size in the sky. The shadow of a black hole is a dark area over a bright background in the direct image of its accretion flow. The boundary of the shadow is expected to depend on the spacetime metric around the compact object and on the viewing angle of the observer. If the accretion flow is geometrically thick and optically thin, the boundary of the shadow corresponds to the apparent photon capture sphere. Very-long baseline interferometric (VLBI) observations at sub-millimeter wavelengths could image the shadow of SgrA* within the next few years.

Gravitational waves are now opening a new window for the study of astrophysical black holes. In February 2016, the LIGO/Virgo collaboration announced the detection of gravitational waves in September 2015 from the coalescence of two black holes, each of them of about 30 $M_\odot$ [1]. The detection of another event was reported in [2].

The sensitivity of current ground-based gravitational wave antennas is increasing and more detections are expected for the future. These experiments will provide unprecedented new data to test general relativity in the strong gravity regime.

1.4 Open Problems

In the past 10–15 years, the quality and the amount of observational data of black holes have significantly increased. Many phenomena are now quite understood, while others are not yet clear. Some techniques have been developed to probe the region very close to these objects and measure their properties. Gravitational waves have been detected. Future observational facilities promise to make further progresses. There are still a number of open questions, ranging from fundamental physics to astrophysics. This is a non-complete list of relevant puzzles:

1. Are astrophysical black holes the Kerr black holes predicted by general relativity? In the past 60 years, Einstein's theory of gravity has been quite accurately tested in weak gravitational fields, mainly with experiments in the Solar System and with observations of binary pulsars. There is today an increasing interest in the possibility of testing the theory in the strong gravity regime and verifying the actual nature of astrophysical black holes.
2. What is the spin distribution of stellar-mass and supermassive black holes? What is the natal spin distribution and how the spin evolves with time? If black holes are the Kerr black hole of general relativity, they should be characterized by only two parameters, namely the mass and the spin. The mass can be measured with dynamical methods. Spin measurements are more challenging and are a hot topic today. They can help us to understand the process of collapse, the evolution of a binary system, the history of the merger of galaxies, etc.
3. What is the mechanism responsible for the formation of the jets observed in these systems? There are several proposals in the literature, but we do not know which of them, if any, is correct.
4. What is the mechanism responsible for the QPOs observed in the X-ray power spectrum of black holes? If properly understood, QPOs may become a new approach to get accurate measurements of the fundamental properties of these objects, like determining the black hole spin or testing the Kerr metric.

References

1. B.P. Abbott et al., LIGO Scientific and Virgo Collaborations. Phys. Rev. Lett. **116**, 061102 (2016), arXiv:1602.03837 [gr-qc]
2. B.P. Abbott et al., LIGO Scientific and Virgo Collaborations. Phys. Rev. Lett. **116**, 241103 (2016), arXiv:1606.04855 [gr-qc]
3. A. Ashtekar, M. Bojowald, Class. Quant. Grav. **22**, 3349 (2005), arXiv:gr-qc/0504029

4. C. Bambi, D. Malafarina, L. Modesto, Eur. Phys. J. C **74**, 2767 (2014), arXiv:1306.1668 [gr-qc]
5. C. Bambi, D. Malafarina, L. Modesto, JHEP **1604**, 147 (2016), arXiv:1603.09592 [gr-qc]
6. J.M. Bardeen, B. Carter, S.W. Hawking, Commun. Math. Phys. **31**, 161 (1973)
7. J.D. Bekenstein, Phys. Rev. D **7**, 2333 (1973)
8. J.D. Bekenstein, Phys. Rev. D **9**, 3292 (1974)
9. C.T. Bolton, Nature **235**, 271 (1972)
10. S. Bowyer, E.T. Byram, T.A. Chubb, H. Friedman, Science **147**, 394 (1965)
11. L.W. Brenneman, C.S. Reynolds, Astrophys. J. **652**, 1028 (2006), arXiv:astro-ph/0608502
12. B. Carter, Phys. Rev. Lett. **26**, 331 (1971)
13. S. Chandrasekhar, Astrophys. J. **74**, 81 (1931)
14. M. Coleman Miller, E.J.M. Colbert, Int. J. Mod. Phys. D **13**, 1 (2004), arXiv:astro-ph/0308402
15. S. Doeleman et al., Nature **455**, 78 (2008), arXiv:0809.2442 [astro-ph]
16. G. Dvali, C. Gomez, Fortsch. Phys. **61**, 742 (2013), arXiv:1112.3359 [hep-th]
17. A. Einstein, Ann. Phys. **49**, 769 (1916) (Ann. Phys. **14**, 517 (2005))
18. H. Falcke, F. Melia, E. Agol, Astrophys. J. **528**, L13 (2000), arXiv:astro-ph/9912263
19. D. Finkelstein, Phys. Rev. **110**, 965 (1958)
20. V.P. Frolov, G.A. Vilkovisky, Phys. Lett. B **106**, 307 (1981)
21. A. Gould, Astrophys. J. **535**, 928 (2000), arXiv:astro-ph/9906472
22. S.W. Hawking, Commun. Math. Phys. **43**, 199 (1975) (Commun. Math. Phys. **46**, 206 (1976))
23. S.W. Hawking, Phys. Rev. D **14**, 2460 (1976)
24. S.W. Hawking, Phys. Rev. D **72**, 084013 (2005), arXiv:hep-th/0507171
25. S.W. Hawking, R. Penrose, Proc. Roy. Soc. Lond. A **314**, 529 (1970)
26. S.A. Hayward, Phys. Rev. Lett. **96**, 031103 (2006), arXiv:gr-qc/0506126
27. W. Israel, Phys. Rev. **164**, 1776 (1967)
28. P.S. Joshi, D. Malafarina, Int. J. Mod. Phys. D **20**, 2641 (2011), arXiv:1201.3660 [gr-qc]
29. R.P. Kerr, Phys. Rev. Lett. **11**, 237 (1963)
30. J. Kormendy, D. Richstone, Ann. Rev. Astron. Astrophys. **33**, 581 (1995)
31. M.D. Kruskal, Phys. Rev. **119**, 1743 (1960)
32. S.D. Mathur, Fortsch. Phys. **53**, 793 (2005), arXiv:hep-th/0502050
33. E.T. Newman, R. Couch, K. Chinnapared, A. Exton, A. Prakash, R. Torrence, J. Math. Phys. **6**, 918 (1965)
34. G. Nordström, Proc. Kon. Ned. Akad. Wet. **20**, 1238 (1918)
35. J.R. Oppenheimer, G.M. Volkoff, Phys. Rev. **55**, 374 (1939)
36. R. Penrose, Phys. Rev. Lett. **14**, 57 (1965)
37. R. Penrose, Riv. Nuovo Cim. **1**, 252 (1969) (Gen. Rel. Grav. **34**, 1141 (2002))
38. H. Reissner, Ann. Phys. **59**, 106 (1916)
39. R.A. Remillard, J.E. McClintock, Ann. Rev. Astron. Astrophys. **44**, 49 (2006), arXiv:stro-ph/0606352
40. D.C. Robinson, Phys. Rev. Lett. **34**, 905 (1975)
41. E.E. Salpeter, Astrophys. J. **140**, 796 (1964)
42. K. Schwarzschild, Sitzungsber. Preuss. Akad. Wiss. Berlin (Math. Phys.) **1916**, 189 (1916), arXiv:physics/9905030
43. B.L. Webster, P. Murdin, Nature **235**, 37 (1972)
44. L. Wyrzykowski et al., Mon. Not. Roy. Astron. Soc. **458**, 3012 (2016), arXiv:1509.04899 [astro-ph.SR]
45. Y.B. Zeldovich, Dokl. Akad. Nauk **155**, 67 (1964) (Sov. Phys. Dokl. **9**, 195 (1964))
46. S.N. Zhang, W. Cui, W. Chen, Astrophys. J. **482**, L155 (1997), arXiv:astro-ph/9704072

Chapter 2
Black Hole Solutions

In 4-dimensional general relativity and in the absence of exotic fields, black holes are completely described by three parameters: the mass M, the spin angular momentum J, and the electric charge Q. This is the conclusion of the no-hair theorem, which holds under specific assumptions. Violations of the no-hair theorem are possible. For instance, hairy black holes naturally emerge in the presence of non-Abelian gauge fields or of fields non-minimally coupled to gravity.

Starting from the Oppenheimer-Snyder model, which describes the gravitational collapse of a homogeneous ball of dust, we know how the complete collapse of a body forms a black hole with a central spacetime singularity. In extensions of general relativity, the picture of the collapse of a massive body may be somewhat different. It is possible that, strictly speaking, black holes cannot form, but only temporary apparent horizons can be created. The latter can however be interpreted as event horizons if the observation time is much shorter than the lifetime of the apparent horizon.

2.1 Definition of Black Hole

Roughly speaking, a black hole is a region of the spacetime in which gravity is so strong that it is impossible to escape or send information to the exterior region. A more technical definition is the following:

A *black hole* in an asymptotically flat spacetime $\mathcal{M}$ is the set of events that do not belong to the causal past of the future null infinity $J^-(\mathscr{I}^+)$, namely

$$\mathscr{B} = \mathcal{M} - J^-(\mathscr{I}^+) \neq \emptyset. \tag{2.1}$$

The *event horizon* is the boundary of the region $\mathscr{B}$.

© Springer Nature Singapore Pte Ltd. 2017
C. Bambi, *Black Holes: A Laboratory for Testing Strong Gravity*,
DOI 10.1007/978-981-10-4524-0_2

This definition of black hole uses the concepts developed by Penrose for study-ing causality and asymptotic properties of asymptotically flat spacetimes [49]. See also [42]. $\mathscr{I}^+$ indicates the future null infinity,[1] namely the region toward which out-going null lines extend. Heuristically, in spherical-like coordinates, it is the region $t + r \to \infty$ at finite $t - r$. $J^-(\mathscr{P})$ is called the causal past of the region $\mathscr{P}$ and is the set of all events that causally precede $\mathscr{P}$; that is, there exists at least one smooth future-directed time-like curve extending from any element in $J^-(\mathscr{P})$ to $\mathscr{P}$. All future-directed curves (either time-like or null) starting from the region $\mathscr{B}$ fail to reach null infinity $\mathscr{I}^+$. A black hole is thus an actual one-way membrane: if something crosses the event horizon it can no longer send any signal to the exterior region.

The event horizon is a global property of an entire spacetime and it can be only determined if we know the whole spacetime, including the faraway future. It has thus no direct observational implications and, for this reason, it is often not a very useful concept in many studies. The apparent horizon is instead a local property of the spacetime and is slicing-dependent. Here we do not discuss all the details, which can be found, for instance, in [7, 42, 50, 60].

Let us consider a $3 + 1$ foliation of the spacetime. A *trapped surface* is a smooth closed 2-dimensional surface in a 3-dimensional space-like slice such that all null geodesics emanating from this surface are pointing inwards. The *trapped region* is the union of all the trapped surfaces of the slice. The *apparent horizon* is the outer boundary of the trapped region. Figure 2.1 shows a simple sketch illustrating these concepts. Outward-pointing light rays behind an apparent horizon thus move inwards and therefore they cannot cross the apparent horizon. In the case of an event horizon, the light rays may initially move outwards and then inwards at some later time. Under certain conditions, the existence of an apparent horizon implies that the slice contains an event horizon; however, the converse may not be true [61].

As it will be discussed in Sect. 2.5, it is possible that black holes cannot form in the Universe, but only apparent horizons can be created. Nevertheless, human observations may be completely unable to check the actual nature of the horizon of astrophysical black holes because any real observation only lasts for a finite time, which may be much shorter than the timescale necessary to distinguish the two scenarios.

An event horizon is a null surface in spacetime. Let us introduce a scalar function f such that at the event horizon $f = 0$. The normal to the event horizon is $n^v = \partial^\mu f$ and is a null vector. The condition for the surface $f = 0$ to be null is [12, 55]

$$g^{\mu\nu} \left(\partial_\mu f \right) \left(\partial_\nu f \right) = 0 \,, \tag{2.2}$$

where $g_{\mu\nu}$ is the metric of the spacetime. In general, one can find the event horizon of a spacetime by integrating null geodesics backwards in time, see [12, 55] for the

---

[1]The symbol $\mathscr{I}$ is usually pronounced "scri".

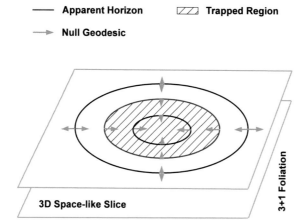

Fig. 2.1 Sketch of a 3-dimensional space-like slice with a trapped region (hatched area). Inside the trapped region, there are trapped surfaces. The outer boundary of the trapped region is the apparent horizon (*red closed curve*). The *green arrows* represent null geodesics. All the null geodesics emanating from the surfaces inside the trapped region move inwards. In the case of the surface outside the trapped region (*black closed curve* outside the hatched area), we see that the arrows can point either outwards or inwards. See the text for more details

details. In the case of a stationary and axisymmetric spacetime,[2] the procedure can significantly simplify. In a coordinate system adapted to the two Killing isometries (stationarity and axisymmetry), and such that f is also compatible with the Killing isometries, Eq. (2.2) reduces to

$$g^{rr} (\partial_r f)^2 + 2g^{r\theta} (\partial_r f) (\partial_\theta f) + g^{\theta\theta} (\partial_\theta f)^2 = 0 \qquad (2.3)$$

in spherical-like coordinates (t, r, θ, ϕ). The surface must be closed and non-singular (namely geodesically complete) in order to be an event horizon and not just a null surface.

If we assume that there is a unique horizon radius for any angle θ (Strahlkörper assumption), we can write f as $f = r - H(\theta)$, where $H(\theta)$ is a function of θ only and the event horizon is $r_H = H(\theta)$; see [12, 55] for the details and the limitations of the Strahlkörper assumption. The problem is thus reduced to finding the solution of the differential equation

$$g^{rr} + 2g^{r\theta} \left(\frac{dH}{d\theta}\right) + g^{\theta\theta} \left(\frac{dH}{d\theta}\right)^2 = 0 \,. \qquad (2.4)$$

---

[2] A space-time is *stationary* if it posses a time-like Killing vector field, and is *static* if it posses a hypersurface-orthogonal time-like Killing vector field. It is always possible to choose coordinates in which the time-like Killing vector field is ∂_t. In this case, in a stationary spacetime the metric is invariant under translations in t, namely the metric coefficients are independent of t. In a static spacetime, the time-like Killing vector is orthogonal to the hypersurfaces of constant t, which implies $g_{0i} = 0$.

Reference [30] argues that the event horizon equation $g^{rr} = 0$ (valid, for instance, in the Kerr spacetime in Boyer–Lindquist coordinates) would hold whenever the surfaces $r = const.$ have a well-defined causal structure, in the sense that any surface $r = const.$ must be null, space-like, or time-like. In such a case, the surface $r = H$ would be null and would correspond to the event horizon, the surfaces $r = const.$ at larger radii would be all space-like, and the surfaces $r = const.$ at smaller radii would be all time-like.

A *Killing horizon* is a null hypersurface on which there is a null Killing vector field. In a stationary and axisymmetric spacetime and employing a coordinate system adapted to the two Killing isometries, the Killing horizon is given by the largest root of

$$g_{tt}g_{\phi\phi} - g_{t\phi}^2 = 0 . \tag{2.5}$$

In general relativity, the Hawking rigidity theorem shows that the event and the Killing horizons coincide [22], so Eqs. (2.4) and (2.5) provide the same result. In alternative theories of gravity, this is at least not guaranteed [30].

In general relativity, the event horizon must have $S^2 \times \mathbf{R}$ topology, and even this property is regulated by certain theorems [22, 29]. For instance, toroidal horizons can form, but they can only exist for a short time, in agreement with these theorems [27].

2.2 Black Holes in General Relativity

2.2.1 Schwarzschild Solution

In general relativity, the simplest black hole metric is the *Schwarzschild solution*, which describes the spacetime of a non-rotating uncharged black hole in a vacuum and asymptotically flat spacetime. In the Schwarzschild coordinates (t, r, θ, ϕ), the line element is

$$ds^2 = -\left(1 - \frac{2M}{r}\right)dt^2 + \left(1 - \frac{2M}{r}\right)^{-1}dr^2 + r^2d\theta^2 + r^2\sin^2\theta d\phi^2 , \tag{2.6}$$

where M is the mass of the black hole. The Schwarzschild metric is relatively straightforward to find and its derivation is presented in many textbooks.

As a consequence of Birkhoff's theorem, the Schwarzschild metric is the only spherically symmetric vacuum solution of the Einstein equations. This means it describes the exterior region of any spherically symmetric body, independently of its interior, which may also change in time (but maintaining its spherical symmetry).

The line element in Eq. (2.6) is singular at the surface $r = 2M$, in the sense that g_{tt} vanishes and g_{rr} diverges. This is the event horizon, as it can be seen from the procedure explained in the previous section. Here $H = 2M$ and is independent of θ.

The singularity is due to the choice of the coordinates, but the spacetime is regular there. For instance, the Kretschmann scalar $\mathscr{K}$ is

$$\mathscr{K} = R^{\mu\nu\rho\sigma} R_{\mu\nu\rho\sigma} = \frac{48M^2}{r^6} . \tag{2.7}$$

The Kretschmann scalar only diverges at $r = 0$, which is the real singularity of the spacetime, both in the sense of curvature singularity (curvature invariants diverge) and in the sense that the spacetime is geodetically incomplete (any geodesic reaching $r = 0$ stops there).

The singularity at the event horizon can be removed by a change of coordinates. For instance, the Lemaitre coordinates (T, R, θ, ϕ) are related to the Schwarzschild coordinates by

$$dT = dt + \left(\frac{2M}{r}\right)^{1/2} \left(1 - \frac{2M}{r}\right)^{-1} dr ,$$
$$dR = dt + \left(\frac{r}{2M}\right)^{1/2} \left(1 - \frac{2M}{r}\right)^{-1} dr , \tag{2.8}$$

and the line element of the Schwarzschild solution becomes

$$ds^2 = -dT^2 + \frac{2M}{r} dR^2 + r^2 d\theta^2 + r^2 \sin^2 \theta d\phi^2 , \tag{2.9}$$

where

$$r = (2M)^{1/3} \left[\frac{3}{2} (R - T)\right]^{2/3} . \tag{2.10}$$

The metric is now regular at the event horizon $r = 2M$, corresponding to the points $4M = 3(R - T)$, and it is still singular at $r = 0$, $R = T$, which is indeed the true singularity in this spacetime.

Other common coordinates to write the Schwarzschild solution are the Kruskal–Szekeres coordinates, the Eddington–Finkelstein coordinates, and the Gullstrand–Painlevé coordinates. The Kruskal–Szekeres coordinates are employed to write the maximal analytic extension of the Schwarzschild solution. In this case, the line element reads

$$ds^2 = -\frac{32M^3}{r} e^{-r/(2M)} dV dU + r^2 d\theta^2 + r^2 \sin^2 \theta d\phi^2 , \tag{2.11}$$

where $U = \tau - \rho$ and $V = \tau + \rho$ are light-cone coordinates and

$$\tau = \begin{cases} \left(\frac{r}{2M} - 1\right)^{1/2} e^{r/(4M)} \sinh\left(\frac{t}{4M}\right) & \text{if } r > 2M, \\ \left(1 - \frac{r}{2M}\right)^{1/2} e^{r/(4M)} \cosh\left(\frac{t}{4M}\right) & \text{if } 0 < r < 2M. \end{cases}$$

$$\rho = \begin{cases} \left(\frac{r}{2M} - 1\right)^{1/2} e^{r/(4M)} \cosh\left(\frac{t}{4M}\right) & \text{if } r > 2M, \\ \left(1 - \frac{r}{2M}\right)^{1/2} e^{r/(4M)} \sinh\left(\frac{t}{4M}\right) & \text{if } 0 < r < 2M. \end{cases} \tag{2.12}$$

As briefly discussed in Sect. 2.6, the maximal extension of the Schwarzschild solution includes also a white hole and a parallel universe, which are not present in the Schwarzschild spacetime in Schwarzschild coordinates.

2.2.2 Reissner–Nordström Solution

If a non-rotating black hole has a non-vanishing electric charge, the metric is described by the *Reissner–Nordström solution*. As a useful recipe to remember, the Reissner–Nordström line element can be obtained from the Schwarzschild one in Eq. (2.6) with the substitution $M \to M - Q^2/(2r)$, where Q is the electric charge of the black hole. The result is[3]

$$ds^2 = -\left(1 - \frac{2M}{r} + \frac{Q^2}{r^2}\right) dt^2 + \left(1 - \frac{2M}{r} + \frac{Q^2}{r^2}\right)^{-1} dr^2 + r^2 d\theta^2$$
$$+ r^2 \sin^2\theta d\phi^2. \tag{2.14}$$

The solution of $g^{rr} = 0$ is

$$r_\pm = M \pm \sqrt{M^2 - Q^2}, \tag{2.15}$$

where the larger root, r_+, is the event horizon, while the smaller root, r_-, is the inner horizon. The latter is a Cauchy horizon and is unstable [13, 51]. The horizons only exist for $|Q| \leq M$. For $|Q| > M$, there is no horizon, the singularity at $r = 0$ is naked, and the Reissner–Nordström solution describes the spacetime of a naked singularity rather than that of a black hole.

---

[3]In international system units, the line element reads (reintroducing also G_N and c)

$$ds^2 = -\left(1 - \frac{2G_N M}{c^2 r} + \frac{G_N Q^2}{4\pi\varepsilon_0 c^4 r^2}\right) dt^2 + \left(1 - \frac{2G_N M}{c^2 r} + \frac{G_N Q^2}{4\pi\varepsilon_0 c^4 r^2}\right)^{-1} dr^2 + r^2 d\theta^2$$
$$+ r^2 \sin^2\theta d\phi^2, \tag{2.13}$$

where $1/(4\pi\varepsilon_0)$ is the Coulomb force constant.

2.2.3 Kerr Solution

A rotating uncharged black hole in 4-dimensional general relativity is described by the *Kerr solution*. In Boyer–Lindquist coordinates, the line element is

$$ds^2 = -\left(1 - \frac{2Mr}{\Sigma}\right) dt^2 - \frac{4aMr\sin^2\theta}{\Sigma} dt\,d\phi + \frac{\Sigma}{\Delta} dr^2 + \Sigma\, d\theta^2$$
$$+ \left(r^2 + a^2 + \frac{2a^2Mr\sin^2\theta}{\Sigma}\right)\sin^2\theta\, d\phi^2 , \tag{2.16}$$

where $\Sigma = r^2 + a^2\cos^2\theta$, $\Delta = r^2 - 2Mr + a^2$, $a = J/M$, and J is the spin angular momentum of the black hole. It is often convenient to introduce the dimensionless spin parameter $a_* = a/M = J/M^2$.

As in the Reissner–Nordström metric, there are two solutions for the equation $g^{rr} = 0$; that is

$$r_\pm = M \pm \sqrt{M^2 - a^2} . \tag{2.17}$$

r_+ is the radius of the event horizon, which requires $|a| \leq M$. For $|a| > M$ there is no horizon and the spacetime has a naked singularity at $r = 0$. It is worth noting that the topology of the spacetime singularity in the Kerr solution is different from that in the Schwarzschild and Reissner–Nordström spacetimes. It is still a curvature singularity and a singularity in the sense of geodesics incompleteness, but this is true only in the equatorial plane. In particular, geodesics outside the equatorial plane can reach the singularity and extend to another universe. The Kretschmann scalar $\mathscr{K}$ is

$$\mathscr{K} = \frac{48M^2}{\Sigma^6}\left(r^6 - 15a^2r^4\cos^2\theta + 15a^4r^2\cos^4\theta - a^6\cos^6\theta\right) , \tag{2.18}$$

and we can see that $\mathscr{K}$ diverges at $r = 0$ only for $\theta = \pi/2$.

Let us consider the Kerr–Schild coordinates (t', x, y, z), which are related to the Boyer–Lindquist ones by

$$x + iy = (r + ia)\sin\theta \exp\left[i\int d\phi + i\int \frac{a}{\Delta} dr\right] ,$$
$$z = r\cos\theta ,$$
$$t' = \int dt - \int \frac{r^2 + a^2}{\Delta} dr - r , \tag{2.19}$$

where i is the imaginary unit, namely $i^2 = -1$. r is implicitly given by

$$r^4 - \left(x^2 + y^2 + z^2 - a^2\right)r^2 - a^2z^2 = 0 . \tag{2.20}$$

The singularity at $r = 0$ and $\theta = \pi/2$ corresponds to $z = 0$ and $x^2 + y^2 = a^2$, namely it is a ring. It is possible to extend the spacetime to negative r, and the ring connects two universes. However, the region $r < 0$ posses closed time-like curves, which means it is possible to go backward in time. More details can be found, for instance, in [9].

As in the Reissner–Nordström case, the inner horizon r_- is likely unstable, but in the Kerr metric there is not a definitive proof. This would make the Kerr solution for $r < r_-$ physically not relevant.

2.2.4 No-Hair Theorem

The most general case of a rotating and electrically charged black hole is described by the *Kerr-Newman solution*. In analogy with the Reissner–Nordström metric, it can be obtained from the line element (2.16) with the substitution $M \to M - Q^2/(2r)$. The horizon is located at

$$r_+ = M + \sqrt{M^2 - Q^2 - a^2}\,, \tag{2.21}$$

and exists for $\sqrt{Q^2 + a^2} \leq M$.

The *no-hair theorem* asserts that black holes have only three asymptotic charges (the mass M, the spin angular momentum J, and the electric charge Q of the compact object) and no more. There are a number of assumptions behind this assertion. The spacetime must be stationary, asymptotically flat, and have 4 dimensions; the exterior region must be regular (no naked singularities and/or closed time-like curves); matter is described by the energy-momentum tensor of the electromagnetic field (but the theorem still holds in the presence of many other fields). For more details, see, e.g., [11]. The no-hair theorem was pioneered in the late 1960s and early 1970s by Israel [28], Carter [8], and Robinson [53], and its final form is still a work in progress. The name no-hair is to indicate that black holes have no features (hairs), although, to be precise, black holes can have three hairs (M, J, and Q). The fact that there is only the Kerr-Newman solution is the result of the *uniqueness theorem*. In the context of tests of the Kerr metric and of general relativity, both theorems are relevant. For instance, as a matter of principle, one may have different classes of black holes (each with characteristic M, J, and Q hairs), thus violating the uniqueness theorem, without any violation of the no-hair theorem.

2.3 Beyond the No-Hair Theorem

The no-hair theorem holds under specific assumptions. For instance, if the spacetime has more than 4 dimensions, there are also other kinds of black holes, e.g. the Myers-Perry black holes [44], as well as other "black objects" [16, 17]; see, for instance, the review article [18]. In an n-dimensional spacetime, a Myers-Perry black hole is characterized by the mass M and other $(n-1)/2$ parameters if n is odd, $n/2$ parameters if n is even, associated to the independent components of the angular momentum. So the number of hairs increases with n. "Hairy" black holes naturally arise also in the presence of non-Abelian gauge fields [58, 59].

In some alternative theories of gravity, the theorem may still holds, and an example is a simple scalar-tensor theory, in which the black hole solutions are the same as in general relativity [54]. Roughly speaking, this is because the Kerr metric is solution of the field equations

$$R_{\mu\nu} = 0\,, \tag{2.22}$$

and even the field equations of other theories of gravity may reduce to this simple form in the vacuum [52].

In other frameworks, the black hole solutions of general relativity may still be solutions of the new field equations, but their uniqueness is not guaranteed. A relevant example of violation of the no-hair theorem is presented in [25], where the authors discovered a family of hairy black holes in 4-dimensional Einstein's gravity minimally coupled to a complex, massive scalar field. Here hairy black holes are possible by introducing a specific harmonic time-dependence in the scalar field, while the spacetime metric and the energy-momentum tensor of the scalar field are still stationary. There are also cases, like dynamical Chern–Simons gravity, in which non-rotating black holes are described by the Schwarzschild solution, but rotating black holes are not those of Kerr [65].

In general, we can distinguish two kinds of hairs, called, respectively, primary and secondary hairs. *Primary hairs* are real hairs of the black hole: if such hairs were to exist, then M, J, and Q would not completely characterize the compact object, and one or more additional parameters would be necessary. An example is a 5-dimensional Myers-Perry black hole: it is the 5-dimensional generalization of Kerr black holes and it has two angular momenta, so one more hair: J' [44].

Secondary hairs are instead related to some new charge that is common to all black holes. For instance, in Einstein-dilaton-Gauss-Bonnet gravity, a black hole has a scalar charge proportional to the volume integral of the Gauss–Bonnet invariant [33, 40]; that is, the scalar charge is determined by the black hole mass and it is not an additional degree of freedom.

In alternative theories of gravity, there may be further complications and the phenomenology can be much richer. For instance, some theories may not have black hole solutions. An example is the model of massive gravity discussed in [1]. The static and spherically symmetric vacuum solution of the corresponding field equations does not describe a black hole but a naked singularity.

In some theories even the definition of black hole may be problematic. For instance, in models with violation of the Lorentz symmetry, null geodesics may depend on the energy of the massless particle. In this case, there are different "event horizons" for photons with different energies, and it is possible that some high energy photons can always escape to infinity, so that there is no real black hole.

It is also possible that some frameworks have black hole solutions, but there is no mechanism to create black holes. In general relativity, black holes emerge as exact solutions of the Einstein equations. However, this is not enough to say that such solutions are physically relevant. In general relativity, we know that black holes can form from gravitational collapse and we have also some simple analytical models that show how this is possible (see Sect. 2.4). In some alternative theories, the gravitational collapse may simply be unable to create a black hole. For instance, black holes may be unstable solutions or would require a set-up impossible to realize. A possible example is represented by the model discussed in [67].

In general, black hole solutions in alternative theories of gravity are known in the non-rotating limit, either in analytic or numerical form, because static and spherically symmetric solutions are relatively easy to find. In some cases, analytic slow-rotating solutions have been found, see e.g. [38, 48, 64, 65]. Exact rotating black hole solutions, especially in analytic form, are more difficult to obtain [34]. This is actually true even in general relativity, and it is proved by the fact that the Kerr metric was discovered more than 45 years after the Schwarzschild solution.

2.4   Gravitational Collapse

When a star exhausts all its nuclear fuel, the gas pressure cannot balance the star own weight, and the body shrinks to find a new equilibrium configuration. For most stars, the pressure of degenerate electrons stops the collapse and the star becomes a white dwarf. However, if the collapsing part of the star is too heavy, the mechanism does not work, matter reaches higher densities, and protons and electrons transform into neutrons. If the pressure of degenerate neutrons stops the collapse, the star becomes a neutron star. If the collapsing core is still too massive and even the neutron pressure cannot stop the process, there is no known mechanism capable of finding a new equilibrium configuration, and the body should undergo a complete collapse. In this case, the final product is a black hole.

The aim of this section is to present the simplest and the next-to-simplest gravitational collapse models. These solutions are analytic and nicely show how the gravitational collapse of a spherically symmetric cloud of dust creates a spacetime singularity and an event horizon. For a review, see e.g. [32]. Numerical simulations can treat more realistic models, where the final product is still a black hole [2, 3].

We want to consider a spherically symmetric collapse, so the spacetime must be spherically symmetric. This means the metric is invariant under the group of spatial rotations $SO(3)$ and we can define the 2-dimensional metric induced on the unit 2-sphere as

$$d\Omega^2 = d\theta^2 + \sin^2\theta d\phi^2 \,. \tag{2.23}$$

We introduce the function R such that $4\pi R^2$ represents the area of each 2-sphere in the spacetime. The 4-dimensional line element reads

$$ds^2 = g_{ab}dx^a dx^b + R^2 d\Omega \,, \tag{2.24}$$

where $a, b = 0, 1$ and R is a function of x^0 and x^1. Choosing $x^0 = t$ and $x^1 = r$ and diagonalizing g_{ab}, we find that the most general line element of a spherically symmetric spacetime can be written as

$$ds^2 = -e^{2\lambda}dt^2 + e^{2\psi}dr^2 + R^2 d\Omega^2 \,, \tag{2.25}$$

where λ, ψ, and R are functions of t and r only.

Let us assume that the collapsing body can be described by a perfect fluid. The coordinate system of the line element in Eq. (2.25) is called comoving because the coordinates t and r are "attached" to every collapsing particle. This is the rest-frame of the collapsing fluid and therefore the fluid 4-velocity is $u^\mu = (e^{-\lambda}, 0, 0, 0)$. The energy momentum tensor is

$$T^\mu_\nu = \mathrm{diag}\left(\rho, P, P, P\right) \,, \tag{2.26}$$

where ρ and P are, respectively, the energy density and the pressure of the fluid.

With the line element in Eq. (2.25), the Einstein tensor reads

$$G^t_t = -\frac{F'}{R^2 R'} + \frac{2\dot{R}e^{-2\lambda}}{RR'}\left(\dot{R}' - \dot{R}\lambda' - \dot{\psi}R'\right) \,, \tag{2.27}$$

$$G^r_r = -\frac{\dot{F}}{R^2 \dot{R}} - \frac{2R'e^{-2\psi}}{R\dot{R}}\left(\dot{R}' - \dot{R}\lambda' - \dot{\psi}R'\right) \,, \tag{2.28}$$

$$G^t_r = -e^{2\psi - 2\lambda}G^r_t = \frac{2e^{-2\lambda}}{R}\left(\dot{R}' - \dot{R}\lambda' - \dot{\psi}R'\right) \,, \tag{2.29}$$

$$G^\theta_\theta = G^\phi_\phi = \frac{e^{-2\psi}}{R}\left[\left(\lambda'' + \lambda'^2 - \lambda'\psi'\right)R + R'' + R'\lambda' - R'\psi'\right] + \tag{2.30}$$

$$-\frac{e^{-2\lambda}}{R}\left[\left(\ddot{\psi} + \dot{\psi}^2 - \lambda\dot{\psi}\right)R + \ddot{R} + \dot{R}\dot{\psi} - \dot{R}\lambda\right] \,. \tag{2.31}$$

From the Einstein equations, we can get the following equations

$$G^t_t = 8\pi T^t_t \Rightarrow \frac{F'}{R^2 R'} = 8\pi\rho \,, \tag{2.32}$$

$$G^r_r = 8\pi T^r_r \Rightarrow \frac{\dot{F}}{R^2 \dot{R}} = -8\pi P \,, \tag{2.33}$$

$$G^t_r = 0 \Rightarrow \dot{R}' - \dot{R}\lambda' - \dot{\psi}R' = 0 \,, \tag{2.34}$$

where the prime $'$ and the dot $\dot{}$ denote, respectively, the derivative with respect to r and t. F is the Misner-Sharp mass [41]

$$F = R\left(1 - e^{-2\psi}R'^2 + e^{-2\lambda}\dot{R}^2\right) \,, \tag{2.35}$$

which is defined by the relation

$$1 - \frac{F}{R} = g_{\mu\nu} \left(\partial^\mu R\right) \left(\partial^\nu R\right) . \tag{2.36}$$

From Eq. (2.32), we can see that the Misner-Sharp mass is proportional to the gravitational mass within the radius r at the time t

$$F(r) = \int_0^r F' d\tilde{r} = 8\pi \int_0^r \rho R^2 R' d\tilde{r} = 2M(r) . \tag{2.37}$$

It is worth noting that $n^\mu = \partial^\mu R$ is the normal to the surface $R = const.$ Therefore, as seen in Sect. 2.1, when $1 - F/R = 0$, the surface $R = const.$ is a null surface (and defines the location of the apparent horizon in the dust collapse models in the next subsections).

A fourth relation can be obtained from the covariant conservation of the matter energy-momentum tensor

$$\nabla_\mu T^\mu_\nu = 0 \Rightarrow \lambda' = -\frac{P'}{\rho + P} . \tag{2.38}$$

2.4.1 Dust Collapse

For dust, $P = 0$, and Eqs. (2.32)–(2.34), and (2.38) become

$$\frac{F'}{R^2 R'} = 8\pi\rho , \tag{2.39}$$

$$\frac{\dot{F}}{R^2 \dot{R}} = 0 , \tag{2.40}$$

$$\dot{R}' - \dot{R}\lambda' - \dot{\psi} R' = 0 , \tag{2.41}$$

$$\lambda' = 0 . \tag{2.42}$$

Equation (2.40) shows that, in the case of dust, F is independent of t, namely there is no inflow or outflow through any spherically symmetric shell with radial coordinate r. This means that the exterior spacetime is described by the Schwarzschild solution. In the general case with $P \neq 0$, this may not be true, and the interior region must be matched with a non-vacuum Vaidya spacetime. If r_b is the comoving radial coordinate of the boundary of the cloud of dust, $F(r_b) = 2M$, where M is the Schwarzschild mass of the vacuum exterior.

Equation (2.42) implies that $\lambda = \lambda(t)$ and permits one to choose the time gauge in such a way that $\lambda = 0$. It is indeed always possible to define a new time coordinate $\tilde{t}$ such that $d\tilde{t} = e^\lambda dt$ and therefore $g_{\tilde{t}\tilde{t}} = -1$.

Equation (2.41) becomes $\dot{R}' - \dot{\psi} R' = 0$ and we can write

$$R' = e^{g(r)+\psi} .$$

(2.43)

We introduce the function $f(r) = e^{2g(r)} - 1$ and Eq. (2.35) becomes

$$\dot{R}^2 = \frac{F}{R} + f .$$

(2.44)

The line element can now be written as

$$ds^2 = -dt^2 + \frac{R'^2}{1+f} dr^2 + R^2 d\Omega^2 .$$

(2.45)

This is the Lemaitre–Tolman-Bondi, or LTB, metric [6, 36, 56].
The Kretschmann scalar of the line element in (2.45) is

$$\mathscr{K} = 12 \frac{F'^2}{R^4 R'^2} - 32 \frac{FF'}{R^5 R'} + 48 \frac{F^2}{R^6} ,$$

(2.46)

and diverges if $R = 0$. The system has a gauge degree of freedom that can be fixed by setting the scale at a certain time. It is common to set the area radius $R(t, r)$ to the comoving radius r at the initial time $t_i = 0$, namely $R(0, r) = r$, and introduce the scale factor a

$$R(t, r) = ra(t, r) .$$

(2.47)

We have thus $a = 1$ at $t = t_i$ and $a = 0$ at the time of the formation of the singularity. The condition for collapse is $\dot{a} < 0$. From Eq. (2.39), the regularity of the energy density at the initial time t_i requires to write the Misner-Sharp mass as $F(r) = r^3 m(r)$, where $m(r)$ is a sufficiently regular function of r in the interval $[0, r_b]$. Equation (2.39) becomes

$$\rho = \frac{3m + rm'}{a^2 (a + ra')} .$$

(2.48)

The function $m(r)$ is usually written as a polynomial expansion around $r = 0$

$$m(r) = \sum_{k=0}^{\infty} m_k r^k ,$$

(2.49)

where $\{m_k\}$ are constants. Requiring that the energy density ρ has no cusps at $r = 0$, $m_1 = 0$.

From Eq. (2.46), we see that the Kretschmann scalar diverges even when $R' = 0$ if $m' \neq 0$. However, the nature of these singularities is different: they arise from the overlapping of radial shells and are called shell crossing singularities [24]. Here the radial geodesic distance between shells with radial coordinates r and $r + dr$

vanishes, but the spacetime may be extended through the singularity by a suitable redefinition of the coordinates. To avoid any problem, it is common to impose that the collapse model has no shell crossing singularities, for instance requiring that $R' \neq 0$ or that m'/R' does not diverge.

At the initial time t_i, Eq. (2.44) becomes

$$\dot{a}(t_i, r) = -\sqrt{m + \frac{f}{r^2}} , \qquad (2.50)$$

and we can see that the choice of f corresponds to the choice of the initial velocity profile of the particles in the cloud. In order to have a finite velocity at all radii, it is necessary to impose some conditions on f. It is common to write $f(r) = r^2 b(r)$ and $b(r)$ as a polynomial expansion around $r = 0$:

$$b(r) = \sum_{k=0}^{\infty} b_k r^k . \qquad (2.51)$$

2.4.2 Homogeneous Dust Collapse

The simplest model of gravitational collapse is the Oppenheimer-Snyder model [46]. It describes the collapse of a homogeneous and spherically symmetric cloud of dust. In this case, $\rho = \rho(t)$ is independent of r, so $m = m_0$ and $b = b_0$. The interior metric is the time reversal of the Friedmann–Robertson–Walker solution

$$ds^2 = -dt^2 + a^2 \left(\frac{dr^2}{1 + b_0 r^2} + r^2 d\Omega^2 \right) . \qquad (2.52)$$

$b_0 = 0$ is the counterpart of a flat universe and corresponds to a marginally bound collapse, namely the scenario in which the falling particles have vanishing velocity at infinity. Equation (2.44) becomes

$$\dot{a} = -\sqrt{\frac{m_0}{a} + b_0} . \qquad (2.53)$$

For $b_0 = 0$, the solution is

$$a(t) = \left(1 - \frac{3\sqrt{m_0}}{2} t \right)^{2/3} . \qquad (2.54)$$

The formation of the singularity occurs at the time

$$t_s = \frac{2}{3\sqrt{m_0}} . \qquad (2.55)$$

The curve $t_{ah}(r)$ describing the time at which the shell r crosses the apparent horizon can be obtained from

$$1 - \frac{F}{R} = 1 - \frac{r^2 m_0}{a} = 0. \tag{2.56}$$

For $b_0 = 0$, the solution is

$$t_{ah}(r) = t_s - \frac{2}{3}F = \frac{2}{3\sqrt{m_0}} - \frac{2}{3}r^3 m_0. \tag{2.57}$$

The Finkelstein diagram of the gravitational collapse of a homogeneous and spherically symmetric cloud of dust is sketched in Fig. 2.2. At the time $t = t_0$, the radius of the surface of the cloud crosses the Schwarzschild radius. We have the formation of both the event horizon in the exterior region and the apparent horizon at the boundary $r = r_b$, i.e. $t_0 = t_{ah}(r_b)$. As shown in Fig. 2.2, the exterior region is now settled down to the static Schwarzschild spacetime, while the radius of the apparent horizon propagates to smaller radii and reaches $r = 0$ at the time of the formation of the singularity t_s.

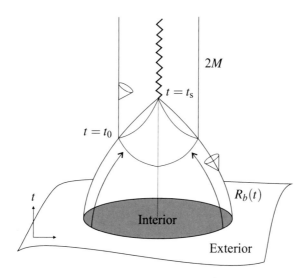

Fig. 2.2 Finkelstein diagram for the gravitational collapse of a homogeneous and spherically symmetric cloud of dust. $R_b(t)$ is the radius of the cloud (in the Schwarzschild coordinates of the exterior region) and separates the interior from the vacuum exterior. The cloud collapses as t increases and at the time $t = t_0$ the horizon forms at the boundary when $R_b(t_0) = 2M$. In the interior, the apparent horizon propagates inwards and reaches the center of symmetry at the time of the formation of the singularity $t = t_s$. For $t > t_s$, the spacetime has settled down to the usual Schwarzschild solution. Figure courtesy of Daniele Malafarina

2.4.3 Inhomogeneous Dust Collapse

In the inhomogeneous dust collapse scenario, ρ depends on both t and r, so we have $m = m(r)$, $b = b(r)$, and $a = a(t, r)$. While the Hawking-Penrose singularity theorems assure that, under certain conditions, the formation of a singularity is inevitable after the formation of an apparent horizon [23], the picture of the collapse may be different from the simple Oppenheimer-Snyder model. In particular, it is possible that the singularity is naked for an infinitesimal time, namely there may be null geodesics that start from the singularity and go to null infinity, see e.g. [10, 31, 45, 63].

If we want to impose that the energy density has no cusps at the center, $m_1 = 0$ and the simplest form of the function m is

$$m(r) = m_0 + m_2 r^2 . \tag{2.58}$$

Now the density profile is described by two parameters, m_0 and m_2. Imposing the condition that ρ must be a decreasing function of r (the density of the cloud is higher at the center and lower at larger radii), $m_2 < 0$.

Equation (2.44) is now

$$\dot{a} = -\sqrt{\frac{m}{a} + b} . \tag{2.59}$$

In the marginally bound case $b = 0$, the solution is

$$a(t, r) = \left(1 - \frac{3\sqrt{m(r)}}{2} t \right)^{2/3} , \tag{2.60}$$

and we see that each shell collapses with a different scale factor and a different velocity. The singularity and the apparent horizon are now described by the curves

$$t_s(r) = \frac{2}{3\sqrt{m}} \tag{2.61}$$

$$t_{ah}(r) = \frac{2}{3\sqrt{m}} - \frac{2}{3} r^3 m . \tag{2.62}$$

If $m = m_0 + m_2 r^2$, we find

$$t_s(r) = \frac{2}{3\sqrt{m_0 + m_2 r^2}} \tag{2.63}$$

$$t_{ah}(r) = \frac{2}{3\sqrt{m_0 + m_2 r^2}} - \frac{2}{3} r^3 \left(m_0 + m_2 r^2 \right) . \tag{2.64}$$

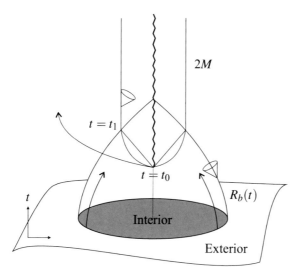

Fig. 2.3 Finkelstein diagram for the gravitational collapse of an inhomogeneous and spherically symmetric cloud of dust. $R_b(t)$ is the radius of the cloud (in the Schwarzschild coordinates of the exterior region) and separates the interior from the vacuum exterior. The cloud collapses as t increases. At the time $t = t_0$, a singularity and a horizon form at the center of the cloud at the same time. Null geodesics from the singularity can reach distant observers and therefore the singularity is temporarily naked. In the interior, the apparent horizon propagates outwards and reaches the boundary at the time $t_1 > t_0$. The exterior spacetime settles down to the usual Schwarzschild solution when the whole star is inside the event horizon. Figure courtesy of Daniele Malafarina

In the dust case, the boundary of the cloud r_b is arbitrary and, for $m_2 < 0$, it is possible to find null geodesics that begin at $r = 0$ at the time of the formation of the singularity and reach observers at infinity [10, 31, 45, 63]. The actual picture depends on the parameters m_0, m_2, and r_b. Figure 2.3 shows the case in which a naked singularity forms at the time $t = t_0$ at $r = 0$ together with the apparent horizon. The singularity is immediately covered by the apparent horizon, which propagates to larger radii. When it reaches the boundary $r = r_b$, the exterior spacetime settles down to the Schwarzschild solution.

2.4.4 Gravitational Collapse for a Distant Observer

In the previous subsections, we adopted comoving coordinates. With such a choice, the black hole and the central singularity are created in a finite time. As already pointed out in the seminal paper by Oppenheimer and Snyder [46], an external observer sees the surface of the collapsing body asymptotically shrinking to the radius of the event horizon of the black hole, without really seeing the formation of the black hole due to the asymptotically increasing gravitational redshift.

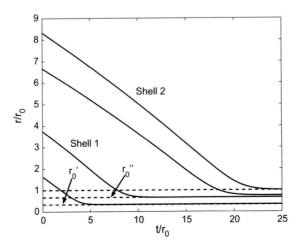

Fig. 2.4 Evolution of the radial coordinates of the inner and of the outer boundaries of shell 1 and shell 2 (*solid lines*) with respect to the time t of a distant observer. The three *horizontal dashed lines* indicate the coordinate of the horizon before the collapse of shell 1 ($r_0' = 2M$, *lower line*), after the collapse of shell 1 and before the collapse of shell 2 ($r_0'' = 2M + 2M_1$, *central line*), and after the collapse of shell 2 ($r_0 = 2M + 2M_1 + 2M_2$, *upper line*). M, M_1, and M_2 are, respectively, the mass of the pre-existing *black hole*, of shell 1, and of shell 2. See the text for more details. From [37] under the terms of the Creative Commons Attribution License

The picture of gravitational collapse as seen by a distant observer can be understood within the toy-models discussed in [37, 66]. Let us consider a pre-existing black hole of mass M and two spherically symmetric shells of matter collapsing onto the black hole and with the mass, respectively, M_1 and M_2. The two shells have a finite thickness, so each shell has inner and outer boundaries. The evolution of such a system is shown in Fig. 2.4.

At the beginning, we have the pre-existing black hole and the two shells at large radii. The event horizon of the black hole is at $r_0' = 2M$. After the collapse of shell 1, the horizon of the new black hole is at $r_0'' = 2M + 2M_1$. However, the distant observer does not see (for the moment) the outer boundary of shell 1 crossing the horizon r_0''. The radius of the outer boundary of shell 1 seems to asymptotically approach r_0'', due to the infinite gravitational redshift. Then we have the collapse of shell 2. Now the horizon of the black hole is at $r_0 = 2M + 2M_1 + 2M_2$ and the distant observer does not see the outer boundary of shell 2 crossing the surface at the radial coordinate r_0.

Let us notice that the radiation emitted by the most outer boundary of any collapsing configuration is more and more redshifted, so that any distant observer eventually sees a black hole for any practical purpose (even if the analytical formula predicts a non-zero emission of radiation at any time). The calculations of the radiation emitted by a collapsing body within some simple toy-models are presented in [35]. The light curves for a homogeneous and inhomogeneous, spherically symmetric, collapsing balls of dust are shown in Fig. 2.5. At late times, the emission of radiation is exponentially suppressed.

Fig. 2.5 Light curves of a homogeneous (*red solid curve*) and inhomogeneous (*blue dashed curve*), spherically symmetric, collapsing balls of dust assuming a simple model of emission. Time T in units $2M = 1$; luminosity in arbitrary units. From [35] under the terms of the Creative Commons Attribution License

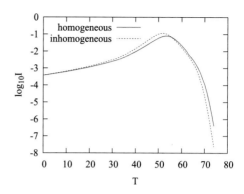

2.5 Beyond the Standard Picture

It is natural to expect that the singularity at the center of black holes is due to the breakdown of the classical theory and that it should be removed when unknown quantum gravity effects are taken into account. If we believe that the energy scale of quantum gravity is the Planck mass M_{Pl}, we may expect that the Kretschmann scalar is bounded by

$$\mathcal{K} \lesssim \frac{1}{M_{Pl}^4} . \tag{2.65}$$

From Eq. (2.7), we see that new physics should show up at the radius

$$r \approx \left(M M_{Pl}^2 \right)^{1/3} . \tag{2.66}$$

However, as already mentioned in Sect. 2.2, charged or rotating black holes have the inner horizon at $r = r_-$, which is unstable. This means that deviations from these metrics should be at least at r_-. In the case of extremal or almost extremal objects, r_- approaches r_+, and therefore new physics may be not far from the event horizon.

There are also arguments suggesting that black holes may be macroscopic quantum objects. Roughly speaking, the identification of the Planck scale as the fundamental energy scale of quantum gravity arises when we quantize general relativity and find the problems of unitarity and renormalizability. The theory looks like a good effective theory at low energies, namely for energies $E \ll M_{Pl}$, but it breaks down when E approaches M_{Pl}. However, this conclusion is obtained by considering the scattering of two particles. In a system made of many components, it is possible that the actual scale of quantum gravity is given by the gravitational radius $\sim M$, where M is the mass/energy of the system, see e.g. [14, 15, 21, 39]. In these scenarios, the metric description typically breaks down inside the black hole.

If the metric description holds at the center of black holes, we can imagine two ways to solve the central singularity, in the sense that geodesics do not stop at $r = 0$

and predictability is not lost. One of these possibilities is a wormhole-like solution. In this scenario, the singularity is replace by a "throat" connecting two universes. For a static/stationary black hole solution, we can have a wormhole, namely a topologically non-trivial structure connecting two asymptotically flat universes. In the case of a black hole formed from gravitational collapse, the throat can connect the original universe to a baby universe generated by the gravitational collapse inside the black hole. The alternative possibility to solve the singularity is to have a spacetime in which the center $r = 0$ can never be reached in a finite time (with a finite value of the affine parameter in the case of null geodesics). These are the two natural scenarios to solve the central singularity, and there are not more, because one has to preserve the property that, roughly speaking, inside a black hole everything must fall to its center. The problem of a geodesically incomplete spacetime can thus be fixed either postulating the extension of the spacetime beyond $r = 0$ or the impossibility of reaching $r = 0$ in a finite time. Other solutions seem at least to require more exotic physics.

The quantum gravity inspired models studied in [4, 5, 20, 67] are characterized by the fact that gravity becomes repulsive at very high densities. The result is that the singularity is replaced by a bounce, after which the collapsing matter starts expanding. In principle, we may have two scenarios: (i) there is a bounce and the creation of a baby universe inside the black hole, or (ii) there is no black hole, in the sense that the collapse only creates an apparent horizon, which can be interpreted as a black hole for a while by the exterior observer if the observational time is shorter than the time scale of the evolution of the process.

In general, both scenarios may be possible, and it depends on the specific model. The scenario (i) is not easy to realize. A similar spacetime can be obtained with a cut-and-paste procedure, in which a singular manifold is extended beyond the singularity by removing the singularity and sewing the spacetime to a new non-singular manifold describing an expanding baby universe. However, this is possible only in very simple examples. Matching of the two manifolds involves the continuity of the first and second fundamental forms across some hypersurface, which is not possible in general due to the lack of a sufficient number of free parameters.

In the case of the scenario (ii) and under certain conditions, the lifetime of the apparent horizon might exactly scale as the Hawking evaporation time. It is thus possible to argue a link between instability of this kind of objects and Hawking radiation [5].

2.6 Penrose Diagrams

Penrose diagrams are 2-dimensional spacetime diagrams used to figure out the global properties and the causal structure of asymptotically flat spacetimes. Since they are 2-dimensional diagrams, every point represents a 2-dimensional sphere of the original 4-dimensional spacetime. Penrose diagrams are obtained by a conformal transformation of the original coordinates such that the entire spacetime is transformed into a compact region. Since the transformation is conformal, angles are preserved, and null geodesics remain lines at 45°. Time-like geodesics are inside the light-cone, space-like geodesics are outside. A more detailed discussion on the topic can be found, for instance, in [57, 62].

It is probably easier to start from the simplest example, namely the Penrose diagram for the Minkowski spacetime. In spherical coordinates (t, r, θ, ϕ), the line elements is

$$ds^2 = -dt^2 + dr^2 + r^2 d\theta^2 + r^2 \sin^2 \theta d\phi^2 . \tag{2.67}$$

We perform the following conformal transformation

$$
\begin{aligned}
t &= \frac{1}{2} \tan \frac{T+R}{2} + \frac{1}{2} \tan \frac{T-R}{2} , \\
r &= \frac{1}{2} \tan \frac{T+R}{2} - \frac{1}{2} \tan \frac{T-R}{2} ,
\end{aligned}
\tag{2.68}
$$

where the use of the tangent function tan is to bring points at infinity to points at a finite value in the new coordinates. The line element now reads

$$
\begin{aligned}
ds^2 = &\left(4\cos^2 \frac{T+R}{2} \cos^2 \frac{T-R}{2} \right)^{-1} \left(-dT^2 + dR^2 \right) \\
&+ r^2 d\theta^2 + r^2 \sin^2 \theta d\phi^2 .
\end{aligned}
\tag{2.69}
$$

The Penrose diagram for the Minkowski spacetime is shown in Fig. 2.6. The semi-infinite (t, r) plane is now a triangle. The dashed vertical line is the origin $r = 0$. Every point corresponds to the 2-sphere (θ, ϕ). There are five different asymptotic regions. Without a rigorous treatment, they can be defined as follows:

Fig. 2.6 Penrose diagram for the Minkowski spacetime. See the text for the details

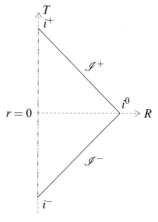

Future time-like infinity i^+: the region toward which time-like geodesics extend. It corresponds to the points at $t \to \infty$ with finite r.

Past time-like infinity i^-: the region from which time-like geodesics come. It corresponds to the points at $t \to -\infty$ with finite r.

Spatial infinity i^0: the region toward which space-like slices extend. It corresponds to the points at $r \to \infty$ with finite t.

Future null infinity $\mathscr{I}^+$: the region toward which outgoing null geodesics extend. It corresponds to the points at $t + r \to \infty$ with finite $t - r$.

Past null infinity $\mathscr{I}^-$: the region from which ingoing null geodesics come. It corresponds to the points at $t - r \to -\infty$ with finite $t + r$.

Such asymptotic regions are points or segments in the Penrose diagram and their T and R coordinates are:

$$
\begin{aligned}
i^+ \quad & T = \pi, \quad R = 0. \\
i^- \quad & T = -\pi, \quad R = 0. \\
i^0 \quad & T = 0, \quad R = \pi,
\end{aligned}
\tag{2.70}
$$

and

$$
\begin{aligned}
\mathscr{I}^+ \quad & T + R = \pi, \quad T - R \in (-\pi; \pi). \\
\mathscr{I}^- \quad & T - R = -\pi, \quad T + R \in (-\pi; \pi).
\end{aligned}
\tag{2.71}
$$

Penrose diagrams become a powerful tool to explore the global properties and the causal structure of more complicated spacetimes. The simplest non-trivial example is the Schwarzschild spacetime. If we consider its maximal extension in Kruskal–Szekeres coordinates and we perform the following coordinate transformation

$$
\begin{aligned}
V &= \frac{1}{2} \tan \frac{T + R}{2} + \frac{1}{2} \tan \frac{T - R}{2}, \\
U &= \frac{1}{2} \tan \frac{T + R}{2} - \frac{1}{2} \tan \frac{T - R}{2},
\end{aligned}
\tag{2.72}
$$

the line element becomes

$$
ds^2 = \frac{32M^3}{r} e^{-r/(2M)} \left(4 \cos^2 \frac{T + R}{2} \cos^2 \frac{T - R}{2} \right)^{-1} \left(-dT^2 + dR^2 \right)
$$
$$
+ r^2 d\theta^2 + r^2 \sin^2 \theta d\phi^2.
\tag{2.73}
$$

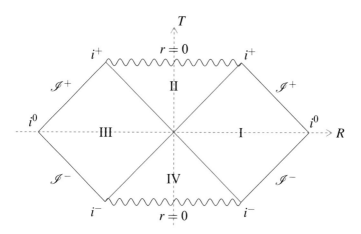

Fig. 2.7 Penrose diagram for the maximal extension of the Schwarzschild spacetime. See the text for the details

The Penrose diagram of the maximal extension of the Schwarzschild spacetime is shown in Fig. 2.7. The asymptotic regions i^+, i^-, i^0, $\mathscr{I}^+$, and $\mathscr{I}^-$ are those already found in the Minkowski case. We can distinguish four regions, indicated, respectively, by I, II, II, and IV in the figure.

The region I corresponds to our universe, namely the exterior region of the Schwarzschild spacetime in Schwarzschild coordinates. The region II is the black hole, so the Schwarzschild spacetime in Schwarzschild coordinates has only the regions I and II. The central singularity of the black hole at $r = 0$ is represented by the line with wiggles above the region II and the diagram clearly shows it is a space-like singularity.[4] The event horizon of the black hole at $r = 2M$ is the red line at 45° (it is indeed a null surface) separating the regions I and II. Any ingoing light ray in the region I is captured by the black hole, while any outgoing light ray in the region I reaches future null infinity $\mathscr{I}^+$. Null and time-like geodesics in the region II cannot exit the black hole and they necessarily fall to the singularity at $r = 0$.

The regions III and IV emerge from the extension of the Schwarzschild spacetime. The region III corresponds to another universe. The red line at 45° separating the regions II and III is the event horizon of the black hole at $r = 2M$. Like in the region I, any light ray in the region III can either cross the event horizon or escape to infinity. No future-oriented null or time-like geodesics can escape from the region II. Our universe in the region I and the other universe in the region III cannot communicate: no null or time-like geodesic can go from one region to another.

---

[4]Singularities, like trajectories, can be space-like, null, or time-like, depending on their causal properties. Space-like singularities are represented by lines with an inclination lower than 45° (like space-like trajectories). Time-like singularities are represented by lines with an inclination higher than 45° (like time-like trajectories). The singularity at $r = 0$ in the Penrose diagram of the Schwarzschild solution is represented by a horizontal line and is thus space-like. For more details, see e.g. [19] and references therein.

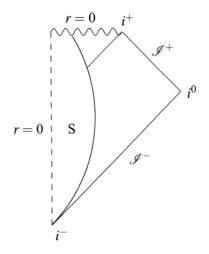

Fig. 2.8 Penrose diagram for the complete collapse of a homogeneous cloud of dust, corresponding to the situation in Fig. 2.2. The letter S indicates the interior region of the collapsing star and the black arc extending from i^- to $r = 0$ is its boundary. See the text for the details

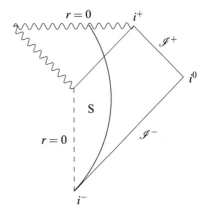

Fig. 2.9 Penrose diagram for the complete collapse of an inhomogeneous cloud of dust with a temporary naked singularity, namely the situation in Fig. 2.3. The letter S indicates the interior region of the collapsing star and the *black arc* extending from i^- to $r = 0$ is its boundary. See the text and [43, 47] for the details.

The region IV is a *white hole*. If a black hole is a region of the spacetime where null and time-like geodesics can only enter and never exit, a white hole is a region where null and time-like geodesics can only exit and never enter. The red lines at 45° separating the region IV from the regions I and III are the horizons at $r = 2M$ of the white hole.

Figure 2.7 is the Penrose diagram for the maximal extension of the Schwarzschild solution, which is static. However, black holes in the Universe should be created from gravitational collapse of massive bodies. The corresponding Penrose diagrams are shown in Figs. 2.8 and 2.9 and are substantially different from the diagram in Fig. 2.7. Figure 2.8 is for the case of the collapse of a homogeneous cloud of dust (the Penrose diagram corresponding to the Finkelstein diagram in Fig. 2.2): the singularity is created at the same time $t = t_s$ for all shells. Figure 2.9 shows the collapse of an inhomogeneous cloud of dust with the formation of a temporary naked singularity (the counterpart of the Finkelstein diagram in Fig. 2.3) [43, 47]: here the singularity is created first at the center and, for an infinitesimal time, is naked.

The Penrose diagram for a static spacetime with a massive body would be equivalent to that for the Minkowski spacetime in Fig. 2.6. In the case of a collapsing body, the diagram changes when the radius of the body crosses the corresponding Schwarzschild radius at $r = 2M$. At this point, we have the formation of the event horizon, represented by the red line at 45° in Figs. 2.8 and 2.9. Now the exterior region looks like the region I in the Penrose diagram of the Schwarzschild spacetime with the future null infinity $\mathscr{I}^+$. In the interior region, the radius of the body goes to the space-like singularity at $r = 0$. There is no white hole or parallel universe.

Figures 2.8 and 2.9 show the Penrose diagrams of two examples of complete collapse in classical general relativity. The picture changes when we consider "quantum" effects, broadly defined. The left diagram in Fig. 2.10 is the Penrose diagram for the formation of a black hole from the collapse of a star and its "complete evaporation" due to Hawking radiation. Even in this case, the event horizon is represented by a red line at 45°. While the evaporation process progressively reduces the radius of the horizon and the object emits radiation moving along null geodesics, as an artifact of the conformal transformation the horizon is still a line at 45° and it seems like all Hawking radiation is emitted together at once. In the case of complete evaporation, the upper part of the Penrose diagram is like that of the Minkowski spacetime.

The right diagram in Fig. 2.10 is one of the Penrose diagrams for the gravitational collapse with bounce studied in [4, 20]. The diagram is slightly different if the evaporation of the apparent horizon is associated to some Hawking-like process. In these scenarios, there is neither formation of black hole nor formation of spacetime

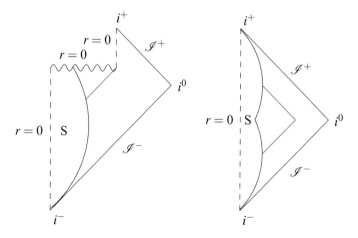

Fig. 2.10 Penrose diagrams for the formation and the Hawking evaporation of a *black hole* (*left*) and for the gravitational collapse with bounce and without formation of singularities (*right*). The letter S indicates the interior region of the collapsing star and the *black arc* extending from i^- is its boundary. See the text for the details

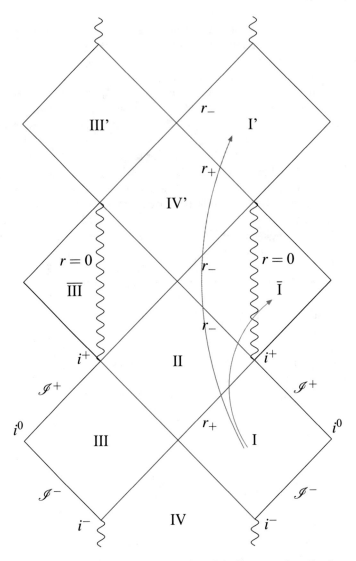

Fig. 2.11 Penrose diagram for the maximal extension of the Kerr spacetime. See the text for the details

singularity. When the radius of the collapsing object crosses the corresponding Schwarzschild radius, we have an apparent horizon and the object first looks like a black hole for a finite time and then a white hole for a finite time. Eventually the apparent horizon disappears.

   More details on Penrose diagrams associated to scenarios of gravitational collapse beyond classical general relativity can be found in [26].

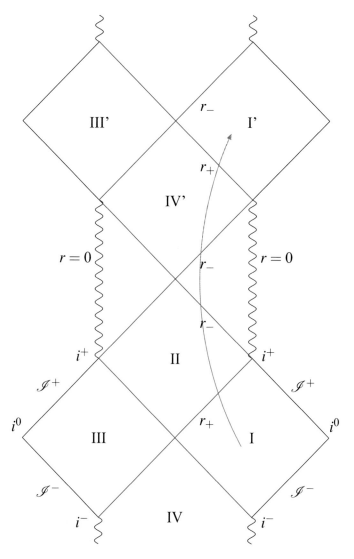

Fig. 2.12 Penrose diagram for the maximal extension of the Reissner–Nordström spacetime. See the text for the details

Figure 2.11 shows the Penrose diagram for the maximal extension of the Kerr spacetime. The region I is our universe outside the black hole and the red line at 45° separating the regions I and II and labeled by r_+ is the event horizon. As in the Schwarzschild spacetime, the region III is another universe and the region IV is a white hole. The region II is the black hole interior between the event horizon r_+ and the inner horizon r_-. The singularity at $r = 0$ is time-like (not space-like as in Schwarzschild) and therefore it is a vertical line and can be avoided by a test-particle.

A test-particle in the region I can cross the event horizon, cross the inner horizon, and then enter the ring singularity outside the equatorial plane (green trajectory). The latter is a "gate" to the region $\bar{\text{I}}$, which is an anti-universe at $r < 0$. Alternatively, the test-particle can cross the event horizon, cross the inner horizon, and then cross again the inner horizon to reach the region IV', which is a white hole, to eventually enter the region I' representing another universe (orange trajectory).

The Penrose diagram for the maximal extension of the Reissner–Nordström space-time is shown in Fig. 2.12. It is very similar to the diagram in Fig. 2.11 for the Kerr solution, with the remarkable difference that the singularity at $r = 0$ is not a ring singularity and therefore the region $\bar{\text{I}}$ and $\overline{\text{III}}$ in Fig. 2.11 do not exist in the Reissner–Nordström spacetime.

References

1. S.V. Babak, L.P. Grishchuk, Int. J. Mod. Phys. D **12**, 1905 (2003), arXiv:gr-qc/0209006
2. L. Baiotti, I. Hawke, P.J. Montero, F. Loffler, L. Rezzolla, N. Stergioulas, J.A. Font, E. Seidel, Phys. Rev. D **71**, 024035 (2005), arXiv:gr-qc/0403029
3. L. Baiotti, L. Rezzolla, Phys. Rev. Lett. **97**, 141101 (2006), arXiv:gr-qc/0608113
4. C. Bambi, D. Malafarina, L. Modesto, Eur. Phys. J. C **74**, 2767 (2014), arXiv:1306.1668 [gr-qc]
5. C. Bambi, D. Malafarina, L. Modesto, JHEP **1604**, 147 (2016), arXiv:1603.09592 [gr-qc]
6. H. Bondi, Mon. Not. Roy. Astron. Soc. **107**, 410 (1947)
7. I. Booth, Can. J. Phys. **83**, 1073 (2005), arXiv:gr-qc/0508107
8. B. Carter, Phys. Rev. Lett. **26**, 331 (1971)
9. S. Chandrasekhar, *The Mathematical Theory of Black Holes* (Clarendon Press, Oxford, 1998)
10. D. Christodoulou, Commun. Math. Phys. **93**, 171 (1984)
11. P.T. Chrusciel, J.L. Costa, M. Heusler, Living Rev. Rel. **15**, 7 (2012), arXiv:1205.6112 [gr-qc]
12. P. Diener, Class. Quant. Grav. **20**, 4901 (2003), arXiv:gr-qc/0305039
13. S. Droz, Phys. Rev. D **55**, 3575 (1997)
14. G. Dvali, C. Gomez, Fortsch. Phys. **61**, 742 (2013), arXiv:1112.3359 [hep-th]
15. G. Dvali, C. Gomez, Phys. Lett. B **719**, 419 (2013), arXiv:1203.6575 [hep-th]
16. R. Emparan, JHEP **0403**, 064 (2004), arXiv:hep-th/0402149
17. R. Emparan, H.S. Reall, Phys. Rev. Lett. **88**, 101101 (2002), arXiv:hep-th/0110260
18. R. Emparan, H.S. Reall, Living Rev. Rel. **11**, 6 (2008), arXiv:0801.3471 [hep-th]
19. F. Fayos, R. Torres, Class. Quant. Grav. **28**, 215023 (2011), arXiv:1204.4651 [gr-qc]
20. V.P. Frolov, G.A. Vilkovisky, Phys. Lett. B **106**, 307 (1981)
21. S.B. Giddings, Phys. Rev. D **90**, 124033 (2014), arXiv:1406.7001 [hep-th]
22. S.W. Hawking, Commun. Math. Phys. **25**, 152 (1972)
23. S.W. Hawking, R. Penrose, Proc. Roy. Soc. Lond. A **314**, 529 (1970)
24. C. Hellaby, K. Lake, Astrophys. J. **290**, 381 (1985)
25. C.A.R. Herdeiro, E. Radu, Phys. Rev. Lett. **112**, 221101 (2014), arXiv:1403.2757 [gr-qc]
26. S. Hossenfelder, L. Smolin, Phys. Rev. D **81**, 064009 (2010), arXiv:0901.3156 [gr-qc]
27. S.A. Hughes, C.R. Keeton, P. Walker, K.T. Walsh, S.L. Shapiro, S.A. Teukolsky, Phys. Rev. D **49**, 4004 (1994)
28. W. Israel, Phys. Rev. **164**, 1776 (1967)
29. T. Jacobson, S. Venkataramani, Class. Quant. Grav. **12**, 1055 (1995), arXiv:gr-qc/9410023
30. T. Johannsen, Phys. Rev. D **87**, 124017 (2013), arXiv:1304.7786 [gr-qc]
31. P.S. Joshi, I.H. Dwivedi, Phys. Rev. D **47**, 5357 (1993), arXiv:gr-qc/9303037
32. P.S. Joshi, D. Malafarina, Int. J. Mod. Phys. D **20**, 2641 (2011), arXiv:1201.3660 [gr-qc]

33. P. Kanti, N.E. Mavromatos, J. Rizos, K. Tamvakis, E. Winstanley, Phys. Rev. D **54**, 5049 (1996), arXiv:hep-th/9511071
34. B. Kleihaus, J. Kunz, E. Radu, Phys. Rev. Lett. **106**, 151104 (2011), arXiv:1101.2868 [gr-qc]
35. L. Kong, D. Malafarina, C. Bambi, Eur. Phys. J. C **74**, 2983 (2014), arXiv:1310.8376 [gr-qc]
36. G. Lemaitre, Annales Soc. Sci. Brux. I A **53**, 51 (1933) [Gen. Rel. Grav. **29**, 641 (1997)]
37. Y. Liu, S.N. Zhang, Phys. Lett. B **679**, 88 (2009), arXiv:0907.2574 [gr-qc]
38. A. Maselli, P. Pani, L. Gualtieri, V. Ferrari, Phys. Rev. D **92**, 083014 (2015), arXiv:1507.00680 [gr-qc]
39. S.D. Mathur, Fortsch. Phys. **53**, 793 (2005), arXiv:hep-th/0502050
40. S. Mignemi, N.R. Stewart, Phys. Rev. D **47**, 5259 (1993), arXiv:hep-th/9212146
41. C.W. Misner, D.H. Sharp, Phys. Rev. **136**, B571 (1964)
42. C.W. Misner, K.S. Thorne, J.A. Wheeler, *Gravitation* (W. H. Freeman and Company, San Francisco, 1973)
43. U. Miyamoto, S. Jhingan and T. Harada, PTEP **2013**, 053E01 (2013), arXiv:1108.0248 [gr-qc]
44. R.C. Myers, M.J. Perry, Ann. Phys. **172**, 304 (1986)
45. R.P.A.C. Newman, Class. Quant. Grav. **3**, 527 (1986)
46. J.R. Oppenheimer, H. Snyder, Phys. Rev. **56**, 455 (1939)
47. N. Ortiz, O. Sarbach, Class. Quant. Grav. **28**, 235001 (2011), arXiv:1106.2504 [gr-qc]
48. P. Pani, C.F.B. Macedo, L.C.B. Crispino, V. Cardoso, Phys. Rev. D **84**, 087501 (2011), arXiv:1109.3996 [gr-qc]
49. R. Penrose, Proc. Roy. Soc. Lond. A **284**, 159 (1965)
50. E. Poisson, *A Relativists Toolkit: The Mathematics of Black-Hole Mechanics* (Cambridge University Press, Cambridge, 2004)
51. E. Poisson, W. Israel, Phys. Rev. D **41**, 1796 (1990)
52. D. Psaltis, D. Perrodin, K. R. Dienes and I. Mocioiu, Phys. Rev. Lett. **100**, 091101 (2008) [Erratum: Phys. Rev. Lett. **100**, 119902 (2008)] arXiv:0710.4564 [astro-ph]
53. D.C. Robinson, Phys. Rev. Lett. **34**, 905 (1975)
54. T.P. Sotiriou, V. Faraoni, Phys. Rev. Lett. **108**, 081103 (2012), arXiv:1109.6324 [gr-qc]
55. J. Thornburg, Living Rev. Rel. **10**, 3 (2007), arXiv:gr-qc/0512169
56. R.C. Tolman, Proc. Nat. Acad. Sci. **20**, 169 (1934) [Gen. Rel. Grav. **29**, 935 (1997)]
57. P.K. Townsend, arXiv:gr-qc/9707012
58. M.S. Volkov, D.V. Galtsov, JETP Lett. **50**, 346 (1989) [Pisma Zh. Eksp. Teor. Fiz. **50**, 312 (1989)]
59. M.S. Volkov, D.V. Gal'tsov, Phys. Rept. **319**, 1 (1999), arXiv:hep-th/9810070
60. R.M. Wald, *General Relativity* (University of Chicago Press, Chicago, 1984)
61. R.M. Wald, V. Iyer, Phys. Rev. D **44**, 3719 (1991)
62. M. Walker, J. Math. Phys. **11**, 2280 (1970)
63. B. Waugh, K. Lake, Phys. Rev. D **38**, 1315 (1988)
64. K. Yagi, N. Yunes and T. Tanaka, Phys. Rev. D **86**, 044037 (2012) [Phys. Rev. D **89**, 049902 (2014)] arXiv:1206.6130 [gr-qc]
65. N. Yunes, F. Pretorius, Phys. Rev. D **79**, 084043 (2009), arXiv:0902.4669 [gr-qc]
66. S.N. Zhang, Int. J. Mod. Phys. D **20**, 1891 (2011), arXiv:1003.1359 [gr-qc]
67. Y. Zhang, Y. Zhu, L. Modesto, C. Bambi, Eur. Phys. J. C **75**, 96 (2015), arXiv:1404.4770 [gr-qc]

Chapter 3
Motion Around Black Holes

The electromagnetic spectrum of an astrophysical black hole results from the radiation emitted by the particles in the accretion disk and in possible jets/outflows. The spacetime metric determines the motion of these particles and the propagation of the radiation from the point of emission to the point of detection in the flat far-away region. The electromagnetic spectrum of a black hole has features related to relativistic effects, like Doppler boosting and gravitational redshift, occurring in the strong gravity region. The study of geodesic motion is the first step to understand these effects and to use the electromagnetic spectrum for testing strong gravity.

The study of circular orbits in the equatorial plane is particularly important. In the Novikov-Thorne model, the accretion disk is in the plane perpendicular to the black hole spin. The particles[1] of the gas move on nearly geodesic circular orbits in the equatorial plane. The inner edge of the disk is at the radius of the innermost stable circular orbit (ISCO). The latter assumption is crucial in current techniques of spin measurements.

3.1 Orbits in the Equatorial Plane

The aim of this section is to study the main properties of equatorial circular orbits around a compact object in a generic stationary, axisymmetric, and asymptotically flat spacetime. These orbits are interesting because the particles of the gas in a thin disk are assumed to follow nearly geodesic circular orbits in the equatorial plane (see Chap. 6).

The geodesic motion of a test-particle in a spacetime with the metric $g_{\mu\nu}$ is governed by the Lagrangian

$$\mathcal{L} = \frac{1}{2} g_{\mu\nu} \dot{x}^\mu \dot{x}^\nu, \tag{3.1}$$

---

[1]In this context, the term "particle" is used to indicate a parcel of gas.

© Springer Nature Singapore Pte Ltd. 2017
C. Bambi, *Black Holes: A Laboratory for Testing Strong Gravity*,
DOI 10.1007/978-981-10-4524-0_3

where $\dot{} = d/d\tau$ and τ is an affine parameter of the geodesic. When we consider time-like geodesics, here and in the rest of the book, τ will be the proper time of the test-particle.

The line element of a generic stationary and axisymmetric spacetime can always be written in the following form[2]

$$ds^2 = g_{tt}dt^2 + 2g_{t\phi}dt d\phi + g_{rr}dr^2 + g_{\theta\theta}d\theta^2 + g_{\phi\phi}d\phi^2, \qquad (3.2)$$

where the metric coefficients are independent of t and ϕ. This is often called the *canonical form* of the line element for a stationary and axisymmetric spacetime and is obtained when the coordinate system is adapted to the two Killing isometries.

Since the metric is independent of the coordinates t and ϕ, we have two constants of motion, namely the specific energy at infinity E and the axial component of the specific angular momentum at infinity L_z:

$$\frac{d}{d\tau}\frac{\partial\mathscr{L}}{\partial\dot{t}} - \frac{\partial\mathscr{L}}{\partial t} = 0 \Rightarrow p_t \equiv \frac{\partial\mathscr{L}}{\partial\dot{t}} = g_{tt}\dot{t} + g_{t\phi}\dot{\phi} = -E, \qquad (3.3)$$

$$\frac{d}{d\tau}\frac{\partial\mathscr{L}}{\partial\dot{\phi}} - \frac{\partial\mathscr{L}}{\partial\phi} = 0 \Rightarrow p_\phi \equiv \frac{\partial\mathscr{L}}{\partial\dot{\phi}} = g_{t\phi}\dot{t} + g_{\phi\phi}\dot{\phi} = L_z. \qquad (3.4)$$

The term "specific" is used to indicate that E and L_z are, respectively, the energy and the angular momentum per unit rest-mass.[3] This is automatic when the affine parameter τ is the proper time. In such a case, the conservation of the rest-mass reads $g_{\mu\nu}\dot{x}^\mu\dot{x}^\nu = -1$. We remind the reader that here we use the convention of a metric with signature $(-++ +)$. For a metric with signature $(+ - - -)$, we would have $p_t = E$ and $p_\phi = -L_z$. Equations (3.3) and (3.4) can be solved to find the t- and the ϕ-component of the 4-velocity of the test-particle

$$\dot{t} = \frac{E g_{\phi\phi} + L_z g_{t\phi}}{g_{t\phi}^2 - g_{tt}g_{\phi\phi}}, \qquad (3.5)$$

$$\dot{\phi} = -\frac{E g_{t\phi} + L_z g_{tt}}{g_{t\phi}^2 - g_{tt}g_{\phi\phi}}. \qquad (3.6)$$

From the conservation of the rest-mass, $g_{\mu\nu}\dot{x}^\mu\dot{x}^\nu = -1$, we can write the equation

$$g_{rr}\dot{r}^2 + g_{\theta\theta}^2\dot{\theta}^2 = V_{\text{eff}}(r, \theta), \qquad (3.7)$$

---

[2]Here and in the rest of the book we will only consider stationary and axisymmetric spacetimes whose line element can be written as in Eq. (3.2). However, this is not "the most general" stationary and axisymmetric spacetime. In principle, even the g_{tr} metric coefficient may be non-vanishing, see Sect. 7.1 in [12] for more details. Nevertheless, black hole solutions in general relativity and in other theories of gravity seem to have $g_{tr} = 0$.

[3]In this and in the next sections about equatorial orbits, we will mainly focus on time-like geodesics, because we have in mind the motion of the gas in a thin accretion disk around a black hole.

where V_{eff} is the effective potential of the test-particle with specific energy E and axial component of the specific angular momentum L_z

$$V_{\text{eff}} = \frac{E^2 g_{\phi\phi} + 2EL_z g_{t\phi} + L_z^2 g_{tt}}{g_{t\phi}^2 - g_{tt} g_{\phi\phi}} - 1. \tag{3.8}$$

Circular orbits in the equatorial plane are located at the zeros and the turning points of the effective potential: $\dot{r} = \dot{\theta} = 0$, which implies $V_{\text{eff}} = 0$, and $\ddot{r} = \ddot{\theta} = 0$, which require, respectively, $\partial_r V_{\text{eff}} = 0$ and $\partial_\theta V_{\text{eff}} = 0$. From these conditions, we could obtain the specific energy E and the axial component of the specific angular momentum L_z of a test-particle in equatorial circular orbits. However, it is faster to proceed in the following way. We write the geodesic equations as

$$\frac{d}{d\tau}\left(g_{\mu\nu}\dot{x}^\nu\right) = \frac{1}{2}\left(\partial_\mu g_{\nu\rho}\right)\dot{x}^\nu\dot{x}^\rho. \tag{3.9}$$

Since $\dot{r} = \dot{\theta} = \ddot{r} = 0$ for equatorial circular orbits, the radial component of Eq. (3.9) reduces to

$$\left(\partial_r g_{tt}\right)\dot{t}^2 + 2\left(\partial_r g_{t\phi}\right)\dot{t}\dot{\phi} + \left(\partial_r g_{\phi\phi}\right)\dot{\phi}^2 = 0. \tag{3.10}$$

The angular velocity $\Omega = \dot{\phi}/\dot{t}$ is

$$\Omega_\pm = \frac{-\partial_r g_{t\phi} \pm \sqrt{\left(\partial_r g_{t\phi}\right)^2 - \left(\partial_r g_{tt}\right)\left(\partial_r g_{\phi\phi}\right)}}{\partial_r g_{\phi\phi}}, \tag{3.11}$$

where the upper (lower) sign refers to corotating (counterrotating) orbits, namely orbits with angular momentum parallel (antiparallel) to the spin of the central object. From $g_{\mu\nu}\dot{x}^\mu\dot{x}^\nu = -1$ with $\dot{r} = \dot{\theta} = 0$, we can write

$$\dot{t} = \frac{1}{\sqrt{-g_{tt} - 2\Omega g_{t\phi} - \Omega^2 g_{\phi\phi}}}. \tag{3.12}$$

Equation (3.3) becomes

$$\begin{aligned} E &= -\left(g_{tt} + \Omega g_{t\phi}\right)\dot{t} \\ &= -\frac{g_{tt} + \Omega g_{t\phi}}{\sqrt{-g_{tt} - 2\Omega g_{t\phi} - \Omega^2 g_{\phi\phi}}}. \end{aligned} \tag{3.13}$$

Equation (3.4) becomes

$$\begin{aligned} L_z &= \left(g_{t\phi} + \Omega g_{\phi\phi}\right)\dot{t} \\ &= \frac{g_{t\phi} + \Omega g_{\phi\phi}}{\sqrt{-g_{tt} - 2\Omega g_{t\phi} - \Omega^2 g_{\phi\phi}}}. \end{aligned} \tag{3.14}$$

As can be seen from Eqs. (3.13) and (3.14), E and L_z diverge when their denominator vanishes. This happens at the radius of the *photon orbit* r_γ

$$g_{tt} + 2\Omega g_{t\phi} + \Omega^2 g_{\phi\phi} = 0 \quad \Rightarrow \quad r = r_\gamma. \tag{3.15}$$

Usually – e.g. in the Kerr metric, but it is quite a common property even for black holes in alternative theories of gravity – there are circular orbits for $r > r_\gamma$ and there are no circular orbits for $r < r_\gamma$. In such a case, the radius of the photon orbit is the equatorial circular orbit with smallest radius for massless and massive particles, and massive particles can reach the photon orbit in the limit of infinite specific energy E. However, in some particular spacetimes, circular orbits with $r < r_\gamma$ may exist, and it is also possible the presence of two or more photon orbits.

The radius of the *marginally bound orbit* r_{mb} is defined by

$$E = -\frac{g_{tt} + \Omega g_{t\phi}}{\sqrt{-g_{tt} - 2\Omega g_{t\phi} - \Omega^2 g_{\phi\phi}}} = 1 \quad \Rightarrow \quad r = r_{mb}. \tag{3.16}$$

The orbit is marginally bound, which means that the test-particle has the sufficient energy to escape to infinity (the test-particle cannot reach infinity if $E < 1$, and can reach infinity with a finite velocity if $E > 1$).

The radius of the *marginally stable orbit* r_{ms}, more often called the ISCO radius r_{ISCO}, is defined by

$$\partial_r^2 V_{eff} = 0 \quad \text{or} \quad \partial_\theta^2 V_{eff} = 0 \quad \Rightarrow \quad r = r_{ISCO}. \tag{3.17}$$

Since equatorial circular orbits at $r < r_{ISCO}$ are unstable, in the Novikov-Thorne model for thin accretion disks the inner edge of the disk is assumed at the ISCO radius.

In the Kerr metric, equatorial circular orbits are always stable for $r > r_{ISCO}$, while they are radially unstable for $r < r_{ISCO}$. In more general spacetimes, the picture may be more complicated. For instance, it is possible that equatorial circular orbits are stable for $r > \tilde{r}$, while there are some stable "islands" at smaller radii, say for $r_1 < r < r_2$, $r_3 < r < r_4$, etc. (and the orbits are unstable for $r < r_1$, $r_2 < r < r_3$, etc.). See e.g. [2]. In this case, it is possible that an accreting compact object has an accretion disk with the inner edge $r_{in} = \tilde{r}$, and then some annuli of accreting material at smaller radii.

To figure out the origin of the ISCO, which has no counterpart in Newtonian mechanics, we can consider the motion around a Schwarzschild black hole. In the case of motion in the equatorial plane ($\theta = \pi/2$ and $\dot{\theta} = 0$), Eq. (3.7) is

$$\frac{1}{2}\dot{r}^2 = \frac{E^2 - 1}{2} - U_{eff}, \tag{3.18}$$

where

$$U_{\text{eff}} = -\frac{M}{r} + \frac{L_z^2}{2r^2} - \frac{L_z^2 M}{r^3}. \tag{3.19}$$

Equation (3.18) can be seen as the equation of motion in Newtonian mechanics for a particle in the effective potential U_{eff}. The first term on the right hand side of Eq. (3.19) corresponds to the usual potential of Newtonian gravity for a spherically symmetric massive body. The second term is the centrifugal potential, which is repulsive and in Newtonian gravity prevents that a test-particle with non-vanishing angular momentum falls to the center. The third term, which gets important at small radii and is attractive, is absent in Newtonian mechanics and is responsible for the existence of the ISCO radius. The correction of the Schwarzschild solution to Newtonian gravity makes the gravitational force stronger than the centrifugal force. Figure 3.1 shows the effective potential around a massive spherically symmetric body in Newtonian gravity and around a Schwarzschild black hole.

While the ISCO is a relativistic concept and has no Newtonian counterpart, this does not mean that the ISCO must always exist. Within the description of a Newtonian motion in an effective potential, we can say that the ISCO exists when at a certain point gravity becomes so strong that stable orbits are not possible any more. However, if the gravitational force is never strong enough, there is no ISCO. This may happen, for instance, in the spacetime around a massive body made of non-interacting particles (where, indeed, there is no event horizon), see e.g. [3, 9]. The gravitational mass within the shell of radius r decreases as r decreases, and therefore it is possible that gravity is never strong enough to have an ISCO.

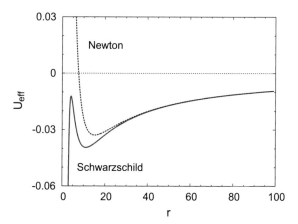

Fig. 3.1 Effective potential $U_{\text{eff}}(r)$ for a test-particle moving in the gravitational field of a Schwarzschild *black hole (red solid curve)* and of a point-like mass in Newtonian gravity (*blue dashed curve*). Here $L_z = 3.9\,M$ and $M = 1$. See the text for more details

3.2 Orbits in the Equatorial Plane in the Kerr Metric

The spacetime around astrophysical black holes is thought to be well approximated by the Kerr solution, and therefore it is useful to revise the results of the previous section in the case of the Kerr metric (2.16). More details can be found in the seminal paper [4]. The conserved specific energy E and the conserved specific axial component of the angular momentum L_z of a test-particle in equatorial circular orbits, Eqs. (3.13) and (3.14), reduce to

$$E = \frac{r^{3/2} - 2Mr^{1/2} \pm aM^{1/2}}{r^{3/4}\sqrt{r^{3/2} - 3Mr^{1/2} \pm 2aM^{1/2}}}, \tag{3.20}$$

$$L_z = \pm \frac{M^{1/2}\left(r^2 \mp 2aM^{1/2}r^{1/2} + a^2\right)}{r^{3/4}\sqrt{r^{3/2} - 3Mr^{1/2} \pm 2aM^{1/2}}}, \tag{3.21}$$

where the angular velocity of the test-particle is now

$$\Omega_\pm = \pm \frac{M^{1/2}}{r^{3/2} \pm aM^{1/2}} \tag{3.22}$$

and, again, the upper sign refers to corotating orbits, the lower sign to counterrotating ones. Figures 3.2 and 3.3 show, respectively, E and L_z as a function of r for different values of the spin parameter of the black hole.

From the vanishing of the denominator in the expression of E and L_z, we get the equation for the radius of the photon orbit

$$r^{3/2} - 3Mr^{1/2} \pm 2aM^{1/2} = 0. \tag{3.23}$$

Fig. 3.2 Specific energy E of a test-particle moving in a Kerr spacetime in a circular equatorial orbit as a function of the radial coordinate r. Every curve corresponds to the case of motion in a spacetime with a *black hole* with a different value of the spin parameter. The values are $a/M = 0, 0.5, 0.8, 0.9, 0.95, 0.99,$ and 0.999

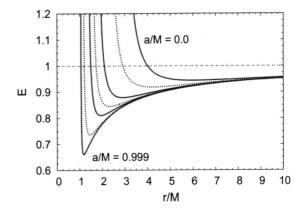

Fig. 3.3 As in Fig. 3.2 for the axial component of the specific angular momentum L_z

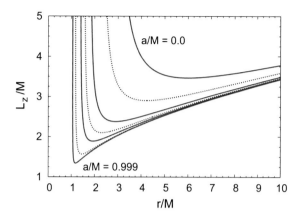

The solution is

$$r_{\rm ph} = 2M \left\{ 1 + \cos \left[\frac{2}{3} \arccos \left(\mp \frac{a}{M} \right) \right] \right\}. \tag{3.24}$$

For $a = 0$, $r_\gamma = 3M$. For $a = M$, $r_\gamma = M$ for corotating orbits and $4M$ for counterrotating orbits.

From the condition $E = 1$, we find the radius of the marginally bound orbit in the Kerr metric

$$r_{\rm mb} = 2M \mp a + 2\sqrt{M\,(M \mp a)}. \tag{3.25}$$

For $a = 0$, $r_{\rm mb} = 4M$. For $a = M$, $r_{\rm mb} = M$ for a corotating orbit and

$$r_{\rm mb} = \left(3 + 2\sqrt{2} \right) M \approx 5.83M \tag{3.26}$$

for counterrotating orbits.

Lastly, the ISCO radius in the Kerr metric is given by

$$r_{\rm ISCO} = 3M + Z_2 \mp \sqrt{(3M - Z_1)(3M + Z_1 + 2Z_2)},$$
$$Z_1 = M + \left(M^2 - a^2 \right)^{1/3} \left[(M + a)^{1/3} + (M - a)^{1/3} \right],$$
$$Z_2 = \sqrt{3a^2 + Z_1^2}. \tag{3.27}$$

For $a = 0$, $r_{\rm ISCO} = 6M$. For $a = M$, $r_{\rm ISCO} = M$ for corotating orbits and $r_{\rm ISCO} = 9M$ for counterrotating orbits. The derivation can be found, for instance, in [6].

It is worth noting that, in the Kerr metric but also in many other black hole metrics in alternative theories of gravity, the ISCO radius is located at the minimum of the energy E. It can thus be obtained also from the equation $dE/dr = 0$. After some manipulation, one finds

$$\frac{dE}{dr} = \frac{r^2 - 6Mr \pm 8aM^{1/2}r^{1/2} - 3a^2}{2r^{7/4}\left(r^{3/2} - 3Mr^{1/2} \pm aM^{1/2}\right)^{3/2}}M, \tag{3.28}$$

and $r^2 - 6Mr \pm 8aM^{1/2}r^{1/2} - 3a^2 = 0$ is the same equation as that we can obtain from Eq. (3.17), see [6]. Moreover, the minimum of the energy E and of the axial component of the angular momentum L_z are located at the same radius. After some manipulation, dL_z/dr is

$$\frac{dL_z}{dr} = \frac{r^2 - 6Mr \pm aM^{1/2}r^{1/2} - 3a^2}{2r^{7/4}\left(r^{3/2} - 3Mr^{1/2} \pm aM^{1/2}\right)^{3/2}}\left(M^{1/2}r^{3/2} \pm aM\right), \tag{3.29}$$

and we see that $dE/dr = 0$ and $dL_z/dr = 0$ have the same solution.

Figure 3.4 shows the radius of the event horizon r_+, of the photon orbit r_γ, of the marginally bound circular orbit $r_{\rm mb}$, and of the ISCO $r_{\rm ISCO}$ in the Kerr metric in Boyer–Lindquist coordinates as functions of a/M.

It is worth noting that, in the case of an extremal Kerr black hole with $a = M$, one finds $r_+ = r_\gamma = r_{\rm mb} = r_{\rm ISCO} = M$ for corotating orbits. However, the Boyer–Lindquist coordinates are not well-defined at the event horizon and these special orbits do not coincide [4]. If we write $a = M - \epsilon$ with $\epsilon \to 0$, we find

$$r_+ = M + \sqrt{2\epsilon} + \cdots, \qquad r_\gamma = M + 2\sqrt{\frac{2\epsilon}{3}} + \cdots,$$

$$r_{\rm mb} = M + 2\sqrt{\epsilon} + \cdots, \qquad r_{\rm ISCO} = M + (4\epsilon)^{1/3} + \cdots. \tag{3.30}$$

Fig. 3.4 Radial coordinates of the event horizon r_+, of the photon orbit r_γ, of the marginally bound circular orbit $r_{\rm mb}$, and of the ISCO $r_{\rm ISCO}$ in the Kerr metric in Boyer–Lindquist coordinates as functions of a/M. For every radius, the upper curve refers to the counterrotating orbit, the lower curve to the corotating orbit

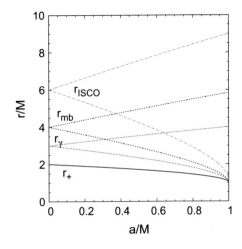

We can then evaluate the proper radial distance between r_+ and the other radii [4]. Sending ϵ to zero, the result is

$$\int_{r_+}^{r_\gamma} \frac{r'\,dr'}{\sqrt{\Delta}} \to \frac{1}{2} M \ln 3,$$

$$\int_{r_+}^{r_{mb}} \frac{r'\,dr'}{\sqrt{\Delta}} \to M \ln\left(1 + \sqrt{2}\right),$$

$$\int_{r_+}^{r_{ISCO}} \frac{r'\,dr'}{\sqrt{\Delta}} \to \frac{1}{6} M \ln\left(\frac{2^7}{\epsilon}\right), \tag{3.31}$$

which clearly shows that these orbits do not coincide even if in Boyer–Lindquist coordinates they have the same value. The energy E and the axial component of the angular momentum L_z are

$$\begin{aligned}
r &= r_\gamma & E &\to \infty, & L_z &\to 2EM, \\
r &= r_{mb} & E &\to 1, & L_z &\to 2M, \\
r &= r_{ISCO} & E &\to \tfrac{1}{\sqrt{3}}, & L_z &\to \tfrac{2}{\sqrt{3}} M.
\end{aligned} \tag{3.32}$$

3.3 Geodesics in the Kerr Metric

The motion of massive and massless test-particles in a background metric is determined by the geodesic equations, which are second order partial differential equations in the coordinates $\{x^\mu\}$

$$\ddot{x}^\mu + \Gamma^\mu_{\rho\sigma} \dot{x}^\rho \dot{x}^\sigma = 0, \tag{3.33}$$

where here the dot ˙ denotes the derivative with respect to some affine parameter of the geodesic and $\Gamma^\mu_{\rho\sigma}$ s are the Christoffel symbols. If the spacetime is stationary and axisymmetric, it is possible to exploit the conservation of the energy and of the axial component of the angular momentum and slightly simplify the equations of motion.

The Kerr metric is a special case, because there are four constants of motion (mass, energy, axial component of the angular momentum, and the so-called Carter constant) instead of three [5]. This implies that, in some coordinate system (for example, in Boyer–Lindquist coordinates), the equations of motion are separable and of first order, which significantly simplifies the calculations. Such a property is not only of the Kerr metric, but of any Petrov type D spacetime. However, in the Kerr metric, and not in any Petrov type D spacetime, these equations can be reduced to elliptic integrals. This section reviews how we can obtain the master equations of motion. More details can be found, for instance, in [6] and in Appendix B.

The Hamilton–Jacobi equation for the geodesic motion of a test-particle is

$$2\frac{\partial S}{\partial \tau} = g^{\mu\nu} \frac{\partial S}{\partial x^\mu} \frac{\partial S}{\partial x^\nu}, \tag{3.34}$$

where S is the Hamilton's principal function. In Boyer–Lindquist coordinates, the metric of the Kerr solution is given by

$$\left(\frac{\partial}{\partial s}\right)^2 = -\frac{A}{\Sigma \Delta} \left(\frac{\partial}{\partial t}\right)^2 - \frac{4aMr}{\Sigma \Delta} \left(\frac{\partial}{\partial t}\right)\left(\frac{\partial}{\partial \phi}\right) + \frac{\Delta}{\Sigma} \left(\frac{\partial}{\partial r}\right)^2 + \frac{1}{\Sigma} \left(\frac{\partial}{\partial \theta}\right)^2$$
$$+ \frac{\Delta - a^2 \sin^2 \theta}{\Sigma \Delta \sin^2 \theta} \left(\frac{\partial}{\partial \phi}\right)^2, \tag{3.35}$$

where $A = \left(r^2 + a^2\right)^2 - a^2 \Delta \sin^2 \theta$. Equation (3.34) becomes

$$2\frac{\partial S}{\partial \tau} = -\frac{A}{\Sigma \Delta} \left(\frac{\partial S}{\partial t}\right)^2 - \frac{4aMr}{\Sigma \Delta} \frac{\partial S}{\partial t} \frac{\partial S}{\partial \phi} + \frac{\Delta}{\Sigma} \left(\frac{\partial S}{\partial r}\right)^2 + \frac{1}{\Sigma} \left(\frac{\partial S}{\partial \theta}\right)^2$$
$$+ \frac{\Delta - a^2 \sin^2 \theta}{\Sigma \Delta \sin^2 \theta} \left(\frac{\partial S}{\partial \phi}\right)^2$$
$$= -\frac{1}{\Sigma \Delta} \left[\left(r^2 + a^2\right) \frac{\partial S}{\partial t} + a\frac{\partial S}{\partial \phi}\right]^2 + \frac{1}{\Sigma \sin^2 \theta} \left[a \sin^2 \theta \frac{\partial S}{\partial t} + \frac{\partial S}{\partial \phi}\right]^2$$
$$+ \frac{\Delta}{\Sigma} \left(\frac{\partial S}{\partial r}\right)^2 + \frac{1}{\Sigma} \left(\frac{\partial S}{\partial \theta}\right)^2. \tag{3.36}$$

At this point, we look for a solution of the Hamilton–Jacobi equation of the form

$$S = \frac{1}{2}\delta \tau - Et + L_z \phi + S_r(r) + S_\theta(\theta). \tag{3.37}$$

where $\delta = -1$ ($\delta = 0$) for time-like (null) geodesics and S_r and S_θ are, respectively, functions of r and θ only. Equation (3.36) becomes

$$\delta \Sigma = \frac{1}{\Delta} \left[\left(r^2 + a^2\right) E - aL_z\right]^2 - \frac{1}{\sin^2 \theta} \left(aE \sin^2 \theta - L_z\right)^2 - \Delta \left(\frac{\partial S}{\partial r}\right)^2 - \left(\frac{\partial S}{\partial \theta}\right)^2$$
$$= \frac{1}{\Delta} \left[\left(r^2 + a^2\right) E - aL_z\right]^2 - \left(\frac{L_z^2}{\sin^2 \theta} - a^2 E^2\right) \cos^2 \theta - (L_z - aE)^2$$
$$- \Delta \left(\frac{\partial S}{\partial r}\right)^2 - \left(\frac{\partial S}{\partial \theta}\right)^2, \tag{3.38}$$

which can be rewritten as

$$\Delta \left(\frac{\partial S}{\partial r}\right)^2 - \frac{1}{\Delta} \left[(r^2 + a^2)\, E - a L_z\right]^2 + (L_z - a E)^2 + \delta r^2$$

$$= -\left(\frac{\partial S}{\partial \theta}\right)^2 - \left(\frac{L_z^2}{\sin^2 \theta} - a^2 E^2 + \delta a^2\right) \cos^2 \theta. \qquad (3.39)$$

In Eq. (3.39), the left hand side depends on r only, while the right hand side depends on θ only. So they must be separately equal to a constant $\mathcal{Q}$, which is the so-called *Carter constant*. We can now write two equations, one for r and one for θ

$$\Delta \left(\frac{\partial S}{\partial r}\right)^2 = \frac{1}{\Delta} \left[(r^2 + a^2)\, E - a L_z\right]^2 - \left[\mathcal{Q} + (L_z - a E)^2 + \delta r^2\right], \quad (3.40)$$

$$\left(\frac{\partial S}{\partial \theta}\right)^2 = \mathcal{Q} - \left(\frac{L_z^2}{\sin^2 \theta} - a^2 E^2 + \delta a^2\right) \cos^2 \theta. \qquad (3.41)$$

The solution for S is thus

$$S = \frac{1}{2}\delta \tau - E t + L_z \phi + \int^r dr' \frac{\sqrt{R(r')}}{\Delta} + \int^\theta d\theta' \sqrt{\Theta(\theta')}, \qquad (3.42)$$

where we have introduced the functions $R(r)$ and $\Theta(\theta)$

$$R(r) = \left[(r^2 + a^2)\, E - a L_z\right]^2 - \Delta \left[\mathcal{Q} + (L_z - a E)^2 + \delta r^2\right], \qquad (3.43)$$

$$\Theta(\theta) = \mathcal{Q} - \left[a^2 (\delta - E^2) + \frac{L_z^2}{\sin^2 \theta}\right] \cos^2 \theta. \qquad (3.44)$$

Since

$$p_\mu = \frac{\partial S}{\partial x^\mu}, \qquad (3.45)$$

and $p_\phi = L_z$, Eq. (3.41) provides the Carter constant $\mathcal{Q}$

$$\mathcal{Q} = p_\theta^2 + p_\phi^2 \cot^2 \theta + a^2 (\delta - E^2) \cos^2 \theta. \qquad (3.46)$$

For $\mathcal{Q} = 0$, the motion is in the equatorial plane at any time. In the Schwarzschild limit $a = 0$, the Carter constant reduces to

$$\mathcal{Q} = \left(p_\theta^2 + \frac{p_\phi^2}{\sin^2 \theta}\right) - p_\phi^2 = L^2 - L_z^2, \qquad (3.47)$$

where $L = p_\theta^2 + p_\phi^2 / \sin^2 \theta$ is the total angular momentum. In the general case with $a \neq 0$, there is not a direct physical interpretation of $\mathcal{Q}$.

The equations of motion can be obtained by setting to zero the partial derivatives of S with respect to the four constants of motion, δ, E, L_z, and $\mathcal{Q}$. From $\partial S/\partial \mathcal{Q} = 0$, we get

$$\int^r \frac{dr'}{\sqrt{R}} = \int^\theta \frac{d\theta'}{\sqrt{\Theta}}. \tag{3.48}$$

From $\partial S/\partial \delta = 0$, $\partial S/\partial E = 0$, and $\partial S/\partial L_z = 0$, we find, respectively,

$$\tau = \int^r dr' \frac{r'^2}{\sqrt{R}} + a^2 \int^\theta d\theta' \frac{\cos^2 \theta'}{\sqrt{\Theta}}, \tag{3.49}$$

$$t = \tau E + 2M \int^r \frac{dr'}{\Delta\sqrt{R}} \left[r'^2 E - a \left(L_z - a E \right) \right] r', \tag{3.50}$$

$$\phi = a \int^r \frac{dr'}{\Delta\sqrt{R}} \left[\left(r'^2 + a^2 \right) E - a L_z \right] + \int^\theta \frac{d\theta'}{\sqrt{\Theta}} \left(\frac{L_z}{\sin^2 \theta'} - a E \right). \tag{3.51}$$

An important property of the Kerr metric (not of any Petrov type D spacetime) is that the integrals in Eq. (3.48) are elliptic integrals (see Appendix B). Such a property is very useful in numerical calculations and is commonly exploited in astrophysical codes.

Since $\dot{x}^\mu = p^\mu = g^{\mu\nu} p_\nu$, Eq. (3.45) provides the expression for the first order equations of motion

$$\Sigma^2 \dot{r}^2 = R, \tag{3.52}$$

$$\Sigma^2 \dot{\theta}^2 = \Theta, \tag{3.53}$$

$$\Sigma \dot{\phi} = \frac{1}{\Delta} \left[2aMrE + (\Sigma - 2Mr) \frac{L_z}{\sin^2 \theta} \right], \tag{3.54}$$

$$\Sigma \dot{t} = \frac{1}{\Delta} \left(AE - 2aMrL_z \right), \tag{3.55}$$

where still $A = \left(r^2 + a^2 \right)^2 - a^2 \Delta \sin^2 \theta$.

3.4   Image Plane of a Distant Observer

In the calculations of the electromagnetic spectrum of black holes, it is usually necessary to compute the image of the disk in the plane of the distant observer or, in other words, to relate the point of the photon emission in the disk to the detection point in the plane of the distant observer. In general, one has to consider a grid in the plane of the distant observer and photons with momentum orthogonal to the plane. The photon trajectories are then integrated backwards in time and one finds the position of the emission point in the disk. In the Kerr metric, there are some simplifications due to the existence of the Carter constant.

3.4.1 General Case

We have a black hole surrounded by an accretion disk and an observer at the distant D from the black hole and with the viewing angle i, as sketched in Fig. 3.5 [8]. The image plane of the distant observer is provided with a system of Cartesian coordinates (X, Y, Z). Another system of Cartesian coordinates (x, y, z) is centered at the black hole. The two Cartesian coordinates are related by

$$
\begin{aligned}
x &= D \sin i - Y \cos i + Z \sin i, \\
y &= X, \\
z &= D \cos i + Y \sin i + Z \cos i.
\end{aligned}
\tag{3.56}
$$

Let us assume the black hole metric is expressed in spherical-like coordinates. Far from the compact object, the spatial coordinates reduce to the usual spherical coordinates in flat spacetime and they are related to (x, y, z) by

$$
\begin{aligned}
r &= \sqrt{x^2 + y^2 + z^2}, \\
\theta &= \arccos\left(\frac{z}{r}\right), \\
\phi &= \arctan\left(\frac{y}{x}\right).
\end{aligned}
\tag{3.57}
$$

Fig. 3.5 The Cartesian coordinates (x, y, z) are centered at the black hole, while the Cartesian coordinates (X, Y, Z) are for the image plane of the distant observer, who is located at the distant D from the *black hole* and with the inclination angle i. From [1]

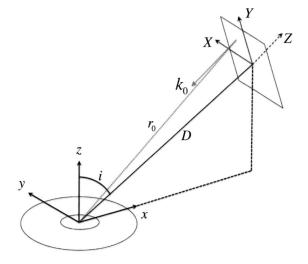

Let us consider a photon at the position $(X_0, Y_0, 0)$ and with 3-momentum $\mathbf{k}_0 = -k_0 \hat{Z}$ perpendicular to the image plane. The initial conditions for the photon position are

$$
\begin{aligned}
t_0 &= 0, \\
r_0 &= \sqrt{X_0^2 + Y_0^2 + D^2}, \\
\theta_0 &= \arccos \frac{Y_0 \sin i + D \cos i}{r_0}, \\
\phi_0 &= \arctan \frac{X_0}{D \sin i - Y_0 \cos i}.
\end{aligned}
\tag{3.58}
$$

For the photon 4-momentum, from $k^\mu = (\partial x^\mu / \partial \tilde{x}^\alpha) \tilde{k}^\alpha$ where $\tilde{k}^\alpha$ is the 4-momentum in Cartesian coordinates, we have

$$
\begin{aligned}
k_0^r &= -\frac{D}{r_0} k_0, \\
k_0^\theta &= \frac{\cos i - (Y_0 \sin i + D \cos i) \frac{D}{r_0^2}}{\sqrt{X_0^2 + (D \sin i - Y_0 \cos i)^2}} k_0, \\
k_0^\phi &= \frac{X_0 \sin i}{X_0^2 + (D \sin i - Y_0 \cos i)^2} k_0,
\end{aligned}
\tag{3.59}
$$

while k_0^t can be obtained from the condition $g_{\mu\nu} k^\mu k^\nu = 0$ with the metric tensor of flat spacetime (the observer is in the flat faraway region), so

$$
k_0^t = \sqrt{\left(k_0^r\right)^2 + r_0^2 \left(k_0^\theta\right)^2 + r_0^2 \sin^2 \theta_0 (k_0^\phi)^2}.
\tag{3.60}
$$

With the photon initial conditions (3.58), (3.59), and (3.60), one can compute the trajectory of any photon with position $(X_0, Y_0, 0)$ in the image plane and compute the transfer function of the accretion disk (see Sect. 6.3.1).

3.4.2 Kerr Spacetime

In the Kerr metric, it is more convenient to follow a different approach, which can be generalized to other spacetimes of Petrov type D.

Let us consider the non-coordinate basis, or orthonormal tetrad and its covariant counterpart, $\{\mathbf{E}_{(\alpha)}\}$ and $\{\mathbf{E}^{(\alpha)}\}$, defined by

$$
\mathbf{E}_{(\alpha)} = E_{(\alpha)}^\mu \partial_\mu, \quad \eta_{(\alpha)(\beta)} = E_{(\alpha)}^\mu g_{\mu\nu} E_{(\beta)}^\nu,
\tag{3.61}
$$

$$
\mathbf{E}^{(a)} = E_\mu^{(a)} dx^\mu, \quad \eta^{(\alpha)(\beta)} = E_\mu^{(\alpha)} g^{\mu\nu} E_\nu^{(\beta)},
\tag{3.62}
$$

where $\det|E^{\mu}_{(\alpha)}| > 0$ to preserve the orientation, $\eta_{(\alpha)(\beta)}$ is the Minkowski metric, and $\eta^{(\alpha)(\beta)}$ is its inverse. The components of a generic vector V^{μ} and of its dual-vector V_{μ} in the non-coordinate basis are

$$V^{(\alpha)} = E^{(\alpha)}_{\mu} V^{\mu}, \quad V_{(\alpha)} = E^{\mu}_{(\alpha)} V_{\mu}. \tag{3.63}$$

More details on non-coordinate basis can be found in textbooks like [10].

Let us consider a stationary, axisymmetric, and asymptotically flat spacetime whose line element can be written in the following form

$$ds^2 = -e^{2\nu} dt^2 + e^{2\mu} dr^2 + e^{2\rho} d\theta^2 + e^{2\sigma} (d\phi - \omega dt)^2, \tag{3.64}$$

where the functions μ, ν, ρ, σ, and ω are independent of t and r. One can identify a special class of observers, the so-called *locally non-rotating observers*, whose world lines are $r = \mathrm{const.}$, $\theta = \mathrm{const.}$, and $\phi = \omega t + \mathrm{const.}$ These observers somewhat rotate with the geometry of the spacetime (see the next section for more details). The non-coordinate basis of these observers are given by [4]

$$E^{\mu}_{(t)} = \left(e^{-\nu}, 0, 0, \omega e^{-\nu}\right), \quad E^{\mu}_{(r)} = \left(0, e^{-\mu}, 0, 0\right),$$
$$E^{\mu}_{(\theta)} = \left(0, 0, e^{-\rho}, 0\right), \quad E^{\mu}_{(\phi)} = \left(0, 0, 0, e^{-\sigma}\right), \tag{3.65}$$

and

$$E^{(t)}_{\mu} = \left(e^{\nu}, 0, 0, 0\right), \quad E^{(r)}_{\mu} = \left(0, e^{\mu}, 0, 0\right),$$
$$E^{(\theta)}_{\mu} = \left(0, 0, e^{\rho}, 0\right), \quad E^{(\phi)}_{\mu} = \left(-\omega e^{\sigma}, 0, 0, e^{\sigma}\right). \tag{3.66}$$

In the Kerr metric, Eqs. (3.65) and (3.66) becomes, respectively,

$$E^{\mu}_{(t)} = \left(\sqrt{\frac{A}{\Sigma\Delta}}, 0, 0, \frac{2Mar}{\sqrt{A\Sigma\Delta}}\right), \quad E^{\mu}_{(r)} = \left(0, \sqrt{\frac{\Delta}{\Sigma}}, 0, 0\right),$$
$$E^{\mu}_{(\theta)} = \left(0, 0, \frac{1}{\sqrt{\Sigma}}, 0\right), \quad E^{\mu}_{(\phi)} = \left(0, 0, 0, \frac{1}{\sin\theta}\sqrt{\frac{\Sigma}{A}}\right), \tag{3.67}$$

and

$$E^{(t)}_{\mu} = \left(\sqrt{\frac{\Sigma\Delta}{A}}, 0, 0, 0\right), \quad E^{(r)}_{\mu} = \left(0, \sqrt{\frac{\Sigma}{\Delta}}, 0, 0\right),$$
$$E^{(\theta)}_{\mu} = \left(0, 0, \sqrt{\Sigma}, 0\right), \quad E^{(\phi)}_{\mu} = \left(-\frac{2Mar\sin\theta}{\sqrt{\Sigma A}}, 0, 0, \sin\theta\sqrt{\frac{A}{\Sigma}}\right). \tag{3.68}$$

The conjugate momenta $p_\mu = \partial \mathscr{L}/\partial \dot{x}^\mu$ are (writing $\dot{x}^\mu$ as p^μ)

$$p_t = -\left(1 - \frac{2Mr}{\Sigma}\right) p^t - \frac{2aMr\sin^2\theta}{\Sigma} p^\phi = -E,$$

$$p_r = \frac{\Sigma}{\Delta} p^r = \frac{\sqrt{R}}{\Delta},$$

$$p_\theta = \Sigma p^\theta = \sqrt{\Theta},$$

$$p_\phi = -\frac{2aMr\sin^2\theta}{\Sigma} p^t + \sin^2\theta \left(r^2 + a^2 + \frac{2a^2 Mr\sin^2\theta}{\Sigma}\right) p^\phi = L_z. \quad (3.69)$$

The tetrad components can be quickly obtained by using $E_\mu^{(\alpha)}$ or $E_{(\alpha)}^\mu$

$$p^{(t)} = E_\mu^{(t)} p^\mu = -E_{(t)}^\mu p_\mu,$$

$$p^{(r)} = E_\mu^{(r)} p^\mu = E_{(r)}^\mu p_\mu,$$

$$p^{(\theta)} = E_\mu^{(\theta)} p^\mu = E_{(\theta)}^\mu p_\mu,$$

$$p^{(\phi)} = E_\mu^{(\phi)} p^\mu = E_{(\phi)}^\mu p_\mu. \quad (3.70)$$

Now the position of a photon in the image plane of the distant observer is given by [6]

$$X_0 = \lim_{r \to \infty} \left(\frac{r p^{(\phi)}}{p^{(t)}}\right) = \frac{\lambda}{\sin i}, \quad (3.71)$$

$$Y_0 = \lim_{r \to \infty} \left(\frac{r p^{(\theta)}}{p^{(t)}}\right) = \pm\sqrt{q^2 + a^2\cos^2 i - \lambda^2\cot^2 i}, \quad (3.72)$$

where i is again the viewing angle of the distant observer and λ and q are defined by

$$\lambda = \frac{L_z}{E}, \quad q^2 = \frac{\mathscr{Q}}{E^2}, \quad (3.73)$$

and are constants of motion along the photon path. In other words, the photon at the position $(X_0, Y_0, 0)$ is characterized by the constants of motion λ and q, and it is then possible to integrate the photon trajectories to relate the photon position in the image plane to the point of emission in the accretion disk (this is discussed in Sect. 6.3.2).

3.5   Frame Dragging

In the framework of Newtonian gravity, only the mass of a body is responsible for the gravitational force. On the contrary, its angular momentum has no gravitational effects. In general relativity, even the angular momentum alters the geometry of the spacetime. *Frame dragging* refers to the capability of a spinning massive body of "dragging" the spacetime.

In Eq. (3.4), L_z is the specific angular momentum of a test-particle as measured at infinity. However, even if $L_z = 0$, the angular velocity of a test-particle $\Omega = \dot{\phi}/\dot{t}$ may be non-vanishing if $g_{t\phi} \neq 0$. In other words, the spinning body generating the gravitational field forces the test-particle to orbit with a non-vanishing angular velocity. If $L_z = 0$, from Eq. (3.4) we find

$$\Omega = -\frac{g_{t\phi}}{g_{\phi\phi}}. \tag{3.74}$$

In the case of the Kerr metric, we have

$$\Omega = \frac{2Mar}{\left(r^2 + a^2\right)\Sigma + 2Ma^2 r \sin^2\theta}. \tag{3.75}$$

The *angular velocity of the event horizon* is the angular velocity at the event horizon of a test-particle with vanishing angular momentum at infinity:

$$\Omega_{\mathrm{H}} = -\left(\frac{g_{t\phi}}{g_{\phi\phi}}\right)_{r=r_+}, \tag{3.76}$$

where r_+ is the radius of the event horizon. In the case of the Kerr metric, $\Omega_{\mathrm{H}} = a_*/(2r_+)$.

The phenomenon of frame dragging is particularly strong in the *ergoregion*, which is the exterior region of the spacetime in which $g_{tt} > 0$. In the Schwarzschild spacetime, there is no ergoregion. In the Kerr spacetime, the ergoregion is between the event horizon and the *static limit*; that is

$$r_+ < r < r_{\mathrm{sl}}, \tag{3.77}$$

where r_{sl} the radius of the static limit and in Boyer–Lindquist coordinates is

$$r_{\mathrm{sl}} = M + \sqrt{M^2 - a^2 \cos^2\theta}. \tag{3.78}$$

The static limit is simply the surface $g_{tt} = 0$, while $g_{tt} < 0 (> 0)$ at $r > r_{\mathrm{sl}} (< r_{\mathrm{sl}})$. Ergoregions may exist even inside spinning compact bodies like neutron stars, so it is not a concept related to black holes only. However, spacetimes with an ergoregion and without a horizon are usually unstable (even if the timescale of the instability might be long enough that we cannot exclude the possibility that ergoregions exist inside compact stars) [7, 11].

In the ergoregion, frame dragging is so strong that static test-particles are not possible, namely test-particles with constant spatial coordinates. Everything must rotate. This can be easily seen by the line element of a static test-particle ($dr = d\theta = d\phi = 0$)

$$ds^2 = g_{tt}dt^2. \tag{3.79}$$

Outside the ergoregion, $g_{tt} < 0$, and therefore a static test-particle follows a time-like geodesic. Inside the ergoregion, $g_{tt} > 0$, and a static test-particle would correspond to a space-like trajectory, which is not allowed.

References

1. C. Bambi, A. Cardenas-Avendano, T. Dauser, J.A. Garcia, S. Nampalliwar, arXiv:1607.00596 [gr-qc]
2. C. Bambi, G. Lukes-Gerakopoulos, Phys. Rev. D **87**, 083006 (2013), arXiv:1302.0565 [gr-qc]
3. C. Bambi, D. Malafarina, Phys. Rev. D **88**, 064022 (2013), arXiv:1307.2106 [gr-qc]
4. J.M. Bardeen, W.H. Press, S.A. Teukolsky, Astrophys. J. **178**, 347 (1972)
5. B. Carter, Phys. Rev. **174**, 1559 (1968)
6. S. Chandrasekhar, *The Mathematical Theory of Black Holes* (Clarendon Press, Oxford, 1998)
7. J.L. Friedman, Commun. Math. Phys. **62**, 247 (1978)
8. T. Johannsen, D. Psaltis, Astrophys. J. **718**, 446 (2010), arXiv:1005.1931 [astro-ph.HE]
9. C.F.B. Macedo, P. Pani, V. Cardoso, L.C.B. Crispino, Astrophys. J. **774**, 48 (2013), arXiv:1302.2646 [gr-qc]
10. M. Nakahara, *Geometry, Topology, and Physics* (Institute of Physics Publishing, Bristol, 1990)
11. B.F. Schutz, N. Comins, Mon. Not. R. Astron. Soc. **182**, 69 (1978)
12. R.M. Wald, *General Relativity* (University of Chicago Press, Chicago, 1984)

Chapter 4
Astrophysical Black Holes

In general relativity, black holes can form from the complete gravitational collapse of massive bodies. When a star exhausts all its nuclear fuel, it shrinks to find a new equilibrium configuration. In the case of ordinary stars, the degenerate pressure of electrons or of neutrons can eventually stop the collapse and the final product is, respectively, a white dwarf or a neutron star [94]. For very heavy stars, there is no mechanism to stop the collapse and the final product is thought to be a black hole. Black holes may have been formed even in the early Universe from the collapse of primordial inhomogeneities or during phase transitions.

Astrophysical observations have discovered at least two classes of black holes in the Universe. Stellar-mass black holes can have a mass from about $3\ M_\odot$ and up to $\sim 100\ M_\odot$. The known stellar-mass black holes in X-ray binaries[1] have masses in the range 5–20 $M_\odot$, while heavier objects, up to $\sim 60\ M_\odot$, have been observed by gravitational wave detectors. Supermassive black holes are at the center of galaxies and have a mass $M \sim 10^5$–$10^{10}\ M_\odot$. Both object classes can be naturally interpreted as black holes. As it has been already pointed out, it is fundamentally impossible to prove the existence of an event horizon with a real observation, so some authors prefer to adopt a more conservative attitude and call all these objects black hole candidates. Stellar-mass black holes have a mass exceeding $3\ M_\odot$, which is the maximum mass for a relativistic star. Supermassive black holes are too heavy, too compact, and too old to be clusters of neutron stars. The non-detection of any thermal component from the surface of these objects is also consistent with the idea that they have an event

---

[1]X-ray binaries are binary systems of a compact object (black hole or neutron star) and a stellar companion. The X-ray radiation is mainly generated by gas falling from the companion star (donor) to the compact object (accretor). The gas releases energy as it falls into the gravitational potential of the compact object and X-rays are emitted from the inner part of the accretion disk (see Sect. 4.5 for more details).

© Springer Nature Singapore Pte Ltd. 2017
C. Bambi, *Black Holes: A Laboratory for Testing Strong Gravity*,
DOI 10.1007/978-981-10-4524-0_4

horizon. The only possible explanation in the framework of conventional physics is that these objects are the black holes of general relativity, and they could be something else only in the presence of new physics.

Today there is an increasing number of observational data suggesting also the existence of intermediate-mass black holes, objects with a mass $M \sim 10^2$–$10^4 \, M_\odot$ filling the gap between the stellar-mass and the supermassive ones. However, their nature is not clear, even because there are no dynamical measurements of their masses. It is possible that current intermediate-mass black hole candidates are actually a heterogeneous class of objects, including real intermediate-mass black holes and also other sources.

4.1 Stellar-Mass Black Holes

From stellar evolution arguments, we expect that in our Galaxy there are about 10^8–10^9 black holes formed from the collapse of very massive stars [97, 101], and a similar number can be expected in any typical galaxy. However, the identification of these remnants is not easy. In the end, only in quite exceptional cases we can discover a stellar-mass black hole. This can happen, for instance, when the object is in a binary and we can observe the X-ray radiation from its accretion disk. Another way is to observe the gravitational waves produced by the coalescence of two black holes (or of a black hole and a neutron star, even if there has not been yet a similar observation). There are also attempts to exploit other approaches, for instance microlensing events to detect isolated stellar-mass black holes [3].

A *stellar-mass black hole* is a compact object with a mass exceeding the maximum mass for a neutron star [88]. Assuming general relativity and reasonable equations of state for matter above nuclear densities, the maximum mass of a relativistic star cannot be more than $3 \, M_\odot$ [48, 54, 91]. If a compact object exceeds this bound, it cannot be interpreted as a relativistic star made of neutrons, mesons, or quarks: there is no known mechanism capable of balancing its own weight and preventing a complete collapse when the nuclear reactions are off. A similar object is thus classified as a black hole. We have also a body of observational evidence that is consistent with the fact that these objects have some kind of horizon (see Sect. 4.4). The observed gravitational waves from the coalesce of stellar-mass black holes well match with those expected from black holes in general relativity.

Today we have 24 "confirmed" stellar-mass black holes[2] in X-ray binaries (*black hole binaries*), where the term confirmed is sometimes used to indicate that these objects have a dynamical measurement of their mass and that the latter exceeds $3 \, M_\odot$. We know also about 40 "unconfirmed" stellar-mass black holes. The latter have no dynamical measurements of their masses, but they present features that

---

[2]This number does not include the black holes observed with gravitational waves, but the latter could also be considered "confirmed" stellar-mass black holes because we have robust measurements of their masses from the gravitational wave signal.

can be associated to black holes. However, it is likely that at least some of the unconfirmed black holes are actually neutron stars. Some authors call black holes the confirmed black holes, and black hole candidates the unconfirmed ones. The current stellar-mass black holes in X-ray binaries are mainly in our Galaxy and, in a minor number, in nearby galaxies. Their number is clearly much lower than the number of black hole remnants in our Galaxy from stellar evolution considerations. 22 binaries with a confirmed stellar-mass black hole are sketched in Fig. 4.1. The size of these systems can be better understood by the comparison with the Sun and the distance Sun-Mercury, in the top left corner of Fig. 4.1.

Black hole binaries can be grouped into two classes: low-mass X-ray binaries and high-mass X-ray binaries. *Low-mass X-ray binaries* (LMXBs) are systems in which the stellar companion is not more than a few Solar masses ($\lesssim 3\ M_\odot$). The mass transfer from the donor star to the black hole occurs by Roche lobe overflow [53, 62]. These systems are typically *transient X-ray sources*, because the mass transfer is not continuous. For instance, a similar X-ray binary may be bright for a period ranging from some days to a few months and then be in a quiescent state for months or even decades. During the quiescent state, the X-ray luminosity is typically below 10^{32} erg/s. This may allow the optical detection of the stellar companion and the measurement of the mass of the compact object. Stellar evolution models suggest the existence of a population of 10^3–10^4 stellar-mass black holes in LMXBs in the Galaxy [51, 108]. Today we know 18 confirmed stellar-mass black holes in LMXBs, and about 40 more objects are not dynamically confirmed. New black hole candidates in LMXBs are regularly discovered every year, at a rate of about 2 objects/yr, when they pass from a quiescent state to an outburst [15].

The second class of systems is represented by *high-mass X-ray binaries* (HMXBs). Here the stellar companion is massive ($\gtrsim 10\ M_\odot$) and the mass transfer from the companion star to the black hole is usually due to the wind[3] of the former, but even incipient Roche lobe overflow might be possible. These binaries are *persistent X-ray sources*, with typical X-ray luminosities around 10^{37} erg/s. Among the 24 confirmed black holes, 6 are in HMXBs (Cygnus X-1, LMC X-1, LMC X-3, M33 X-7, NGC 300-1, IC 10 X-1). Only Cygnus X-1 is in our Galaxy.

In this classification, GRS1915+105 is quite a special object. It is a LMXB, because the stellar companion has a mass $\sim 0.5\ M_\odot$. However, it is a bright X-ray source since 1992. This may be explained with its large accretion disk (see Fig. 4.1), which can thus provide enough accretion material at any time.

Gravitational wave detectors can observe the gravitational radiation emitted by a binary system of two stellar-mass black holes (*binary black holes*) at the last stage of their coalescence. In September 2015, the two LIGO antennas detected the coalescence of two black holes with a mass, respectively, of $36 \pm 5\ M_\odot$ and $29 \pm 4\ M_\odot$ [1]. The merger produced a black hole with a mass $62 \pm 4\ M_\odot$, while $3.0 \pm 0.5\ M_\odot$ was radiated in gravitational waves. The event was called GW150914

---

[3]Flows of gas ejected from the upper atmosphere of a star are quite common. In the case of HMXBs, the wind can be driven by radiation pressure on the resonance absorption lines of heavy elements [21].

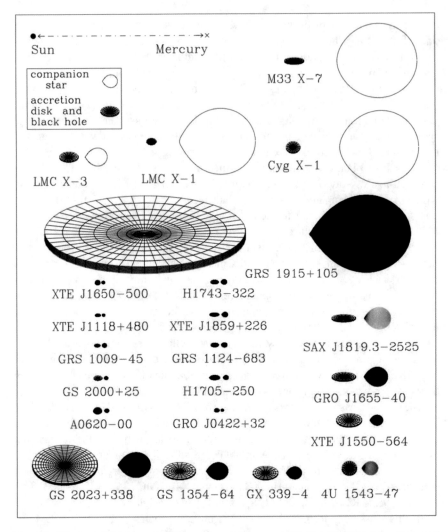

Fig. 4.1 Sketches of 22 X-ray binaries with a stellar-mass black hole confirmed by dynamical measurements. For every system, the black hole accretion disk is on the *left* and the companion star is on the *right*. The orientation of the disks indicates the inclination angles i of the binaries. The distorted shapes of the stellar companions is due to the gravitational fields of the black holes. The size of the latter should be about 50 km, to be compared with the distance Sun-Mercury of about 50 millions km and the radius of the Sun of 0.7 millions km (*top left corner*). Figure courtesy of Jerome Orosz

(because detected on 14 September 2015). The masses of the initial and the final black holes are larger than the masses of the known stellar-mass black holes in X-ray binaries, but it is clear that heavier black holes produce a stronger gravitational wave signal and therefore are easier to detect. On 26 December 2015, the LIGO experiment detected the gravitational waves from the coalescence of another system. The event, called GW151226, was generated by the coalescence of two black holes with a mass, respectively, of about 14 and 7 $M_\odot$ [2].

4.1.1 Dynamical Mass Measurements

Relatively robust mass measurements can be obtained through dynamical methods, by studying the orbital motion of the stellar companion, typically with optical observations. The system can be studied within Newtonian mechanics, because the companion star is relatively far from the black hole.[4]

The key-quantity is the mass function (reintroducing Newton's gravitational constant G_N)

$$f(M) = \frac{K_c^3 P_{orb}}{2\pi G_N} = \frac{M \sin^3 i}{(1+q)^2}, \qquad (4.1)$$

where $K_c = v_c \sin i$ is the maximum line-of-sight Doppler velocity of the companion star, v_c is the velocity of the companion star, i is the inclination angle of the orbital plane with respect to the line of sight of the observer, P_{orb} is the orbital period, M is the mass of the compact object, and $q = M_c/M$ where M_c is the mass of the companion star. Equation (4.1) is valid for circular orbits, which is usually a reasonable assumption considering the short circularization timescale and the age of the X-ray binary [106].

If we can get an independent estimate of i and q and we can measure K_c and P_{orb}, from Eq. (4.1) it is possible to determine the mass of the compact object M. In general relativity, the maximum mass for a compact star made of neutrons, mesons, or quarks for plausible matter equations of state is about 3 $M_\odot$ [48, 54, 91]. If M turns out to be higher than this limit, the object is classified as a black hole because it cannot be explained otherwise in the framework of conventional physics. In alternative theories of gravity, the 3 $M_\odot$ bound may somewhat change, but usually not too much [4, 25]. Of course, it is important to be sure that the object is compact and is not a star. An ordinary star should be observed in the spectrum of the binary. Moreover, the short timescale variability can confirm that the dark object is compact.

The estimate of the inclination angle i is crucial and its uncertainty usually dominates that of the mass measurement. The inclination angle is often obtained by modeling optical/near infrared light curves. Especially in LMXBs, light curves are

---

[4]Systems in which the companion star is very close to the black hole cannot exist, because the strong gravitational field around the compact object would disrupt an ordinary star.

characterized by a specific modulation due to the tidal distortion of the companion star and the amplitude of the modulation depends on the inclination angle i [72]. However, estimates from different groups of the same binary sometimes do not agree because of systematic effects not fully under control. In the case of GRS 1915+105, the estimate of the inclination angle comes from the orientation of the radio jet, assuming that the latter is orthogonal to the plane of the binary [31]. Upper bounds on i are sometimes obtained from the absence of the detection of X-ray eclipses.

The mass ratio q can be obtained from a number of different approaches. The most reliable technique is currently the measurement of the broadening of some photospheric lines from the companion star [103]. Other approaches are based, for instance, on fitting stellar atmosphere models [10] or stellar evolutionary models [109].

Examples of measurements of light curves and of radial velocities are shown in Figs. 4.2 and 4.3. In Fig. 4.2, we have the optical light curves in the U, B, and V bands (top panels) and the radial velocity (bottom panels) of Cygnus X-1. In the left panels, the fitting model has an eccentric orbit with eccentricity $e = 0.018$. In the right panels, the fitting model assumes that the orbit is circular. Figure 4.3 shows the X-ray light curve (top panel) and the radial velocity (bottom panel) of M33 X-7. The low count rate in the top panel in Fig. 4.3 is due to the occultation by the companion star of the accretion disk around the black hole. M33 X-7 is currently the only known black hole binary presenting X-ray eclipses.

Table 4.1 shows some selected measurements of the orbital parameters of 17 LMXBs and of 4 HMXBs with a dynamically confirmed stellar-mass black hole. The mass estimates of the black holes are typically higher than 5 $M_\odot$. In some cases, the same mass function $f(M)$ exceeds 5 $M_\odot$, which means the mass of the compact object is heavier than 5 $M_\odot$ independently of the estimate of the mass of the companion star M_c and the viewing angle of the orbit i, since $M > f(M)$.

Accurate mass measurements are important for a number of reasons. First, it is the crucial estimate to classify a compact object as a black hole or not. Accurate mass measurements are necessary in the continuum-fitting method to measure the black hole spin parameter (see Chap. 7). More in general, accurate mass measurements are required to test black hole formation models. Present data show an unexpected absence, or at least a low number, of objects in the interval 2–5 $M_\odot$, namely between neutron stars and black holes [82]. Numerical simulations of black hole formation typically do not predict a similar feature in the mass distribution of remnants from massive stars. The absence of high mass neutron stars may be attributed to the supernova explosion model (for instance, it could be explained within a delayed supernova explosion scenario), while the absence of low mass black holes may be due to systematic effects (for instance, incorrect estimates of the inclination angles of the orbits).

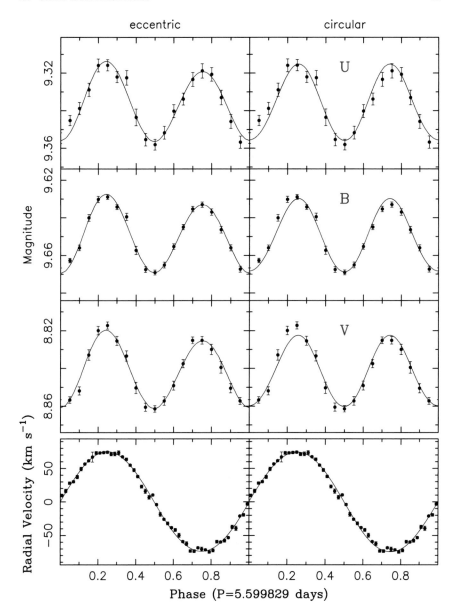

Fig. 4.2 Black hole binary Cygnus X-1. *Top panels* optical light curves in the U, B, and V bands. *Bottom panels* radial velocity. The solid lines are the best-fitting models assuming an eccentric orbit with eccentricity $e = 0.018$ (*left panels*) and a circular orbit (*right panels*). From [73]. © AAS. Reproduced with permission

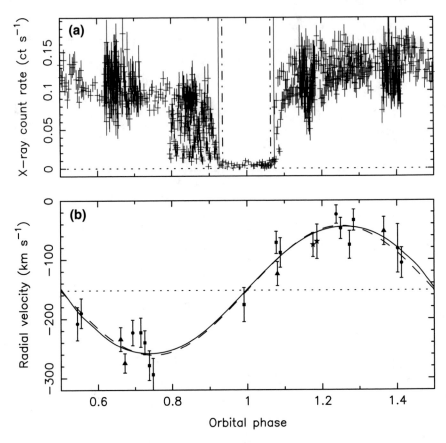

Fig. 4.3 Black hole binary M33 X-7. *Top panel* the Chandra ACIS light curve in the 0.5–5 keV energy band. The low count rate at the *center* is the X-ray eclipse. *Bottom panel* radial velocity curve. The *solid line* corresponds to the best-fitting model, the *dashed line* is the best-fitting sinusoid. The radial velocity shows a nearly sinusoidal variation. From [80]. Reprinted by permission from Macmillan Publishers Ltd

4.2 Supermassive Black Holes

Supermassive black holes have a mass in the range $M \sim 10^5$–10^{10} $M_\odot$ [52]. They are harbored at the center of galaxies, and actually it is thought that every normal galaxy has a supermassive black hole at its center, namely every galaxy that is not too small.[5] Small galaxies usually have no supermassive object in their nucleus.

As in the case of stellar-mass black holes, there is no direct observational evidence that these objects are really black holes. They are supposed to be supermassive black holes because, at least in some cases, they are too massive, too compact, and too old

---

[5]Exceptions might be possible: the galaxy A2261-BCG has a very large mass but it might not have any supermassive black hole at its center [87].

Table 4.1 Summary of the mass measurements of 17 confirmed black holes in LMXBs (top) and 4 confirmed black holes in HMXBs (bottom). Table adapted from [15]

BH Binary	P_{orb} [days]	$f(M)$ [$M_\odot$]	q	i [deg]	M [$M_\odot$]	Principal references
GRS 1915+105	33.85 (16)	7.02 ± 0.17	0.042 ± 0.024	66 ± 2	10.1 ± 0.6	[96]
GS 2023+338	6.47129 (7)	6.07 ± 0.05	0.067 ± 0.005	67^{+3}_{-1}	$9.0^{+0.2}_{-0.6}$	[13, 14, 16, 49]
SAX J1819.3-2525	2.81730 (1)	2.74 ± 0.12	0.63–0.70	60–71	8.7–11.7	[78]
GRO J1655-40	2.62168 (14)	2.73 ± 0.09	0.42 ± 0.03	68.7 ± 1.5	6.6 ± 0.5	[5, 92, 93, 102]
GS 1354-64	2.54451 (8)	5.73 ± 0.29	$0.12^{+0.03}_{-0.04}$	<79	>7.0	[19, 20]
GX 339-4	1.7557 (4)	5.8 ± 0.5	–	–	>6.0	[46, 63]
XTE J1550-564	0.5420333 (24)	7.65 ± 0.38	~ 0.03	74.7 ± 3.8	7.8–15.6	[77, 79]
4U 1543-475	1.123 (8)	0.22 ± 0.02	–	20–40	2.7–7.5	[75]
H1705-250	0.5213 (13)	4.86 ± 0.13	≤ 0.053	60–80	4.9–7.9	[34, 44, 90]
GS 1124-684	0.4326058 (31)	3.01 ± 0.15	0.13 ± 0.04	60^{+5}_{-4}	5.0–7.5	[18, 74, 89]
GS 2000+250	0.3440873 (2)	5.01 ± 0.15	0.042 ± 0.012	58–74	5.5–8.8	[17, 33, 43, 47]
A0620-00	0.32301405 (1)	2.76 ± 0.01	0.067 ± 0.010	51 ± 0.9	6.6 ± 0.3	[12, 41, 57]
XTE J1650-500	0.3205 (7)	2.73 ± 0.56	–	>47	–	[76]
GRS 1009-45	0.285206 (14)	3.17 ± 0.12	–	–	–	[35]
XTE J1859+226	0.274 (2)	4.5 ± 0.6	–	<70	>5.42	[24, 32]
GRO J0422+32	0.2121600 (2)	1.19 ± 0.02	$0.11^{+0.05}_{-0.02}$	45 ± 2	3.97 ± 0.95	[38, 105]
XTE J1118+480	0.1699339 (2)	6.27 ± 0.04	0.024 ± 0.009	68–79	6.9–8.2	[11, 42, 50, 58, 98, 104]
Cyg X-1	5.599829 (16)	0.244 ± 0.006	1.29 ± 0.15	27.1 ± 0.8	14.8 ± 1.0	[73]
LMC X-1	3.90917 (5)	0.149 ± 0.007	4.91 ± 0.53	36.4 ± 1.9	10.9 ± 1.4	[81]
M33 X-7	3.453014 (20)	0.46 ± 0.08	4.47 ± 0.60	74.6 ± 1.0	15.7 ± 1.5	[80]
LMC X-3	1.7048089 (11)	2.77 ± 0.04	–	50–70	9.5–13.6	[95, 100]

to be a cluster of non-luminous bodies, like white dwarfs, neutron stars, or stellar-mass black holes [56]. Within standard physics, the only explanation is that they are supermassive black holes.

Such a conclusion follows from the consideration that a similar cluster would have a lifetime due to evaporation and/or physical collisions of its constituents much shorter than 10 Gyr, making its observation in the Universe at the present time highly unlikely [56]. Assuming favorable conditions for a long cluster lifetime (the lowest possible concentration to minimize collisions, cluster of equal-mass objects to reduce the velocity of the cluster evolution, etc.), one employs some model for the cluster structure and finds the timescale for evaporation and physical collisions.

Evaporation is an inevitable process due to gravitational scattering and makes some bodies escape from the system. Following [56], in the case of a Plummer model for the cluster, the evaporation timescale is

$$t_{evap}(m_0) \approx \frac{4.3 \cdot 10^4 \; x}{\ln (0.8 \; x)} \left(\frac{10^8 \; M_\odot/\text{pc}^3}{\rho_h} \right)^{1/2} \text{yr} , \qquad (4.2)$$

where $x = M_h/m_0$, M_h is the half-mass of the cluster, and m_0 is the mass of the non-luminous bodies the cluster is made of. ρ_h is the half-mass density, namely the mean density within the cluster half-radius R_h (the radius of the spherical surface within which there is the half-mass of the cluster).

The timescale for a body of the cluster to collide with another body is, still assuming a Plummer model, [56]

$$t_{coll}(m_0, r_0) \approx \left[23.8 \; G_N^{1/2} M_h^{1/3} \rho_h^{7/6} \frac{r_0^2}{m_0} \left(1 + \frac{m_0}{2^{1/2} \; M_h^{2/3} \; \rho_h^{1/3} \; r_0} \right) \right]^{-1} \text{s}, \quad (4.3)$$

where r_0 is the radius of the non-luminous bodies and it has been reintroduced Newton's constant G_N.

Considering a cluster of half-mass M_h and half-mass density ρ_h, its lifetime depends on the mass and the radius of its constituents and cannot exceed the shorter timescale between t_{evap} and t_{coll}; that is $\tau(m_0, r_0) < \min(t_{evap}, t_{coll})$. For a certain combination of M_h and ρ_h, the maximum cluster lifetime is

$$\tau_{max}(M_h, \rho_h) = \max [\tau(m_0, r_0)] . \qquad (4.4)$$

Figure 4.4 shows the contour levels of τ_{max} on the plane (M_h, ρ_h) according to the analysis in [56], which considers the possibilities that the cluster is made of brown dwarfs, white dwarfs, neutron stars, and stellar-mass black holes. If the dark mass at the center of the Galaxy and of NGC 4258 were a cluster of these non-luminous bodies, the cluster lifetime would be much shorter than 10 Gyr. This would make its observation today definitely unlikely. If these dark masses cannot be clusters of non-luminous bodies, the simplest interpretation is that they are supermassive black holes.

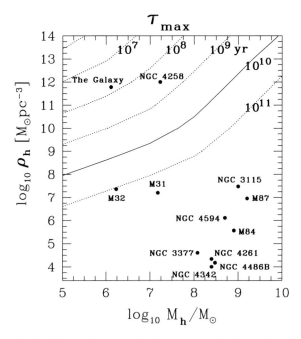

Fig. 4.4 Contour levels of the maximum possible lifetime τ_{max} of a cluster of non-luminous objects with half-mass M_h and half-mass density ρ_h. The processes considered for the cluster lifetime are those of evaporation and physical collisions of its constituents. The *black dots* represent the location of specific supermassive black holes in this plane. If we demand that the possible cluster of non-luminous bodies is older than 10 Gyr, only the observations from the supermassive black holes in our Galaxy and in NGC 4258 can rule out the dark cluster scenario. From [56]. © AAS. Reproduced with permission

Today, the most stringent dynamical constraints on the existence of a supermassive black hole and against the dark cluster scenario are obtained from the observational data of SgrA*, the supermassive object at the center of our Galaxy [39]. From the study of the Newtonian motion of individual stars, we can infer that the mass of the compact object is about $4 \cdot 10^6 \ M_\odot$. An upper bound on the size of the black hole can be obtained from the minimum distance approached by one of these stars, which is less than 45 AU and corresponds to $\sim$1,200 M for a $4 \cdot 10^6 \ M_\odot$ object. Figure 4.5 shows the astrometric positions and orbital fits for seven stars orbiting the supermassive black hole at the center of the Galaxy.

4.3 Intermediate-Mass Black Holes

The initial mass of a stellar-mass black hole should depend on the properties of the progenitor star: on its mass, on its evolution, and on the supernova explosion mechanism [6]. A crucial quantity is the metallicity of the star. For a low-metallicity star, the maximum mass of the black hole remnant should be around 100 $M_\odot$, because

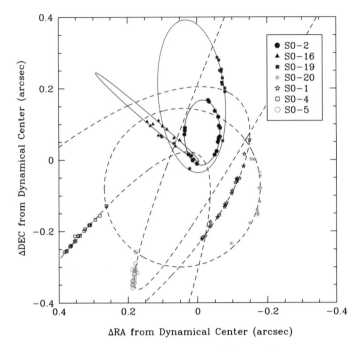

Fig. 4.5 Astrometric positions and orbital fits for seven stars orbiting the supermassive black hole at the *center* of the Galaxy. From [39]. © AAS. Reproduced with permission

stars with $M > 130\,M_\odot$ are supposed to undergo a runaway thermonuclear explosion that completely destroys the system, without leaving any black hole remnant. All the known stellar-mass black holes in X-ray binaries have $M < 20\,M_\odot$ (see Table 4.1). In the event GW150914 detected by LIGO, each of the two black holes in the binary system had a mass $M \approx 30\,M_\odot$ and merged to form a black hole with $M \approx 60\,M_\odot$ [1].

In the case of the supermassive black holes at the center of galaxies, their actual formation mechanism is not clear. However, observations show that their mass ranges from about 10^5 to about $10^{10}\,M_\odot$ [40].

The term intermediate-mass black holes refers to black hole candidates with a mass $M \sim 10^2$–$10^4\,M_\odot$ [23]. They may be the remnants of primordial massive stars of zero-metallicity, the so-called Pop III stars [55]. Intermediate-mass black hole candidates may form at the center of stellar clusters by merger [61, 86]. They may be some unevolved supermassive black holes that, for some reason, did not grow enough [71].

Observational evidence of this class of objects is still controversial, because there are no dynamical measurements of their masses. For instance, intermediate-mass black hole candidates may be associated to some ultraluminous X-ray sources. The latter are X-ray sources in nearby galaxies with an X-ray-luminosity $L_X > 10^{39}$–10^{42} erg/s [22]. If they radiate isotropically, their luminosity would exceed the

Eddington limit for a stellar-mass object.[6] At the same time, they do not look like underluminous supermassive black holes (for instance, they are not at the center of their galaxy and there may be more than one object in a single galaxy). The nature of these objects is not clear. They may be intermediate-mass black holes with mass in the range 10^2–10^4 $M_\odot$. Otherwise, they may be stellar-mass black holes in which the emission is not isotropic and/or their accretion luminosity may exceed the Eddington limit. For instance, numerical simulations show that it is possible to exceed the Eddington luminosity by an order of magnitude in the case of non-isotropic emission and a moderate super-Eddington accretion mass rate [70].

Since the formation of intermediate-mass black holes in dense stellar clusters seems to be quite a likely possibility, several studies have explored the presence of these objects in globular clusters. This can be done by studying the kinematics of the stars in the cluster. Measurements of the radial velocity profile and of the velocity dispersion profile are compared with theoretical models. The presence of an intermediate-mass black hole at the center of the cluster would increase the velocity dispersion in the cluster core. Some studies support the conclusion that at least some globular clusters have an intermediate-mass black hole at the center [36, 37], but there is not yet a common consensus on this.

4.4 Existence of Event Horizons

As discussed in Chap. 2, a back hole is a region of the spacetime causally disconnected to future null infinity. If an observer enters the black hole, it cannot communicate with an observer in the exterior region any more. The fact that this must be "forever" is crucial and clearly implies that no human experiment can really prove the existence of a black hole or of an event horizon, because our observations can only last for a finite time. The existence of an apparent horizon is instead more accessible and fits better with what one would expect to observe from astrophysical black holes. In the special case of a stationary spacetime, the event and the apparent horizons coincide for an observer at infinity, which also means that they cannot be observationally distinguished if the observational timescale is much shorter than that of the evolution of the system.

Bearing in mind it is fundamentally impossible to prove the existence of a black hole with real data, it is anyway very interesting that a number of observations are consistent with the fact that astrophysical black holes have an event horizon and there are no observations that are inconsistent with this hypothesis. The next subsections briefly review these observations. The key-point is that the accreting gas can release radiation if the compact object has a surface, or can at least heat the compact object increasing its temperature. The accretion energy can instead be lost in the presence of an event horizon. The comparison between observations of neutron stars and black

---

[6]The Eddington limit is the maximum luminosity for an object and is reached when the pressure of the radiation luminosity balance the gravitational force. See Appendix G for more details.

holes makes these arguments stronger, because in the neutron star case we observe what we expect for an object with a surface, while if the source is a black hole we observe what we would expect in the present of an event horizon.

4.4.1 Type I X-Ray Bursts

In the case of a neutron star, the gas of accretion can be accumulated on the neutron star surface and eventually develop a thermonuclear instability. This causes a thermonuclear explosion called type I X-ray burst. The phenomenon seems to be well understood and theoretical models well agree with observations. These bursts are observed from binaries in which the compact object is a neutron star, but they have been never observed from binaries in which the compact object is supposed to be a black hole [99]. It seems thus that both neutron stars and black holes behave as we would expect, and the absence of bursts from sources with a black hole can be easily explained with the fact that the gas cannot accumulate on the surface of the compact object. The gas instead crosses the event horizon and then cannot emit radiation to the exterior region any more [66, 107].

4.4.2 X-Ray Binaries in Quiescent State

Most X-ray binaries are transient sources, and they spend a long period in a quiescent state with low mass accretion rate and luminosity. Even supermassive black holes may be observed in quiescent state, and the best example is SgrA*. One may argue that the luminosity of black holes in quiescent state is too low with respect to their mass accretion rate [68]. The interpretation is that the gas crosses the event horizon and is lost, while in the presence of a surface it should stop on the surface and released radiation. However, this depends on the theoretical model and the presence of outflows makes the argument weaker.

The argument gets instead stronger when we compare neutron stars and black holes in quiescent state, because the accretion scenario should be roughly the same and independent of the actual nature of the central object.

It turns out that black holes can be extremely underluminous in comparison with neutron stars. Figure 4.6 shows the luminosity of X-ray transients in which the compact object is supposed to be either a black hole (filled circles) or a neutron star (open circles) as a function of the orbital period P_{orb} [67], since the mass accretion rate should be proportional to P_{orb}. The luminosity of neutron stars is a factor 100 higher than the luminosity of black holes, which becomes a factor 1,000 when measured in Eddington units (black holes have a mass about ten times larger). Once again, the interpretation is that the thermal energy locked in the gas can be completely lost when the gas crosses the event horizon, while in the neutron star case the gas hits the surface of the compact object and releases energy [67].

Fig. 4.6 Eddington scaled
luminosity in the energy
range 0.5–10 keV of X-ray
binaries with black holes
(*filled circles*) and neutron
stars (*open circles*) in
quiescent state. The *diagonal
hatched areas* highlight the
fact that binaries with black
holes and neutron stars
occupied two different
regions in the plane, and the
former are about three orders
of magnitude fainter than the
latter. From [67]

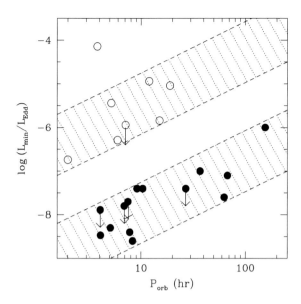

Last, it is quite interesting that, during the quiescent period, neutron star binaries have a thermal blackbody-like component in the X-ray band, which is interpreted as the emission from the neutron star surface. No similar component is found in binaries with a black hole [59]. The non-detection of thermal component may be interpreted as evidence for the absence of a surface and it is consistent with the existence of a horizon.

4.4.3 SgrA*

The strongest constraint on the possible emission of radiation from the surface of black holes is probably reported in [9] (see also [8]). The object under consideration is SgrA*. The basic idea is, again, that in the presence of a surface the energy of the accreting gas must ultimately be radiated, while in the case of a black hole it can be lost after the gas crosses the event horizon.

The argument is based on three assumptions: (i) SgrA* is powered by accretion, (ii) it has already reached a steady state, and (iii) it should emit as a blackbody if it has a surface instead of an event horizon. Relaxing one of these assumptions requires new physics. However, this is not impossible: for instance, the blackbody assumption is clearly violated in scenarios in which black holes are actually Bose-Einstein condensates of gravitons [26, 27]. A similar condensate does not emit as a blackbody because its constituents do not follow the Boltzmann statistics.

Let L_{tot} be the total accretion luminosity, namely the gravitational binding energy released by a particle falling onto the possible surface of SgrA*. L_{tot} is in part converted into radiation, $L_{\text{obs}} = \eta_r L_{\text{tot}}$ where η_r is the radiative efficiency, and in part into the kinetic energy of outflows, $L_k = \eta_k L_{\text{tot}}$ where η_k is the outflow efficiency. If there is a horizon,

$$L_{\text{tot}} - L_{\text{obs}} - L_k \tag{4.5}$$

is advected and cannot be observed any more. If there is a surface, such an energy should be radiated (it cannot heat the object and increase its temperature because we are assuming a steady state). This radiation should be in the form of a blackbody spectrum, so the luminosity observed at infinity can be written as

$$L_{\text{surf}} = 4\pi \sigma R_a^2 T^4 , \tag{4.6}$$

where σ is the Stefan-Boltzmann constant, R_a is the apparent radius of SgrA*, and T the temperature of its putative surface. Since no thermal component is observed in the spectrum of SgrA*, it is possible to put an upper bound on $L_{\text{surf}}/L_{\text{tot}}$ or, equivalently, a lower bound on $\eta_r + \eta_k$. The constraint is shown in Fig. 4.7: the region above the black line is excluded.

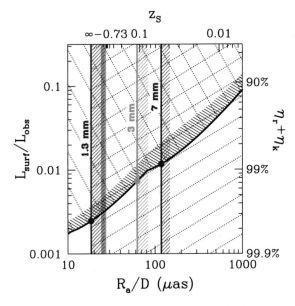

Fig. 4.7 Constraints from SgrA* on the ratio between the surface luminosity and the electromagnetic accretion luminosity as measured at infinity, $L_{\text{surf}}/L_{\text{obs}}$, as a function of the photosphere radius R_a as seen at infinity. $D \approx 8\,\text{kpc}$ is the distance of the source. The constraints from IR measurements exclude the region of the plane above the *black line*. The *red, green,* and *blue horizontal lines* are the constraints on R_a/D from VLBI observations, respectively at 1.3, 3, and 7 mm wavelengths. The region on the right of each line is excluded by the corresponding VLBI observation. In the end, the allowed region is the small corner in the *bottom left* part of the plane. From [9]. © AAS. Reproduced with permission

At the same time, mm-very long baseline interferometric (VLBI) observations can constrain the size of SgrA*. These observations can image the accretion flow around SgrA* and put an upper bound on R_a/D, where D is the distance of SgrA* from us. Figure 4.7 shows the measurements of the size of SgrA* at 1.3, 3, and 7 mm and the region on the right of the corresponding line is excluded. Actually the observation at 1.3 mm is smaller than the apparent diameter of the horizon and is usually interpreted as an orbiting accretion flow. However, it can still provide the correct length scale of the system.

Combining the constraint from the non-detection of any thermal component and the upper bound on the apparent size of SgrA*, the allowed region is the small left bottom corner in Fig. 4.7. Since we do not observe any radiation from the surface of SgrA*, the gravitational energy should be released in the form of radiation and kinetic energy of outflows. Observations eventually require that $\eta_r + \eta_k > 0.996$. However η_r is estimated to be 10^{-4}–10^{-2} and we do not observe any powerful outflow. The interpretation is that such a gravitational binding energy is indeed lost when the gas crosses the event horizon of SgrA*.

4.5 Spectral States

Roughly speaking, the electromagnetic spectrum of accreting black holes has four main components, as sketched in Fig. 4.8. In the radio/IR band, we may observe the synchrotron radiation emitted by particles accelerated in jets. It is typically a flat

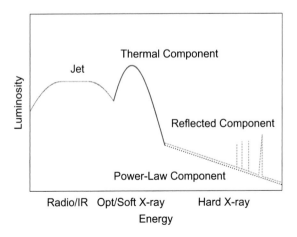

Fig. 4.8 Sketch of the electromagnetic spectrum of a black hole. The thermal component of the accretion disk is in the optical/UV bands for supermassive black holes and in the soft X-ray band in the case of black hole binaries. The hard X-ray spectrum is dominated by the direct radiation from a hot corona (power-law component) and the reflection spectrum of the disk (reflected component). The radio/IR part of the spectrum is due to jets. See Fig. 4.11 and the text for more details

power-law ($S \propto E^\gamma$ with $\gamma \approx 0$) with a break at low frequencies and one at high frequencies. The thermal component of the accretion disk is in the soft X-ray band ($\sim 1\,$keV) for stellar-mass black holes and in the optical/UV bands (~ 1–$10\,$eV) for the supermassive ones. This is because the temperature of the disk roughly scales as $M^{-0.25}$. The hard X-ray spectrum can be usually approximated by a power-law ($S \propto E^{-\Gamma}$ with $\Gamma \approx 1$–3). It is attributed to a corona, a hotter electron cloud ($T \sim 100\,$keV) which enshrouds the central disk and acts as an X-ray source as the result of inverse Compton scattering of the thermal photons from the accretion disk off the electrons in the corona. The reflected component is produced by the illumination of the disk by the corona and presents some emission lines.

The contribution from each component may change from source to source, from different observations of the same source, and some components may not be observed. Black holes can thus be found in different *spectral states*. The transition between different states was discovered in the early 1970s with Cygnus X-1, and then studied in more details with stellar-mass black holes in transient X-ray sources. This section provides a basic review on the topic, without entering the details and without discussing the fast variability, which is also very important for the classification of the spectral states but is beyond the scope of this introductory review. It should be noted that the spectral state classification is still a work in progress, some spectral states and their physical interpretation are not yet well understood, and different authors may use a different nomenclature. More details on the topic can be found, for instance, in [7, 45] and references therein.

4.5.1   Observations

Both transient and persistent X-ray sources exhibit different spectral states. Transient sources may stay in a quiescent state with a very low luminosity for several months or even decades and then have an outburst. The latter typically lasts from some days to a few months. However, exceptions are possible, and an example is GRS1915+105, which started its current outburst in 1992.

During an outburst, the spectrum of the source changes. Black hole transients can be conveniently studied in the *hardness-intensity diagram* (HID), see Fig. 4.9. The x-axis is for the hardness of the source, which is the ratio between its luminosity in the hard and in the soft X-ray bands, for instance between the luminosity in the 6–10 and 2–6 keV bands, but other choices are also common. The y-axis is for the X-ray luminosity, for instance in the 2–10 keV band, but even in this case other options are possible, for instance the count number of the instrument. The HID diagram depends on the source (e.g. the interstellar absorption) and on the instrument (e.g. its effective area at different energies), but, despite that, it turns out to be extremely useful to study transient sources.

The prototype of an outburst in a transient source can be described as follows. The source is initially in the *quiescent state*. In the case of A0620-00, we observed an outburst in 1975 and then the source has always been in a quiescent state. Other

Fig. 4.9 Sketch of the hardness-intensity diagram (HID) of a transient X-ray source with a stellar-mass black hole, bearing in mind that there may be differences between different sources or even for the same source but during different outbursts. See the text for more details

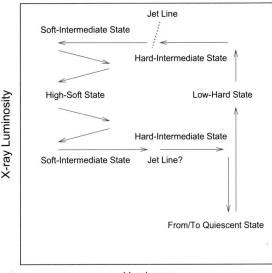

sources, like GX 339-4, are much more active and we have observed several outbursts. The quiescent state is characterized by a very low accretion luminosity, say

$$L_{\text{acc}}^{\text{quiescent}} < 10^{33} \, \text{erg/s} \, ,$$
$$< 10^{-6} \, L_{\text{Edd}} \, , \tag{4.7}$$

where L_{Edd} is the Eddington luminosity of the source. The spectrum is usually hard (but the source may be too faint to detect its spectrum).

At the beginning of the outburst, the hardness of the source is high. The source becomes brighter and brighter, essentially maintaining the same hardness. The initial rise may be so fast in some sources, say a few days, that it may not be observed by instruments. The source is in the so-called *low-hard state*: the flux in the 1–10 keV range is relatively low, and the spectrum is hard with a power-law index around 1.7 (i.e. $S \propto E^{-\Gamma}$ with $\Gamma \approx 1.7$). However, despite its name, at the end of the state the photon flux may not be low any more. The thermal component is subdominant, and the temperature of the inner part of the disk may be low, around 0.1 keV or even lower, but it increases as the luminosity of the source rises. During the low-hard state, compact mildly relativistic steady jets are common.

The source then moves to lower values of hardness. We have first the *hard-intermediate state*, where the power-law index is around 2.4 and the thermal disk component becomes more important, and then the *soft-intermediate state*. These states may be identifies by studying the fast variability of the source. It is worth noting that there exists a *jet line*, not well understood for the moment, in the HID [30]: when the source crosses the jet line, we observe transient jets. The hardness of the source may oscillate from lower to higher values near the jet line, with the result that we can observe several transient jets. Unlike the steady jet of the low-hard state, these jets are typically highly relativistic.

The source then enters the *high-soft state*, which is a rather stable state. The name is to indicate that the flux in the 1–10 keV band is high and the hardness is low. The thermal spectrum of the disk is the dominant component in the spectrum, and it is indeed in the soft X-ray band. The temperature of the inner part of the disk is around 1 keV. In this spectral state, we do not observe any kind of jet. However, strong winds and outflows are common (while they have never been observed in the low-hard state). The luminosity of the source may somewhat decreases and changes hardness, remaining on the left side of the HID, as shown in Fig. 4.9.

At a certain point, the hardness of the source increases. The source re-enters the soft-intermediate state, the hard-intermediate state, then the low-hard state, and eventually, when the hardness is high, the luminosity drops down and the source returns to the quiescent state till the next outburst. Between the soft-intermediate and the hard-intermediate states, we may observe transient jets, but the existence of a jet line is not clear here.

The low-hard and the high-soft states are the two "historical" states. They were the first to be identified and their name was coined to distinguish the case of a low flux and hard spectrum from that of a high flux and soft spectrum. However, as it can be seen from Fig. 4.9, when the source is in the low-hard state it does not necessarily have a flux lower than when it is in the high-soft state.

From the shape of the path of a source on the HID, this diagram is called "q-diagram" or "turtle-head diagram". Any source follows this path counter-clockwise. Figure 4.10 shows the HID for low (left panels) and high (right panels) inclination sources. The gray area in the low hardness region is only reached by sources with a low inclination angle, as a result of inclination dependent relativistic effects on accretion disks [64]. This is one more feature that makes the HID source-dependent.

In the case of persistent sources, the picture is slightly different. The most studied source is Cygnus X-1 (the other persistent sources are in nearby galaxies, so they are fainter and more difficult to study). This object spends most of the time in the low-hard state, but it occasionally moves to a softer state, which is usually interpreted as a high-soft state. LMC X-1 is always in the high-soft state. LMC X-3 is usually observed in the high-soft state, rarely in the low-hard state, and there is no clear evidence that this source can be in an intermediate state.

GRS1915+105 is definitively a special source. The system is a LMXBs, but the source is not transient. Since 1992, it has always been bright without a period of quiescence. The reason may be the large accretion disk around the black hole (see Fig. 4.1), so that there is enough material for the accretion process at any time. This source has never been observed in the typical low-hard state with power-law component with photon index $\Gamma \approx 1.7$ and, on the contrary, the thermal component is always present (again, probably because of the large reservoir of material in the accretion disk).

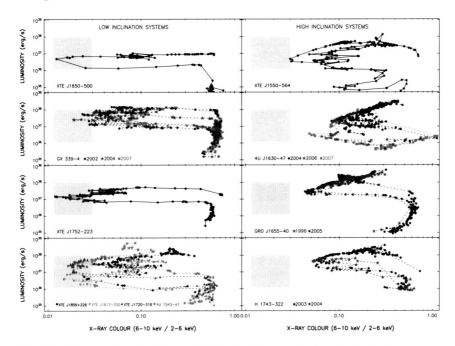

Fig. 4.10 Hardness-intensity diagram (HID) of stellar-mass black holes in transient sources observed from *low* (*left panels*) and *high* (*right panels*) viewing angles. The gray area of very soft spectra is only reached by sources observed from a low inclination angle, because of inclination dependent relativistic effects on accretion disks. Figure 4.1 from [64], reproduced by permission of Oxford University Press

In the case of supermassive black holes, there are at least two important differences. First, the fundamental scale of a black hole is the gravitational radius $r_g = M$; that is, the length and time scales of most processes are proportional to M. 1 day for a stellar-mass black hole corresponds to 3,000 years for a supermassive black hole of $10^7 M_\odot$, which makes impossible to study the evolution of a specific system. Second, the temperature of the disk scales as $T \sim M^{-0.25}$ (see Sect. 6.1). The thermal component of the disk is in the soft X-ray band for a stellar-mass black hole, but in the optical/UV band for a supermassive one. Despite these two issues, it seems that the two object classes behave in a similar way and it is possible to use for supermassive black holes the same spectral states as black hole binaries.

It is worth noting that neutron stars in LMXBs exhibit a similar behavior to that of X-ray transients with a black hole, and one can obtain similar HIDs. This should not be a surprise, because the qualitative behavior of these systems is determined by the accretion process and the presence of a compact object, regardless of its exact nature.

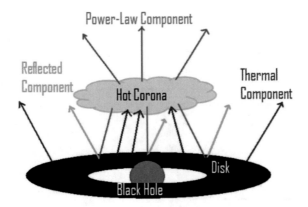

Fig. 4.11 Corona-disk model. The black hole is surrounded by a cold accretion disk, which radiates like a blackbody locally and as a multi-color blackbody when integrated radially. The corona is a hot (∼100 keV) electron cloud acting as an X-ray source, due to inverse Compton scattering of the thermal photons from the accretion disk off the electrons in the corona. Within the lamppost geometry, the corona is compact and located just above the black hole. The corona illuminates also the disk, producing a reflection component with some emission lines. See the corresponding spectrum in Fig. 4.8

4.5.2   Theoretical Models

Figure 4.11 shows an accreting black hole in the framework of the disk-corona model. The compact object is surrounded by a thin accretion disk, which has the inner edge at some radius r_{in}. If the corona is compact and just above the black hole, the set-up is called the lamppost geometry. In such a scenario, the corona could be the base of the jet. As shown in Fig. 4.11, the source has a thermal component from the accretion disk. Through inverse Compton scattering of the thermal photons from the accretion disk off the electrons in the corona, the latter acts as an X-ray source and produces a power-law component in the spectrum of the object. The corona illuminates also the accretion disk, producing a reflection component with some emission lines.

The geometry of the accretion flow changes in different spectral states. Figure 4.12 shows the proposal of [28], in which the key-parameter is the mass accretion rate. There is now a body of observational evidence that this model is too simple and that it is not only the mass accretion rate that determines the spectral state of a source. The HID and the study of the fast variability clearly show that a classification based on the sole mass accretion rate is not adequate. However, the model points out some important features.

In the quiescent state, the accretion rate is low. The inner edge of the disk r_{in} is "truncated", namely it is at some radius larger than the one of the ISCO. The source can exhibit jets (the two black triangles above and below the black dot representing the black hole). Near the black hole, the accretion process is described by models like the advection dominated accretion flow (ADAF) [65], in which the gas is hot

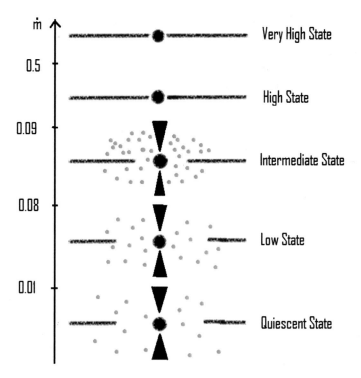

Fig. 4.12 Sketch of the geometry of the accretion flow in different spectral states as a function of the mass accretion rate (in Eddington units) according to the proposal of [28]. This picture seems today too simple to explain observations. See the text for more details

because it cannot efficiently radiate. The accretion luminosity is thus very low and most of the energy is swallowed by the central object.

In the low-hard state, the accretion rate probably ranges from a few percent to something like 70% of the Eddington limit. At low accretion rates, the disk is truncated at some radius larger than the ISCO one, but the inner edge approaches the compact object as the accretion rate/luminosity increases [85]. The temperature of the inner edge of the disk increases accordingly, because the disk is closer to the black hole and the accretion rate is higher (this follows from the properties of thin disks, which will be discussed in Chap. 6). There are observations in the hard state in which the inner edge of the disk may be at the ISCO or at least not far from it [29, 84]. The truncation of the accretion disk is still a controversial and complicated issue. When the disk is truncated, it may coexists with a hot, optically thin and geometrically thick, accretion flow extending from the inner edge of the disk to the ISCO. Even in the low-hard state we observe steady jets.

In the intermediate states, the source present some features common to the low-hard and the high-soft states. The inner edge of the disk might be at the ISCO. Jets are now transient and highly relativistic, so the mechanism responsible for their formation may be different from that of the steady jets in the low-hard and quiescent states. For the moment, however, we do not know the exact origin of both steady and transient jets.

In the high-soft state, the accretion disk is probably well described by the Novikov-Thorne model [69, 83] (see Sect. 6.1 for more details), at least when the accretion luminosity is between 5 and 30% of the Eddington limit [60]. The inner edge is thought to be at the ISCO radius. We normally do not observe any jet.

The very high state of Fig. 4.12 may be reached by some sources in some outbursts. In the HID diagram, it may occurs when the luminosity is very high and the source is switching from the hard to the soft state. It is not used any more as name for a spectral state.

References

1. B.P. Abbott et al., LIGO scientific and virgo collaborations. Phys. Rev. Lett. **116**, 061102 (2016), arXiv:1602.03837 [gr-qc]
2. B.P. Abbott et al., LIGO scientific and virgo collaborations. Phys. Rev. Lett. **116**, 241103 (2016), arXiv:1606.04855 [gr-qc]
3. E. Agol, M. Kamionkowski, L.V.E. Koopmans, R.D. Blandford, Astrophys. J. **576**, L131 (2002), arXiv:astro-ph/0203257
4. A.V. Astashenok, S. Capozziello, S.D. Odintsov, Phys. Rev. D **89**, 103509 (2014), arXiv:1401.4546 [gr-qc]
5. M.E. Beer, P. Podsiadlowski, Mon. Not. R. Astron. Soc. **331**, 351 (2002), arXiv:astro-ph/0109136
6. K. Belczynski, T. Bulik, C.L. Fryer, A. Ruiter, J.S. Vink, J.R. Hurley, Astrophys. J. **714**, 1217 (2010), arXiv:0904.2784 [astro-ph.SR]
7. T.M. Belloni, Lect. Notes Phys. **794**, 53 (2010), arXiv:0909.2474 [astro-ph.HE]
8. A.E. Broderick, R. Narayan, Class. Quant. Grav. **24**, 659 (2007), arXiv:gr-qc/0701154
9. A.E. Broderick, A. Loeb, R. Narayan, Astrophys. J. **701**, 1357 (2009), arXiv:0903.1105 [astro-ph.HE]
10. S.M. Caballero-Nieves et al., Astrophys. J. **701**, 1895 (2009), arXiv:0907.2422 [astro-ph.SR]
11. D.E. Calvelo, S.D. Vrtilek, D. Steeghs, M.A.P. Torres, J. Neilsen, A.V. Filippenko, J.I.G. Hernandez, Mon. Not. R. Astron. Soc. **399**, 539 (2009), arXiv:0905.1491 [astro-ph.SR]
12. A.G. Cantrell et al., Astrophys. J. **710**, 1127 (2010), arXiv:1001.0261 [astro-ph.HE]
13. J. Casares, IAU Colloq. **158**, 395 (1996)
14. J. Casares, P.A. Charles, Mon. Not. R. Astron. Soc. **271**, L5 (1994)
15. J. Casares, P.G. Jonker, Space Sci. Rev. **183**, 223 (2014), arXiv:1311.5118 [astro-ph.HE]
16. J. Casares, P.A. Charles, T. Naylor, Nature **355**, 614 (1992)
17. J. Casares, P.A. Charles, T.R. Marsh, Mon. Not. R. Astron. Soc. **277**, L45 (1995)
18. J. Casares, E.L. Martin, P.A. Charles, P. Molaro, R. Rebolo, New Astron. **1**, 299 (1997)
19. J. Casares, C. Zurita, T. Shahbaz, P.A. Charles, R.P. Fender, Astrophys. J. **613**, L133 (2004), arXiv:astro-ph/0408331
20. J. Casares et al., Astrophys. J. Suppl. **181**, 238 (2009)
21. J.I. Castor, D.C. Abbott, R.I. Klein, Astrophys. J. **195**, 157 (1975)
22. E.J.M. Colbert, R.F. Mushotzky, Astrophys. J. **519**, 89 (1999), arXiv:astro-ph/9901023
23. M. Coleman Miller, E.J.M. Colbert, Int. J. Mod. Phys. D **13**, 1 (2004), arXiv:astro-ph/0308402
24. J.M. Corral-Santana, J. Casares, T. Shahbaz, C. Zurita, I.G. Martinez-Pais, P. Rodriguez-Gil, Mon. Not. R. Astron. Soc. **413**, 15 (2011), arXiv:1102.0654 [astro-ph.SR]

25. D.D. Doneva, S.S. Yazadjiev, N. Stergioulas, K.D. Kokkotas, Phys. Rev. D **88**, 084060 (2013), arXiv:1309.0605 [gr-qc]
26. G. Dvali, C. Gomez, Fortsch. Phys. **61**, 742 (2013), arXiv:1112.3359 [hep-th]
27. G. Dvali, C. Gomez, Phys. Lett. B **719**, 419 (2013), arXiv:1203.6575 [hep-th]
28. A.A. Esin, J.E. McClintock, R. Narayan, Astrophys. J. **489**, 865 (1997), arXiv:astro-ph/9705237
29. A.C. Fabian, A. Lohfink, E. Kara, M.L. Parker, R. Vasudevan, C.S. Reynolds, Mon. Not. R. Astron. Soc. **451**, 4375 (2015), arXiv:1505.07603 [astro-ph.HE]
30. R. Fender, T. Belloni, Ann. Rev. Astron. Astrophys. **42**, 317 (2004), arXiv:astro-ph/0406483
31. R.P. Fender, S.T. Garrington, D.J. McKay, T.W.B. Muxlow, G.G. Pooley, R.E. Spencer, A.M. Stirling, E.B. Waltman, Mon. Not. R. Astron. Soc. **304**, 865 (1999), arXiv:astro-ph/9812150
32. A.V. Filippenko, R. Chornock, IAU Colloq. **7644**, 1 (2001)
33. A.V. Filippenko, T. Matheson, A.J. Barth, Astrophys. J. **455**, L139 (1995)
34. A.V. Filippenko, T. Matheson, D.C. Leonard, A.J. Barth, S.D. Van Dyk, Publ. Astron. Soc. Pac. **109**, 461 (1997)
35. A.V. Filippenko, D.C. Leonard, T. Matheson, W. Li, E.C. Moran, A.G. Riess, Publ. Astron. Soc. Pac. **111**, 969 (1999), arXiv:astro-ph/9904271
36. K. Gebhardt, R.M. Rich, L.C. Ho, Astrophys. J. **578**, L41 (2002), arXiv:astro-ph/0209313
37. K. Gebhardt, R.M. Rich, L.C. Ho, Astrophys. J. **634**, 1093 (2005), arXiv:astro-ph/0508251
38. D.M. Gelino, T.E. Harrison, Astrophys. J. **599**, 1254 (2003), arXiv:astro-ph/0308490
39. A.M. Ghez, S. Salim, S.D. Hornstein, A. Tanner, M. Morris, E.E. Becklin, G. Duchene, Astrophys. J. **620**, 744 (2005), arXiv:astro-ph/0306130
40. J.E. Greene, L.C. Ho, Astrophys. J. **670**, 92 (2007), arXiv:0707.2617 [astro-ph]
41. J.I.G. Hernandez, J. Casares, Astron. Astrophys. **516**, A58 (2010), arXiv:1004.0435 [astro-ph.GA]
42. J.I.G. Hernandez, R. Rebolo, G. Israelian, A.V. Filippenko, R. Chornock, N. Tominaga, H. Umeda, K. Nomoto, Astrophys. J. **679**, 732 (2008), arXiv:0801.4936 [astro-ph]
43. E.T. Harlaftis, K. Horne, A.V. Filippenko, Publ. Astron. Soc. Pac. **108**, 762 (1996)
44. E.T. Harlaftis, D. Steeghs, K. Horne, A.V. Filippenko, Astron. J. **114**, 1170 (1997), arXiv:astro-ph/9707242
45. J. Homan, T. Belloni, Astrophys. Space Sci. **300**, 107 (2005), arXiv:astro-ph/0412597
46. R.I. Hynes, D. Steeghs, J. Casares, P.A. Charles, K. O'Brien, Astrophys. J. **583**, L95 (2003), arXiv:astro-ph/0301127
47. Z. Ioannou, E.L. Robinson, W.F. Welsh, C.A. Haswell, Astrophys. J. **127**, 481 (2004)
48. V. Kalogera, G. Baym, Astrophys. J. **470**, L61 (1996)
49. J. Khargharia, C.S. Froning, E.L. Robinson, Astrophys. J. **716**, 1105 (2010), arXiv:1004.5358 [astro-ph.HE]
50. J. Khargharia, C.S. Froning, E.L. Robinson, D.M. Gelino, Astron. J. **145**, 21 (2013), arXiv:1211.2786 [astro-ph.SR]
51. P.D. Kiel, J.R. Hurley, Mon. Not. R. Astron. Soc. **369**, 1152 (2006), arXiv:astro-ph/0605080
52. J. Kormendy, D. Richstone, Ann. Rev. Astron. Astrophys. **33**, 581 (1995)
53. J.P. Lasota, New Astron. Rev. **45**, 449 (2001), arXiv:astro-ph/0102072
54. J.M. Lattimer, Ann. Rev. Nucl. Part. Sci. **62**, 485 (2012), arXiv:1305.3510 [nucl-th]
55. P. Madau, M.J. Rees, Astrophys. J. **551**, L27 (2001), arXiv:astro-ph/0101223
56. E. Maoz, Astrophys. J. **494**, L181 (1998), arXiv:astro-ph/9710309
57. T.R. Marsh, E.L. Robinson, J.H. Wood, Mon. Not. R. Astron. Soc. **226**, 137 (1994)
58. J.E. McClintock, M.R. Garcia, N. Caldwell, E.E. Falco, P.M. Garnavich, P. Zhao, Astrophys. J. **551**, L147 (2001), arXiv:astro-ph/0101421
59. J.E. McClintock, R. Narayan, G.B. Rybicki, Astrophys. J. **615**, 402 (2004), arXiv:astro-ph/0403251
60. J.E. McClintock et al., Class. Quant. Grav. **28**, 114009 (2011), arXiv:1101.0811 [astro-ph.HE]
61. M.C. Miller, D.P. Hamilton, Mon. Not. R. Astron. Soc. **330**, 232 (2002), arXiv:astro-ph/0106188
62. S. Mineshige, J.C. Wheeler, Astrophys. J. **343**, 241 (1989)

63. T. Munoz-Darias, J. Casares, I.G. Martinez-Pais, Mon. Not. R. Astron. Soc. **385**, 2205 (2008), arXiv:0801.3268 [astro-ph]
64. T. Muoz-Darias, M. Coriat, D.S. Plant, G. Ponti, R.P. Fender, R.J.H. Dunn, Mon. Not. R. Astron. Soc. **432**, 1330 (2013), arXiv:1304.2072 [astro-ph.HE]
65. R. Narayan, I.S. Yi, Astrophys. J. **428**, L13 (1994), arXiv:astro-ph/9403052
66. R. Narayan, J.S. Heyl, Astrophys. J. **574**, L139 (2002), arXiv:astro-ph/0203089
67. R. Narayan, J.E. McClintock, New Astron. Rev. **51**, 733 (2008), arXiv:0803.0322 [astro-ph]
68. R. Narayan, R. Mahadevan, J.E. Grindlay, R.G. Popham, C. Gammie, Astrophys. J. **492**, 554 (1998), arXiv:astro-ph/9706112
69. I.D. Novikov, K.S. Thorne, Astrophysics and black holes, in *Black Holes*, ed. by C. De Witt, B. De Witt (Gordon and Breach, New York, 1973)
70. K. Ohsuga, S. Mineshige, Astrophys. J. **736**, 2 (2011), arXiv:1105.5474 [astro-ph.HE]
71. R.M. O'Leary, A. Loeb, Mon. Not. R. Astron. Soc. **421**, 2737 (2012), arXiv:1102.3695 [astro-ph.CO]
72. J.A. Orosz, P.H. Hauschildt, Astron. Astrophys. **364**, 265 (2000), arXiv:astro-ph/0010114
73. J.A. Orosz, J.E. McClintock, J.P. Aufdenberg, R.A. Remillard, M.J. Reid, R. Narayan, L. Gou, Astrophys. J. **742**, 84 (2011), arXiv:1106.3689 [astro-ph.HE]
74. J.A. Orosz, C.D. Bailyn, J.E. McClintock, R.A. Remillard, Astrophys. J. **468**, 380 (1996)
75. J.A. Orosz, R.K. Jain, C.D. Bailyn, J.E. McClintock, R.A. Remillard, Astrophys. J. **499**, 375 (1998), arXiv:astro-ph/9712018
76. J.A. Orosz, J.E. McClintock, R.A. Remillard, S. Corbel, Astrophys. J. **616**, 376 (2004), arXiv:astro-ph/0404343
77. J.A. Orosz, J.F. Steiner, J.E. McClintock, M.A.P. Torres, R.A. Remillard, C.D. Bailyn, J.M. Miller, Astrophys. J. **730**, 75 (2011), arXiv:1101.2499 [astro-ph.SR]
78. J.A. Orosz et al., Astrophys. J. **555**, 489 (2001), arXiv:astro-ph/0103045
79. J.A. Orosz et al., Astrophys. J. **568**, 845 (2002), arXiv:astro-ph/0112101
80. J.A. Orosz et al., Nature **449**, 872 (2007), arXiv:0710.3165 [astro-ph]
81. J.A. Orosz et al., Astrophys. J. **697**, 573 (2009), arXiv:0810.3447 [astro-ph]
82. F. Ozel, D. Psaltis, R. Narayan, J.E. McClintock, Astrophys. J. **725**, 1918 (2010), arXiv:1006.2834 [astro-ph.GA]
83. D.N. Page, K.S. Thorne, Astrophys. J. **191**, 499 (1974)
84. M.L. Parker et al., Astrophys. J. **808**, 9 (2015), arXiv:1506.00007 [astro-ph.HE]
85. D.S. Plant, R.P. Fender, G. Ponti, T. Munoz-Darias, M. Coriat, Astron. Astrophys. **573**, A120 (2015), arXiv:1309.4781 [astro-ph.HE]
86. S.F. Portegies Zwart, S.L.W. McMillan, Astrophys. J. **576**, 899 (2002), arXiv:astro-ph/0201055
87. M. Postman et al., Astrophys. J. **756**, 159 (2012), arXiv:1205.3839 [astro-ph.CO]
88. R.A. Remillard, J.E. McClintock, Ann. Rev. Astron. Astrophys. **44**, 49 (2006), arXiv:astro-ph/0606352
89. R.A. Remillard, J.E. McClintock, C.D. Bailyn, Astrophys. J. **399**, L145 (1992)
90. R.A. Remillard, J.A. Orosz, J.E. McClintock, C.D. Bailyn, Astrophys. J. **459**, 226 (1996)
91. C.E. Rhoades, R. Ruffini, Phys. Rev. Lett. **32**, 324 (1974)
92. T. Shahbaz, Mon. Not. R. Astron. Soc. **339**, 1031 (2003), arXiv:astro-ph/0211266
93. T. Shahbaz, F. van der Hooft, J. Casares, P.A. Charles, J. van Paradijs, Mon. Not. R. Astron. Soc. **306**, 89 (1999), arXiv:astro-ph/9901334
94. S.L. Shapiro, S.A. Teukolsky, *Black Holes, White Dwarfs and Neutron Stars: The Physics of Compact Objects* (Wiley-VCH, Weinheim, 2004)
95. L. Song et al., Astron. J. **140**, 794 (2010), arXiv:1007.3637 [astro-ph.HE]
96. D. Steeghs, J.E. McClintock, S.G. Parsons, M.J. Reid, S. Littlefair, V.S. Dhillon, Astrophys. J. **768**, 185 (2013), arXiv:1304.1808 [astro-ph.HE]
97. F.X. Timmes, S.E. Woosley, T.A. Weaver, Astrophys. J. **457**, 834 (1996), arXiv:astro-ph/9510136
98. M.A.P. Torres, P.J. Callanan, M.R. Garcia, P. Zhao, S. Laycock, A.K.H. Kong, Astrophys. J. **612**, 1026 (2004), arXiv:astro-ph/0405509

99. D. Tournear et al., Astrophys. J. **595**, 1058 (2003), arXiv:astro-ph/0303480
100. A.K.F. Val Baker, A.J. Norton, I. Negueruela, AIP Conf. Proc. **924**, 530 (2007)
101. E.P.J. van den Heuvel, Endpoints of stellar evolution: the incidence of stellar mass black holes in the galaxy, in *Environment Observation and Climate Modelling Through International Space Projects* (1992), p. 29
102. F. van der Hooft, M. Heemskerk, F. Alberts, J. van Paradijs, Astron. Astrophys. **329**, 538 (1998), arXiv:astro-ph/9709151
103. R.A. Wade, K. Horne, Astrophys. J. **324**, 411 (1988)
104. R.M. Wagner, C.B. Foltz, T. Shahbaz, J. Casares, P.A. Charles, S.G. Starrfield, P.C. Hewett, Astrophys. J. **556**, 42 (2001), arXiv:astro-ph/0104032
105. N.A. Webb, T. Naylor, Z. Ioannou, P.A. Charles, T. Shahbaz, Mon. Not. R. Astron. Soc. **317**, 528 (2000). arXiv:astro-ph/0004235
106. M.G. Witte, G.J. Savonije, Astron. Astrophys. **366**, 840 (2001)
107. Y.F. Yuan, R. Narayan, M.J. Rees, Astrophys. J. **606**, 1112 (2004). arXiv:astro-ph/0401549
108. L.R. Yungelson, J.-P. Lasota, G. Nelemans, G. Dubus, E.P.J. van den Heuvel, J. Dewi, S. Portegies Zwart, Astron. Astrophys. **454**, 559 (2006), arXiv:astro-ph/0604434
109. J. Ziolkowski, Mon. Not. R. Astron. Soc. **358**, 851 (2005), arXiv:astro-ph/0501102

Chapter 5
Observational Facilities

Astrophysical black holes can be studied with electromagnetic radiation and gravitational waves. In the case of electromagnetic radiation, we can study the photons released by the gas in the accretion disk, jet, and outflows, as well as the photons emitted by possible bodies (like stars) orbiting the black hole. In the case of gravitational waves, we can detect the radiation emitted by the inspiral of a compact object into a black hole, the inspiral and merger of two black holes, and the ringdown of a black hole.

The electromagnetic spectrum of a black hole ranges from the radio to the γ-ray band (see Table 5.1 for the list of the bands of the electromagnetic spectrum). The photon energy is determined by the emission mechanism and the environment conditions. Photons with different wavelengths carry different information about the black hole and its environment, and require different observational facilities to be detected. In this chapter, we will briefly review X-ray observatories only, because the X-ray band is the photon energy range of interest for the continuum-fitting method, X-ray reflection spectroscopy, and QPOs discussed in Part II and Part III in this book.

In the case of gravitational waves, the wave frequency depends on the size of the system. In particular, the wavelength roughly scales as the linear size of the system emitting gravitational radiation. Gravitational waves from black holes are expected to range from a few nHz, in the case of the inspiral of supermassive black holes, to a few kHz, for the merger and ringdown of stellar-mass black holes. Like in the case of electromagnetic radiation, gravitational waves with different wavelengths require different observational facilities to be detected.

5.1 X-Ray Observatories

Earth's atmosphere blocks out most of the radiation from space, see Fig. 5.1. If it were not so, life on Earth – at least as we know – would be impossible, because γ-rays, X-rays, and UV photons are harmful for any organism. X-ray observatories must thus be on board of rockets or satellites. The first X-ray observatory can be

© Springer Nature Singapore Pte Ltd. 2017
C. Bambi, *Black Holes: A Laboratory for Testing Strong Gravity*,
DOI 10.1007/978-981-10-4524-0_5

Table 5.1 Bands of the electromagnetic spectrum. Please note that different authors may use slightly different definitions

Band	Wavelength	Frequency	Energy
Radio	>0.1 m	<3 GHz	<12.4 μeV
Microwave	1 mm–0.1 m	3–300 GHz	12.4 μeV–1.24 meV
Infrared (IR)	700 nm–1 mm	300 GHz–430 THz	1.24 meV–1.7 eV
Visible	400–700 nm	430–790 THz	1.7–3.3 eV
Ultraviolet (UV)	10–400 nm	$7.9 \cdot 10^{14}$-$3 \cdot 10^{16}$ Hz	3.3–124 eV
X-Ray	0.01–10 nm	$3 \cdot 10^{16}$-$3 \cdot 10^{19}$ Hz	124 eV–124 keV
γ-Ray	<0.01 nm	$>3 \cdot 10^{19}$ Hz	>124 keV

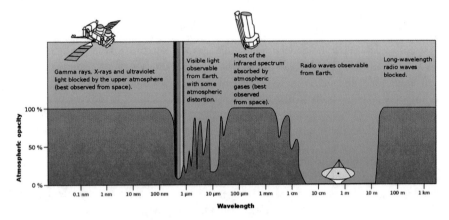

Fig. 5.1 Atmospheric opacity as a function of photon wavelength. Since the atmosphere is opaque at most wavelengths, only optical and radio telescopes can be at ground level on Earth. γ-ray, X-ray, UV, and IR observational facilities are required to be on board of rockets or satellites. Credit: NASA

considered a V2 rocket launched in 1948, which was used to observe the Sun, the brightest X-ray source in the sky. The first extrasolar X-ray source was discovered in 1962 (Scorpius X-1, an X-ray binary with a neutron star of 1.4 $M_\odot$ and a companion of 0.42 $M_\odot$) by a team led by Riccardo Giacconi with an X-ray detector on board of an Aerobee 150 sounding rocket [3]. The Einstein Observatory was the first X-ray observatory with a Wolter type I telescope [4]. It was launched on 13 November 1978 and worked for about 3 years at an altitude of about 500 km. Table 5.2 lists a selection of recent, present, and future X-ray missions.

Table 5.2 Summary of a selection of recent, present, and future X-ray missions. Notes: [1]MAXI is on board of the International Space Station; the launch date is replaced by the initial function checkout on the International Space Station; perigee and apogee are those of the International Space Station. [2]Expected. [3]SRG and ATHENA will not orbit around Earth, but will be located at the L2 point (1.5 millions km from Earth)

Mission	Launch date	End of mission	Perigee (km)	Apogee (km)	Instruments
Advanced Satellite for Cosmology and Astrophysics (ASCA) http://heasarc.gsfc.nasa.gov/docs/asca/	20-2-1993	14-7-2000	500	600	GIS (0.7–10 keV) SIS (0.4–10 keV)
Rossi X-ray Timing Explorer (RXTE) http://heasarc.gsfc.nasa.gov/docs/xte/	30-12-1995	5-1-2012	380.9	384.5	ASM (2–10 keV) PCA (2–60 keV) HEXTE (15–250 keV)
Chandra X-ray Observatory (CXO) http://chandra.harvard.edu	23-7-1999	still active	14,307.9	134,527.6	ACIS (0.2–10 keV) HRC (0.1–10 keV) LETG (0.08–2 keV) HETG (0.4–10 keV)
XMM-Newton http://www.cosmos.esa.int/web/xmm-newton http://sci.esa.int/xmm-newton/	10–12-1999	still active	5,662.7	112,877.6	EPIC-MOS (0.15–15 keV) EPIC-pn (0.15–15 keV) RGS (0.33–2.5 keV) OM (optical/UV band)
International Gamma-Ray Astrophysics Laboratory (INTEGRAL) http://sci.esa.int/integral/	17-10-2002	still active	6,281.9	156,859.1	IBIS (15 keV–10 meV) SPI (18 keV–8 meV) JEM-X (3–35 keV)

(continued)

Table 5.2 (continued)

Mission	Launch date	End of mission	Perigee (km)	Apogee (km)	Instruments
Swift Gamma-Ray Burst Mission (Swift) http://swift.gsfc.nasa.gov	20-11-2004	still active	560.1	576.3	BAT (15–150 keV) XRT (0.2–10 keV) UVOT (optical/UV band)
Suzaku http://global.jaxa.jp/projects/sat/astro_e2/ http://www.astro.isas.jaxa.jp/suzaku/index.html.en	10-7-2005	2-9-2015	550	550	XRS (0.3–12 keV) XIS (0.2–12 keV) HXD (10–600 keV)
Monitor of All-sky X-ray Image (MAXI)[1] http://iss.jaxa.jp/en/kiboexp/theme/first/maxi/	3–8-2009[1]	still active	409[1]	416[1]	SSC (0.5–10 keV) GSC (2–30 keV)
Nuclear Spectroscopic Telescope Array (NuSTAR) http://www.nustar.caltech.edu	13-6-2012	still active	607.5	623.9	FPMA (3–79 keV) FPMB (3–79 keV)
Spektrum-Roentgen-Gamma (SRG) http://hea.iki.rssi.ru/SRG/en/	2017[2]	–	L2 orbit[3]	L2 orbit[3]	eROSITA (0.3–10 keV) ART-XC (0.5–11 keV)
Enhanced X-ray Timing Polarization (eXTP) http://www.isdc.unige.ch/extp/	2022[2]	–	550[2]	550[2]	SFA (0.5–20 keV) LAD (1–30 keV) PFA (2–10 keV) WFM (2–50 keV)
Advanced Telescope for High Energy Astrophysics (ATHENA) http://www.the-athena-x-ray-observatory.eu	2028[2]	–	L2 orbit[3]	L2 orbit[3]	X-IFU (0.2–12 keV) WFI (0.1–15 keV)

5.1.1 X-Ray Missions

An X-ray telescope has two basic elements: the optics and the detector. The *optics* is used to focus the radiation entering the telescope on the detector. In the case of visible light, the optics of a telescope can be based either on lenses or on mirrors. However, in the case of X-ray photons it is not possible to employ the conventional techniques used for visible light. There is no counterpart of lenses for X-ray photons and X-rays are transmitted or absorbed by conventional mirrors, not reflected.

The optics in most X-ray observatories employs *grazing incident mirrors*, in which the angle of incidence of the X-ray photon must be very small (hence the name "grazing"). Snell's law reads $n_1 \cos \alpha_1 = n_2 \cos \alpha_2$, where n_i is the refractive index and α_i is the angle between the propagation direction of the X-ray photon and the boundary of the two materials. Vacuum has $n = 1$. For normal materials, $n < 1$ in the X-ray band. Total reflection occurs at all grazing angles smaller than the critical angle α_c given by

$$\alpha_c = \arccos n \,. \tag{5.1}$$

For 1–10 keV photons (wavelength 0.1–1 nm) and typical materials employed in X-ray mirrors (e.g. gold), the critical angle is about 1°. X-ray mirrors are usually coated with a thin layer of reflective material, like gold. In general, the refractive index can be written as $n = 1 - \delta + i\beta$, where

$$\delta = \frac{n_a r_e \lambda^2}{2\pi} f_1(\lambda) \,, \quad \beta = \frac{n_a r_e \lambda^2}{2\pi} f_2(\lambda) \,, \tag{5.2}$$

n_a is the atomic density of the material, r_e is the classical electron radius, λ is the photon wavelength, and f_1 and f_2 are the atomic scattering functions that depend on λ and the material. n has an imaginary part to take absorption into account. As the photon energy increases, λ decreases, and n approaches 1. This limits the energy range of a telescope, because α_c becomes too small. The 10–15 keV upper bound of the energy range of Chandra and XMM-Newton is determined by the optics. The more modern technology employed in NuSTAR allows to have a telescope covering higher energies.

X-ray mirrors must face other two problems. First, the mirror surface must be very smooth to prevent photon scattering. In typical mirrors, the roughness of the surface should not exceed 1/10 the photon wavelength. This is easy to achieve for visible photons with wavelengths in the range 400–700 nm, but it is technologically much more challenging for X-rays, with wavelengths less than 1 nm. Second, single optical reflection in grazing incidence introduces optical distortions. The problem of optical distortion was solved in the 1950s by Hans Wolter, who proposed three types of mirror configurations, called, respectively, *Wolter telescopes* of type I, II, and III. The three configurations are shown in Fig. 5.2. Current X-ray observatories usually employ some version of the Wolter type I configuration. Wolter telescopes require

Fig. 5.2 Wolter telescopes
of type I (*top picture*),
II (*central picture*), and
III (*bottom picture*). Credit:
NASA's Imagine the
Universe

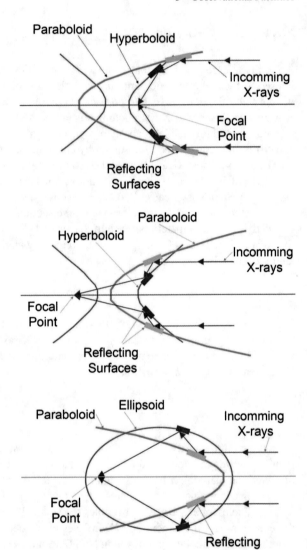

two successive reflections from a pair of mirrors, either paraboloid/hyperboloid or
paraboloid/ellipsoid.

   The *detector* of an X-ray telescope counts X-ray photons and measures their
properties (arrival time, energy, etc.). There are several types of detectors for X-ray
telescopes, but for soft X-ray photons (say 0.1–10 keV) they all work in a similar
way. When an X-ray photon hits the detector, it produces free electrons due to photoelectric effect or photoionization. These electrons then generate an electric current
proportional to the energy of the X-ray photon.

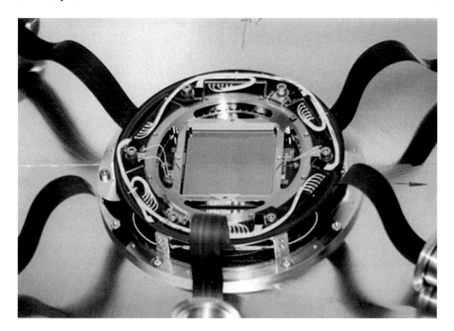

Fig. 5.3 The twelve chips of the EPIC-pn camera of XMM-Newton and the connections to the integrated preamplifiers. Image courtesy of ESA/MPI-semiconductor laboratory/MPE/Institut für Astronomie und Astrophysik Tübingen

CCD (charge-coupled device) detectors are among the most common X-ray detectors in current X-ray missions. While CCDs are properly imaging detectors, X-ray CCDs are also used as X-ray spectrometers. In the case of visible light, a single photon can produce a single electron in a pixel of the camera, and images are built up by detecting many photons. In the case of X-rays, the energy of a single photon is high enough to produce a number of electrons (hundreds or thousands) proportional to its energy, so it is possible to measure the photon energy. XMM-Newton has three CCD detectors, called EPIC (European photon imaging camera): two EPIC-MOS (metal oxide semiconductor), each of them with seven MOS-CCDs, and one EPIC-pn with twelve pn-CCDs (see Fig. 5.3). The energy resolution at 6 keV is around 150 eV (full width half maximum) and is slightly different for EPIC-MOS and EPIC-pn. ACIS on board of Chandra is a CCD detector with an energy resolution of about 170 eV at 6 keV.

NuSTAR has CZT (cadmium zinc telluride, CdZnTe) detectors, which are semiconductor detectors particularly suitable for high energy photons. NuSTARS is indeed designed to detect photons up to 79 keV, see Table 5.2. The energy resolution is lower than CCD detectors, and it is about 400 eV at 6 keV.

X-ray micro-calorimeters promise to reach very good energy resolution. XRS on board of Suzaku had an energy resolution of 6.5 eV, but it was never used because was shut down by a cooling system malfunction immediately after the launch. Hitomi

(ASTRO-H before the launch) was launched in February 2016 and had a micro-calorimeter with an energy resolution of 7 eV in the range 0.3–12 keV. However, Hitomi broke into several parts after the launch due to excessive rotational rate caused by a malfunction of the satellite. The first micro-calorimeter employed in X-ray astronomy will probably be X-IFU, on board of ATHENA, which is expected with an energy resolution of 2.5 eV.

5.1.2   X-Ray Spectrum Analysis

The outcome of an observation is a data set called the *event list*. Roughly speaking, the event list is a table. Every row is an event, namely the "detection of something" by the instrument. It may be an X-ray photon from the target source, but also a background X-ray photon or a cosmic ray. Every column of the event list is reserved for the measurement of a specific quantity of the events, like the detection time, the position coordinates of the event (usually in some detector coordinates), the energy, etc.

The detector does not directly measure the energy of an event, but its PHA (*pulse height amplitude*). The PHA is an engineering unit: it describes the integrated charge per pixel from an event recorded in the detector. The PHA is related to the energy of the event, and such a relationship is determined by the response of the detector. A gain table (which may depend on the energy, the detection point in the detector, etc.) is used to convert the PHA of an event into its energy, say E. A related quantity is the PI (*pulse invariant*), which is an integer. For example, the PI may be given by

$$\mathrm{PI} = \left[\frac{E}{14.6 \text{ eV}} \right] + 1 , \qquad (5.3)$$

where [...] denotes the integer part inside the brackets. With the definition in Eq. (5.3), if $E = 478$ eV we have PI $= 33$. The relation between the PI and the energy E can have any form and, usually, is not linear as in the example in Eq. (5.3).

The spectrum measured by a detector can be written as

$$C(h) = \tau \int dE \, R(h, E) \, A(E) \, s(E) . \qquad (5.4)$$

$C(h)$ is in units of counts per spectral bin, where h is the spectral channel in units of PHA or PI. On the right hand side of Eq. (5.4), we have: the exposure time τ, the energy E, the redistribution matrix $R(h, E)$, the effective area $A(E)$, and the intrinsic spectrum of the source $s(E)$. In general, one has also to take the background into account, so $s(E)$ in Eq. (5.4) is replaced by $s(E) + b(E)$, where $b(E)$ is the (focused) background. Non-focused background $b'(E)$ may show up outside the integral and be insensitive to the redistribution matrix and the effective area.

Fig. 5.4 Response of the
EPIC-pn detector to
monochromatic lines of 0.3,
0.5, 2.0, and 6.0 keV. Figure
courtesy of Matteo
Guainazzi

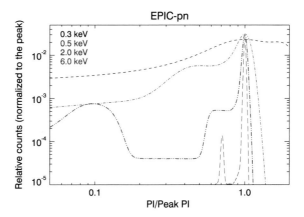

The redistribution matrix is related to the response of the detector: it roughly corresponds to the probability that a photon of energy E is detected in the channel h. For a given detector, it strongly depends on the energy. In the case of an ideal detector, the redistribution matrix would be a δ-function, say $R(h, E) = \delta(h - h_E)$. The width of the curve $R(E)$ defines the instrument resolution at the energy E. Figure 5.4 shows the redistribution matrix of the EPIC-pn camera on board of XMM-Newton for four different photon energies E.

The effective area $A(E)$ depends on the optics, possible filters, and the detector. It is somehow a measurement of the efficiency of these elements and is usually in units of cm^2. A larger effective area increases the photon count for the same exposure time. In the absence of pile-up[1], this is a benefit in general, because it reduces the intrinsic Poisson noise of the source. Figure 5.5 shows the effective area of instruments of XMM-Newton, Chandra, Suzaku, NuSTAR, eXTP, and ATHENA.

Figure 5.6 shows three examples of observed spectrum $C(h)$. The two spectra in the top panels, which look similar, are the AGN Ark 120 (left panel) and the Coma Cluster (right panel) measured with the EPIC-pn camera. The Coma Cluster is a cluster of more than 10^3 galaxies, and is a completely different source with respect to an AGN. The bottom panel is again the observed spectrum of the AGN Ark 120, but now measured by SIS/ASCA. It does not look like the observed spectrum in the top left panel obtained from the EPIC-pn camera. These pictures show that the shape of the count spectra does not have much to do with the intrinsic spectrum of the source $s(E)$, but is instead mainly determined by the characteristics of the instruments, i.e. the redistribution matrix and the effective area.

A detector measures $C(h)$, and we have to determine $s(E)$. We could think of inverting Eq. (5.4) to have $s(E)$ in terms of $C(h)$, but in general this is not possible

---

[1] *Pile-up* occurs in the case of bright sources, when two or more photons are detected as a single event. This clearly causes a distortion in the observed spectrum. For example, the CCD detectors on board of XMM-Newton and Chandra suffer pile-up in the case of bright X-ray binaries, while NuSTAR does not. There are specific procedures to fix the problem of pile-up in an observation, but this inevitably causes a loss of information.

Fig. 5.5 Effective area as a function of energy for a selection of instruments on board of X-ray satellites

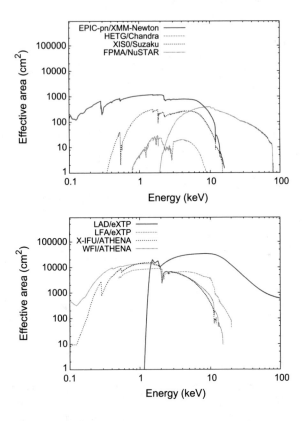

because of cross-correlations among different energies. We can thus employ the so-called *forward-folding approach*, which goes as follows:

1. We consider a theoretical model for the intrinsic spectrum $s(E)$.
2. We convolve the theoretical model with the instrument response, namely redistribution matrix (RMF file of the instrument) and effective area (ARF or ancillary file), and we get the expected $C(h)$ for the input parameters used in the theoretical model. The outcome is called the *folded spectrum*.[2]
3. We compare the observed spectrum with the folded spectrum with some goodness-of-fit statistical test.
4. We find "the best fit", namely we minimize the goodness-of-fit test by changing the input parameters in the theoretical model. The best fit parameters correspond to our measurements.
5. We calculate the confidence intervals on the best fit parameters.

---

[2]The unfolded spectrum is the spectrum obtained by inverting Eq. (5.4). Equation (5.4) cannot be inverted in general, but one can invert this equation under some assumptions/simplifications, and the result is the unfolded spectrum.

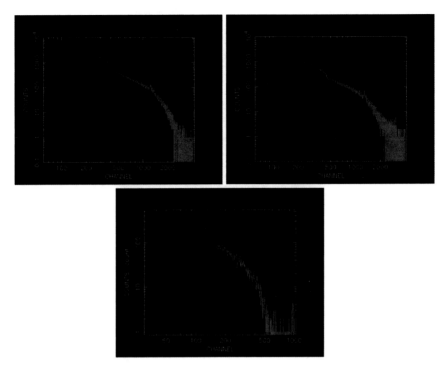

Fig. 5.6 *Top left panel* spectrum of the AGN Ark120 from the EPIC-pn camera on board of XMM-Newton. *Top right panel* spectrum of the Coma Cluster from the EPIC-pn camera. *Bottom panel* spectrum of the AGN Ark120 from the SIS instrument on board of ASCA. The shape of the count spectra is mainly determined by the response of the telescope and of the detector, not by the intrinsic spectrum of the source. Figures courtesy of Matteo Guainazzi

5.2 Gravitational Wave Detectors

Gravitational waves were predicted by Einstein immediately after the formulation of general relativity. However, generally speaking, they are a prediction of any relativistic theory of gravity. Matter makes the spacetime curved and therefore the motion of matter alters the background metric around. Gravitational waves are "ripples" in the curvature of the spacetime propagating at a finite velocity. Nevertheless, for a given source, different gravity theories may predict a different gravitational wave signal, even if the motion of matter may be the same.

The first observational evidence for the existence of gravitational waves followed the discovery of the binary pulsar PSR 1913+16 by Russell Alan Hulse and Joseph Hooton Taylor in 1974. It was the first binary pulsar to be discovered and Hulse and Taylor got the 1993 Noble Prize in physics for it. It is a binary system of two neutron stars, and one of them is seen as a pulsar, which makes this system a perfect laboratory for testing general relativity. Since the discovery of PSR 1913+16, the orbital period of the system has decayed in agreement with the expectations from general relativity

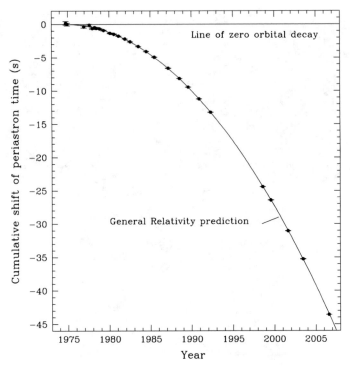

Fig. 5.7 Cumulative shift of the periastron time of PSR 1913+16 over about 30 years, from 1975 to 2005. The dots with the small error bars are the data and the solid curve is the prediction of general relativity for the emission of gravitational waves. From [9]. ©AAS. Reproduced with permission

for the emission of gravitational waves. From the radio data covering about 30 years of observations, we have [9]

$$\frac{\dot{P}_{\text{corrected}}}{\dot{P}_{\text{GR}}} = 1.0013 \pm 0.0021 , \tag{5.5}$$

where $\dot{P}_{\text{corrected}}$ is the (corrected) observed orbital decay[3] and $\dot{P}_{\text{GR}}$ is the orbital decay due to gravitational waves expected in general relativity. Figure 5.7 shows the perfect agreement between the data (the black dots with the small error bars) and the theoretical prediction (the solid line). The study of the orbital decay of binary pulsars can constrain alternative theories of gravity, see e.g. [10, 11] and references therein.

As a ripple in the curvature of the spacetime, the passage of a gravitational wave causes a temporal variation in the proper distance of a set of test-particles or, alternatively, tidal deformations in an extended body. Figure 5.8 shows the deformations

---

[3]One has to remove the effect due to the relative acceleration between us and the pulsar caused by the differential rotation of the Galaxy.

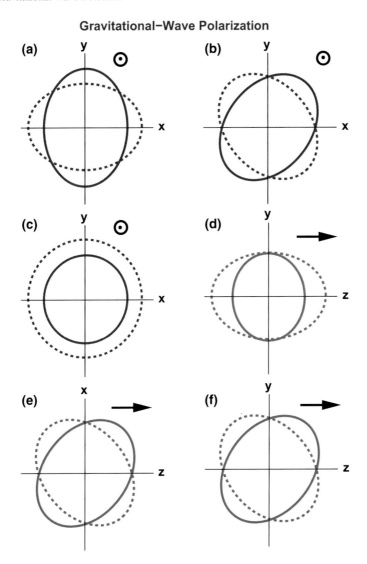

Fig. 5.8 Impact on a ring of test-particles by the six possible polarization modes for gravitational waves in a metric theory of gravity. The wave propagates along the z direction, as indicated in the top right corner of each panel. In general relativity, there are only the polarization modes (**a**) and (**b**). From [10] under the terms of Creative Commons Attribution-Non-Commercial 3.0 Germany License

of a ring of test-particles by the six possible polarization modes in a metric theory of gravity. An observer feels the passage of a gravitational wave as a distortion of the spacetime by the effect of a strain. If two test-particles are at a distance L, a gravitational wave causes a variation ΔL, and the *strain $h = \Delta L/L$* is related to the amplitude of the gravitational wave. h is the quantity measured by a detector and clearly depends on the position of the detector with respect to the propagation direction of the gravitational wave. In general relativity, there are only the polarization modes (a) and (b) of Fig. 5.8, which are usually denoted by h_+ and $h_\times$, respectively.

Direct detection of gravitational waves is very challenging, and indeed the first direct detection has been possible only very recently [1]. This is because the expected amplitude of gravitational waves passing through Earth is extremely small, with h of order 10^{-20}. If r is the distance of the source from the detection point, $h \propto 1/r$. To have a simple idea of the technological difficulties to detect gravitational waves, we can consider that the Earth's radius $R \approx 6{,}000$ km would change by $\Delta R = 60$ fm (1 fm $= 10^{-15}$ m) because of the passage of a gravitational wave with $h = 10^{-20}$. Such a value of ΔR is much smaller than the radius of an atom ($\sim 10^5$ fm).

Table 5.3 lists a selection of recent, present, and future/proposed gravitational wave detectors. There are three main types of detectors, which will be briefly review in the next subsections: resonant detectors, interferometers (either ground-based or space-based), and pulsar timing arrays.

Figure 5.9 shows the sensitivity curves of some gravitational wave detectors and the expected strength of gravitational wave sources. The x-axis is for the frequency ν of the gravitational waves. The y-axis is for the (dimensionless) *characteristic strain* h_c, which is defined as [7]

$$|h_c(\nu)|^2 = 4\nu^2 |\tilde{h}(\nu)|^2 , \qquad (5.6)$$

where $\tilde{h}(\nu)$ is the Fourier transform of $h(t)$. h_c is not directly related to the amplitude of the gravitational wave, while includes the effect of integrating an inspiralling signal. The energy density in gravitational waves is another commonly used quantity [7] (reintroducing the constant c and G_N for clarity)

$$\rho c^2 = \frac{c^2}{16\pi G_N} \int_{-\infty}^{+\infty} (2\pi\nu)^2 \, \tilde{h}(\nu) \, \tilde{h}^*(\nu) \, d\nu = \int_{-\infty}^{+\infty} S_E(\nu) d\nu , \qquad (5.7)$$

where $\tilde{h}^*$ is the complex conjugate of $\tilde{h}$ and $S_E(\nu)$ is the *spectral energy density* (i.e. the energy per unit volume of space and unit frequency)

$$S_E(\nu) = \frac{c^2}{16\pi G_N} (2\pi\nu)^2 \, \tilde{h}(\nu) \, \tilde{h}^*(\nu) . \qquad (5.8)$$

The gravitational wave frequency is determined by the size of the source. In particular, the wavelength of a gravitational wave is usually of the order the size of the source, as one could indeed expect considering that the wave is a ripple in the curvature of the spacetime caused by matter motion. For a system of mass M and size R, the frequency of the gravitational wave is roughly

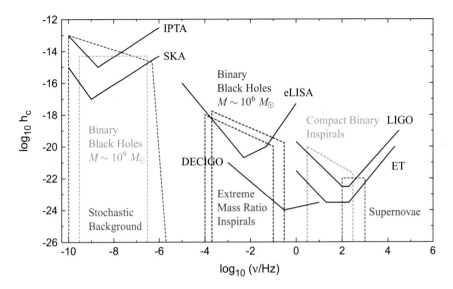

Fig. 5.9 Sketch of the sensitivity curves of some gravitational wave detectors in terms of the characteristic strain h_c and the frequency ν. Resonance spikes in the detector noise curves have been removed for clarity. The figure also shows the expected characteristic strain and frequency of a number of possible sources

$$\nu \sim \left(\frac{M}{R^3}\right)^{1/2} \sim 1 \left(\frac{\rho}{10^8 \text{ g/cm}^3}\right)^{1/2} \text{ Hz} . \tag{5.9}$$

If we have a binary system of two $10\, M_\odot$ black holes separated by 10 gravitational radii ($M/R^3 \sim 10^{13}$ g/cm^3), the frequency of the emitted gravitational waves is about 300 Hz. If the distance between these two black holes is 100 gravitational radii, the frequency is about 10 Hz.

The frequency band of the sensitivity of a detector is also related to the size of the detector. Within the same kind of detectors, roughly speaking a larger/smaller detector may be sensitive to lower/higher frequencies (modulo other complications).

5.2.1 Resonant Detectors

Resonant detectors consist of a large resonant body (bar) which is stretched and squeezed by the passage of a gravitational wave. The sensitivity of the detector is peaked at its mechanical resonance, which corresponds to the first longitudinal mode of the bar and in most detectors is around 1 kHz.

The first gravitational wave detector was the resonant detector constructed by Joseph Weber in the 1960s. It was a 2 m aluminum cylinder held at room temperature and isolated from vibrations in a vacuum chamber. The claim of detection of

Table 5.3 Partial list of gravitational wave detectors. RD = resonant detector; GBLI = ground-based laser interferometer; SBLI = space-based laser interferometer. Notes: [1] While there may not be an official end of the experiment, resonant detectors are not operative any more. [2] Expected/proposed

Project	Activity period	Location	Detector type	Frequency range
EXPLORER	1990–2002 http://www.roma1.infn.it/rog/explorer/	Geneva (Switzerland)	RD	~900 Hz
ALLEGRO	1991–2008	Baton rouge (Louisiana)	RD	~900 Hz
NAUTILUS http://www.roma1.infn.it/rog/nautilus/	1995–2002	Frascati (Italy)	RD	~900 Hz
TAMA 300 http://tamago.mtk.nao.ac.jp/spacetime/tama300_e.html	1995-	Mitaka (Japan)	GBLI	10 Hz–10 kHz
GEO 600 http://www.geo600.org	2001-	Sarstedt (Germany)	GBLI	50 Hz–1.5 kHz
AURIGA http://www.auriga.lnl.infn.it	1997-x[1]	Padua (Italy)	RD	~900 Hz
Gravitational Radiation Antenna in Leiden (MiniGRAIL) http://www.minigrail.nl	2001-x[1]	Leiden (Netherland)	RD	2–4 kHz
Laser Interferometer Gravitational Wave Observatory (LIGO) http://www.ligo.org	2004-	Hanford (Washington) Livingston (Louisiana)	GBLI	30 Hz–7 kHz
Mario Schenberg	2006-x[1]	Sao Paulo (Brazil)	RD	3.0–3.4 kHz
Virgo interferometer http://www.virgo-gw.eu	2007-	Cascina (Italy)	GBLI	10 Hz–10 kHz
Kamioka Gravitational Wave Detector (KAGRA) http://gwcenter.icrr.u-tokyo.ac.jp/en/	From 2018[2]	Kamioka mine (Japan)	GBLI	10 Hz–10 kHz

(continued)

Table 5.3 (continued)

Project	Activity period	Location	Detector type	Frequency range
Indian Initiative in Gravitational Wave Observations (IndiGO) http://gw-indigo.org/tiki-index.php	From 2023[2]	India	GBLI	30 Hz–7 kHz
Deci-hertz Interferometer Gravitational Wave Observatory (DECIGO) http://tamago.mtk.nao.ac.jp/decigo/index_E.html	From 2027[2]	Space	SBLI	1 mHz–10 Hz
Einstein Telescope (ET) http://www.et-gw.eu	From ~2030[2]	To be decided	GBLI	1 Hz–10 kHz
Evolved Laser Interferometer Space Antenna (eLISA) https://www.elisascience.org	From 2034[2]	Space	SBLI	0.1 mHz–1 Hz
TianQin	From ~2040[2]	Space	SBLI	0.1–100 mHz

SUSPENSIONS   RESONANT
                               TRANSDUCER

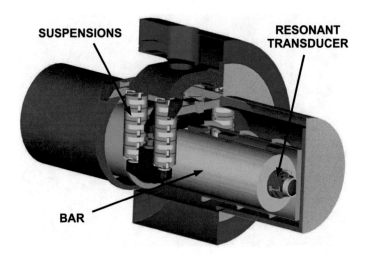

BAR

Fig. 5.10 Cartoon of the AURIGA detector. The mechanical resonator is an aluminum cylinder of 3 m of length, 60 cm of diameter, and with a mass of 2,300 kg. The bar is cooled down to 100 mK to reduce the thermal noise. The acoustic noise of the laboratory is suppressed by placing the bar in two vacuum chambers. Mechanical noise (floor vibrations caused by nearby moving vehicles or walking persons, seismic noise, etc.) is reduced by placing the experiment on the top of a 200 tons heavy concrete platform which sits on a sand layer. The suspensions are also adopted to reduce mechanical noise. Figure adapted from http://www.auriga.lnl.infn.it

gravitational waves was reported in [8]. However, such a detector was only able to measure strains of order 10^{-16}. This should not be enough to detect gravitational waves from astrophysical sources.

Figure 5.10 is a cartoon of the AURIGA detector, which was an INFN (Istituto Nazionale di Fisica Nucleare) experiment installed at the Laboratori Nazionali di Legnaro, near Padua, in Italy. The bar was an aluminum cylinder of 3 m of length, 60 cm of diameter, and with a mass of 2,300 kg. The challenge was to reduce a number of noises caused by the atoms of the bar itself and by the surrounding environment. The bar was cooled down to 100 mK to reduce the thermal noise, namely the motion of the atoms of the bar. The bar was inside two vacuum chambers to reduce the acoustic noise of the laboratory. Mechanical noise due to the vibration of the floor was reduced by placing the detector on the top of a 200 tons heavy concrete platform sitting on a sand layer and by some mechanical suspensions. The vibration of the bar was read out by a smaller mass (resonant transducer) of about 1 kg. The transducer had the same resonant frequency as the bar, so it could pick up resonantly the bar vibrations. Since it was much lighter, the amplitude of its vibrations could be much larger.

EXPLORER, ALLEGRO, NAUTILUS, and AURIGA were all cylindrical bar detectors and worked in a similar way. MiniGRAIL and Mario Schenberg were instead spherical detectors. The mass of MiniGRAIL was 1,150 kg and the diameter was 65 cm. The cryogenic chamber was cooled down to 20 mK. Spherical antennas are technologically more challenging, but they present some advantages in the possibility of detecting gravitational waves; see e.g. [2] and reference therein. In particular, a spherical detector can detect gravitational waves arriving from any direction.

Resonant detectors (both bar and spherical detectors) did not receive funds for some years and their groups have joined gravitational wave laser interferometers.

5.2.2 Interferometers

Gravitational wave laser interferometers are based on a Michelson interferometer. The set-up of the advanced LIGO detectors is sketched in Fig. 5.11. There are two arms, which are orthogonal each other. A beamsplitter splits the original laser beam into two beams, which are reflected by the two mirrors at the end of the two arms and eventually recombine and produce an interference pattern. In general, the passage of a gravitational wave would change the travel time in the two arms in a different way: depending on the propagation direction of a gravitational wave with respect to the orientation of the interferometer, one of the arms can be stretched, while the other can be squeezed. The photodetector at the place of the interference pattern can measure a change in the proper length of the arms. To increase the effective path length of

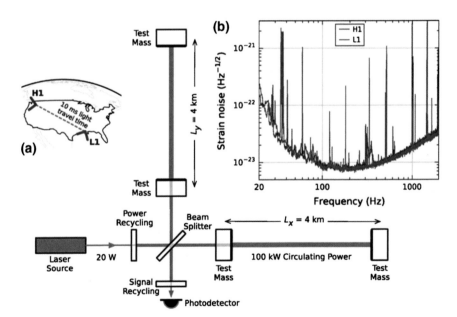

Fig. 5.11 Sketch of the advanced LIGO detectors, location and orientation of the LIGO detectors at Hanford (H1) and Livingston (L1) (**a**), and sensitivity curves in terms of equivalent gravitational wave strain amplitude (**b**). From [1] under the terms of the Creative Commons Attribution 3.0 License

Fig. 5.12 Aerial view of the interferometer detector Virgo (Cascina, Pisa, Italy). From the Virgo collaboration under the Creative Commons CC0 1.0 Universal Public Domain Dedication

the laser light in the arms, there are partially reflecting mirrors, which make the laser light run along the arms many (typically hundreds) times.

TAMA 300 was the first laser interferometer to work, but its sensitivity is limited by its small size. The length of its arms is 300 m. GEO 600 is a laser interferometer with arms of 600 m. In the case of LIGO and Virgo, the length of the arms is, respectively, 4 and 3 km. Figure 5.12 shows the aerial view of the Virgo detector, near Pisa, in Italy. KAGRA and the Einstein Telescope are underground detectors, to reduce the seismic noise. The Einstein Telescope will have a equilateral triangle geometry, with three arms of 10 km and two detectors in each corner.

Space-based laser interferometers work with a constellation of satellites (e.g. three satellites in eLISA and four clusters of three satellites in the case of DECIGO). The working principle is the same as the ground-based laser interferometers, and one wants to monitor the proper distance among mirrors located in different satellites. The distance between satellites is larger than the length of the arms of ground-based laser interferometers, so these experiments are sensitive to gravitational waves at lower frequencies (see Fig. 5.9). This is also possible because the limitation at low frequencies in ground-based experiments is due to seismic noise, but there is no seismic noise in space. In the case of DECIGO, the distance among the satellites should be $\sim 10^3$ km, while the arms of eLISA should be $\sim 10^6$ km.

The sensitivity achievable by a laser interferometer can be understood as follows. We can consider the case of LIGO, where the laser wavelength is $\lambda \sim 1\,\mu\text{m}$ and the interferometer arms have $L = 4\,\text{km}$. If we could measure ΔL only with a precision of order the size of a fringe, i.e. $\Delta L \sim \lambda$, the minimum detectable strain would be $h \sim \lambda/L \sim 3 \cdot 10^{-10}$. Gravitational waves with a strain of order 10^{-20} can be detected if we can measure changes in the arm length much smaller than λ. The photodetector of the experiment can indeed monitor changes in the photon flux and reach a sensitivity $\Delta L \sim \lambda/\sqrt{N}$, where N is the number of photons arriving at the detector and $\sqrt{N}$ is its fluctuation, being a Poisson process. If P is the laser power, $N \sim P/(\nu E_\gamma)$, where ν is the frequency of the gravitational wave (we can collect photons for a time $t \sim 1/\nu$) and E_γ is the photon energy. If we consider $P \sim 1\,\text{W}$, $\nu = 100\,\text{Hz}$, and $\lambda \sim 1\,\mu\text{m}$, we find $N \sim 10^{16}$ and $h \sim 10^{-18}$. Moreover, the two arms of the interferometer are two Fabry–Perot optical cavities, and they can store the light for many round trips. For $\nu = 100\,\text{Hz}$ and $L = 4\,\text{km}$, the light can make about a thousand round trips during the passage of a gravitational wave, which increases the effective arm length by the factor $\sim 10^3$ and the interferometer sensitivity becomes $h \sim 10^{-21}$.

5.2.3 Pulsar Timing Arrays

In a pulsar timing array experiment, one monitors 20–50 well-known millisecond pulsars to find possible variations in the distance of these pulsars. Like in the interferometers in the previous subsection, the passage of gravitational waves contracts the space in one direction and expands the space in the other direction, thus changing the arrival time of the pulsar signals. Since millisecond pulsars can be used as very precise clocks, it is possible to infer variations in the time arrival of the signal of order of some ns. Two review articles on pulsar timing arrays are [5, 6].

The distance between Earth and these pulsars in the Galaxy is 1–10 kpc, which is definitively much larger than the length scale involved in the interferometric detectors. Pulsar timing array experiments can thus detect gravitational waves of very low frequency, in the range 1–100 nHz. There are two main possible sources for such low frequency gravitational waves. (i) Binary systems of two supermassive black holes with an orbital period ranging from a few months to a few years. Even if they are far from us, the power emitted in gravitational waves is huge, and the signal may be strong enough to be detected. (ii) Gravitational waves produced in the early Universe. There is a number of different scenarios predicting a background of low frequency gravitational waves, like decay of cosmic strings, inflationary models, and first order phase transitions. In all these cases, the frequency of the gravitational waves would be very low because of the cosmological redshift.

Table 5.4 lists the present pulsar timing array projects and the near future SKA experiment. With the available pulsar data, it is only possible to get some upper bounds on the amplitude of low frequency gravitational waves. These bounds can be improved with time, because the precision is determined by the observational

Table 5.4 List of pulsar timing array projects. EPTA, PPTA, NANOGrav, and IPTA are already operative. SKA is expected to start in 2020

Project
European Pulsar Timing Array (EPTA)   http://www.epta.eu.org
Parkes Pulsar Timing Array (PPTA)   http://www.atnf.csiro.au/research/pulsar/ppta/
North American Nanohertz Observatory for Gravitational Waves (NANOGrav)   http://nanograv.org
International Pulsar Timing Array (IPTA)   http://www.ipta4gw.org
Square Kilometre Array (SKA)   https://www.skatelescope.org

time of the pulsars. We may also discover new millisecond pulsars suitable for these measurements. This would increase the number of sources monitored, which is also helpful to improve the sensitivity.

References

1. B.P. Abbott et al., LIGO scientific and virgo collaborations. Phys. Rev. Lett. **116**, 061102 (2016), arXiv:1602.03837 [gr-qc]
2. E. Coccia, V. Fafone, G. Frossati, J.A. Lobo, J.A. Ortega, Phys. Rev. D **57**, 2051 (1998), arXiv:gr-qc/9707059
3. R. Giacconi, H. Gursky, F.R. Paolini, B.B. Rossi, Phys. Rev. Lett. **9**, 439 (1962)
4. R. Giacconi et al., Astrophys. J. **230**, 540 (1979)
5. G. Hobbs et al., Class. Quant. Grav. **27**, 084013 (2010), arXiv:0911.5206 [astro-ph.SR]
6. A.N. Lommen, Rept. Prog. Phys. **78**(12), 124901 (2015)
7. C.J. Moore, R.H. Cole, C.P.L. Berry, Class. Quant. Grav. **32**, 015014 (2015), arXiv:1408.0740 [gr-qc]
8. J. Weber, Phys. Rev. Lett. **22**, 1320 (1969)
9. J.M. Weisberg, D.J. Nice, J.H. Taylor, Astrophys. J. **722**, 1030 (2010), arXiv:1011.0718 [astro-ph.GA]
10. C.M. Will, Living Rev. Rel. **17**, 4 (2014), arXiv:1403.7377 [gr-qc]
11. N. Yunes, S.A. Hughes, Phys. Rev. D **82**, 082002 (2010). arXiv:1007.1995 [gr-qc]

Part II
Main Tools for Testing Astrophysical Black Holes

Chapter 6
Thin Accretion Disks

The standard framework for the description of geometrically thin and optically thick accretion disks[1] is the Novikov-Thorne model [38, 40], which is the relativistic generalization of the Shakura–Sunyaev one [47]. The model is relatively simple, and it is formulated in a generic stationary, axisymmetric, and asymptotically flat spacetime. The time-averaged radial structure of the accretion disk follows from the fundamental laws of the conservation of rest-mass, energy, and angular momentum.

In the Kerr metric, the Novikov-Thorne model has four fundamental parameters: the black hole mass M and the black hole spin parameter a_*, both related to the metric of the spacetime, the mass accretion rate $\dot{M}$, and the viscosity parameter α. The latter, however, does not enter the equations for the time-averaged radial structure of the accretion disk.

6.1 Novikov-Thorne Model

The Novikov-Thorne model describes geometrically thin and optically thick accretion disks around black holes. Accretion is possible because viscous magnetic or turbulent stresses and radiation transport energy and angular momentum outward. There are slightly different versions of the model, but the typical assumptions are [40]:

1. The spacetime is stationary, axisymmetric, asymptotically flat, and reflection-symmetric with respect to the equatorial plane.
2. The accretion disk is non-self-gravitating; that is, the impact of the disk's mass on the background metric is ignored.

---

[1] An accretion disk is geometrically thin (thick) if the disk opening angle is $h/r \ll 1$ ($h/r \sim 1$), where h is the semi-thickness of the disk at the radial coordinate r. The disk is optically thick (thin) if the photon mean free path in the disk $l = 1/(\sigma n)$, where σ is the photon scattering cross-section in the disk medium and n is the number density of scattering particles in the disk, is $l \ll h$ ($l \gg h$).

© Springer Nature Singapore Pte Ltd. 2017
C. Bambi, *Black Holes: A Laboratory for Testing Strong Gravity*,
DOI 10.1007/978-981-10-4524-0_6

3. The accretion disk is in the equatorial plane; that is, the disk is perpendicular to the black hole spin.
4. The inner edge of the disk is at the ISCO radius.
5. The accretion disk is geometrically thin, namely the disk opening angle is $h/r \ll 1$, where h is the semi-thickness of the disk at the radial coordinate r.
6. We suppose to average over time scales Δt that are short enough to assume that the spacetime is stationary (for instance, the mass accreted by the central object does not appreciable change the background metric) and large enough to neglect possible inhomogeneities in the accretion fluid.
7. The particles of the gas follow nearly-geodesic circular orbits in the equatorial plane. In this case, we can use the expressions for the 4-velocity, the energy, the axial component of the angular momentum, and the angular velocity found in Sects. 3.1 and 3.2. It is worth noting that here the term "particle" is used to indicate a parcel of gas.
8. Radial heat transport is ignored, and energy and angular momentum are radiated from the disk surface.
9. Magnetic fields are ignored.
10. Energy and angular momentum from the disk's surface are only carried away by photons with wavelength $\lambda \ll M$.
11. The effect of energy and angular momentum transport by photons emitted from the disk and returning to the disk due to strong light bending in the vicinity of the black hole (returning radiation) is neglected.

Some assumptions are sometimes relaxed, while others cannot. For instance, the fact that the inner edge of the disk is assumed at the ISCO radius plays a crucial role in the spin measurements via the continuum-fitting and iron line methods. However, it is straightforward to relax this assumption and set the inner edge of the disk at a larger radius. The time-average radial structure of the disk changes, but not the form of its equations. The fact that the accretion disk must be geometrically thin cannot be relaxed, because otherwise other assumptions are not satisfied and the structure of the disk is completely different. The effect of the returning radiation was neglected in the original paper [40], but it is often taken into account in spin measurements via the continuum-fitting method [29] and is definitively non-negligible in the calculation of the polarization of the spectrum of the disk (see Sect. 7.3).

6.1.1   Validity of the Novikov-Thorne Model

The validity of the Novikov-Thorne model has been explored in a number of studies. The assumptions 1 and 2 are very natural. The Assumption 1 holds in the Kerr metric and in any plausible extension. The disk's mass is usually many orders of magnitude smaller than the mass of the central black hole, so its effects on the metric of the spacetime can indeed be ignored; see Sect. 6.5 for more details.

The validity of the assumption 3 depends on the origin and the evolution of the system. The cases of stellar-mass black holes and of supermassive black holes are somewhat different. Let us first consider stellar-mass black holes. If the object is the final product of the supernova explosion of a heavy star in a binary, its spin should be orthogonal to the orbital plane of the binary in the case of a symmetric explosion without strong shocks and kicks [18]. A misalignment may be introduced by a non-symmetric supernova explosion and/or shocks and kicks, as well as in those systems formed through multi-body interactions (binary capture or replacement), where the orientation of the spin of the black hole and that of the orbital angular momentum of the binary are initially uncorrelated.

The inner part of the disk – which is the one important in the continuum-fitting and the iron line measurements – may be in any case expected to be in the equatorial plane perpendicular to the spin of the compact object as a result of the Bardeen–Petterson effect [12, 28]. This mechanism works for thin disks, because it requires $\alpha > h/r$, where $\alpha \sim 0.01$–0.1 is the viscosity parameter. The combination of the Lense-Thirring precession with the disk viscosity eventually drags the innermost part of the disk into alignment with the black hole spin. Because of the short range of the Lense-Thirring effect, the outer part tends to remain in its original configuration. "Bardeen–Petterson configuration" refers to a system in which the inner part of the disk is flat and perpendicular to the black hole spin, while the outer part is also flat but in the plane perpendicular to the angular momentum vector of the binary.

The alignment timescale of thin disks has been estimated to be in the range 10^6–10^8 years, and therefore the disk should be already adjusted in the black hole equatorial plane for not too young systems [50] (but see [33, 34] for more details). However, the actual timescale depends on some unknown parameters, like the viscosity α, [24, 32] and it should be noted that at least some numerical simulations do not find the adjustment of the disk [16, 57]. If the inner part of the disk is a hot, geometrically thick accretion flow, the picture is different and the inner disk precesses as a solid body [22]. Future X-ray spectropolarimetric measurements of the thermal spectrum of accretion disks will be able to check the validity of the assumption that the disk is in the equatorial plane (see Sect. 7.3).

In the case of supermassive black holes, the orientation of the accretion disk with respect to the black hole spin is expected to change during the evolution of the system, in particular because of galaxy merger processes. However, the Bardeen–Petterson effect should have had the time to make the inner part of the disk orthogonal to the black hole spin in the case of prolonged disk accretion.

The assumption that the inner edge of the disk is at the ISCO radius has a crucial role in the spin measurements via the continuum-fitting and the iron line methods. This is because, assuming the Kerr metric, there is a one-to-one correspondence between the spin parameter a_* and the ISCO radius r_{ISCO} (see Fig. 3.4), and the exact position of the inner edge of the disk has a strong impact on the features of the spectrum. Observations show that the inner edge of the disk does not change appreciably over several years when the source is in the thermal state. The most compelling evidence comes from LMC X-3. The analysis of many spectra collected during eight X-ray missions and spanning 26 years shows that the radius of the inner

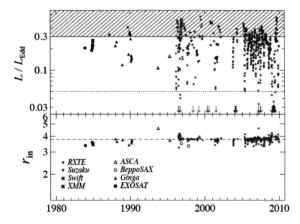

Fig. 6.1 *Top panel* Accretion disk luminosity in Eddington units versus time for 766 spectra of LMC X-3. The shaded region does not satisfy the thin disk selection criterion $L/L_{Edd} < 0.3$, as well as the data below the dotted line, which marks $L/L_{Edd} = 0.05$. *Bottom panel* fitted value of the inner disk radius of the 411 spectra in the *top panel* that can meet the thin disk selection criterion. See the text for more details. From [51]. © AAS. Reproduced with permission

edge of the disk is quite constant [51], see Fig. 6.1. The most natural interpretation is that the inner edge is associated to some intrinsic property of the geometry of the spacetime, namely the radius of the ISCO, and is not affected by variable phenomena like the accretion process.

From the theoretical point of view, the fact that the inner edge of the disk is at the ISCO radius is related to the assumption of the model that the shear stress, driving the accretion at large radii, vanishes at the ISCO. Within a hydrodynamical description, the problem has been studied in [2, 39, 46]. The conclusion of these studies is that deviations from the Novikov-Thorne model decrease monotonically with the disk thickness h/r. So, thin disks with $h/r \ll 1$ should be well described by the Novikov-Thorne model.

The case of magnetized accretion disks is more difficult to address, and studies with general relativistic magnetohydrodynamics (GRMHD) simulations have been reported in [36, 37, 41, 45]. In particular, the impact of deviations from the Novikov-Thorne model in the spin measurements via the continuum-fitting method has been discussed in [27] (see also Sect. 7.2). According to [27], there are deviations from the theoretical model, but they decrease as the disk thickness h/r decreases. This thus requires the Assumption 5 and justifies the Assumption 9. Moreover, current measurements via the continuum-fitting method are dominated by the uncertainties on the black hole mass, distance, and viewing angle, while deviations from the Novikov-Thorne model are subdominant.

The Assumption 6 is necessary because we want that the accretion process is stationary and axisymmetric. Since inhomogeneities in the accretion flow are present in the reality, physical quantities of the gas are averaged over a time interval Δt short enough to have negligible changes in the background metric (which is not a problem for our observations of stellar-mass and supermassive black holes) and long enough to assume that the system is perfectly stationary. With the same spirit, we have to average over the coordinate ϕ to remove deviations from a perfectly axisymmetric configuration. For any "real" quantity $\Psi(t, r, \theta, \phi)$, in what follows we will use the quantity

$$
\langle \Psi(r, \theta) \rangle = \frac{1}{2\pi \, \Delta t} \int_0^{\Delta t} \int_0^{2\pi} \Psi(t, r, \theta, \phi) \, dt \, d\phi .
\tag{6.1}
$$

The notation $\langle . \rangle$ will be omitted for simplicity.

The Assumption 7 requires that the radial acceleration of the gas due to pressure gradients is negligible in comparison with the gravitational acceleration due to the black hole. This requires that, as the gas falls onto the black hole, its potential energy is transported away or radiated away, and only a negligible part is converted to internal energy of the gas [40]. This is true in the case of radiatively efficient accretion flow and it is the case for thin disks.

Concerning the Assumption 8, heat advection scales as $\sim h^2$, so even this assumption requires thin accretion disks.

The condition $\lambda \ll M$ permits us to neglect possible coherent superpositions of the radiation reaction in nearby regions of the disk and superradiance phenomena. In the case of stellar-mass black holes, $M \approx 10$ km, while the radiation emitted by the inner part of the disk is mainly in the X-ray band and the condition $\lambda \ll M$ is surely satisfied. The same conclusion holds for supermassive black holes.

Returning radiation may be neglected in some cases, but not in others. For instance, it is quite important in the case of the spectrum of the polarization, see Sect. 7.3, and it is indeed included in standard calculations.

In spin measurements, it is clearly very important to select the observations and the sources in which the disk is geometrically thin and its inner edge is at the ISCO radius. In the case of the continuum-fitting method, one usually selects sources in the high-soft state with a strong contribution from the thermal disk emission. The luminosity of the source should be between $\sim 5\%$ and 20–30% of the Eddington limit [35]. At lower luminosities, the disk may be truncated. In such a case, the inner edge of the disk would be at a radius larger than the ISCO. For higher accretion rates, the disk is not thin any more: several assumptions of the model are not valid and even the inner edge of the disk may not be at the ISCO radius. Thick disks are briefly reviewed in Appendix E.

6.1.2 General Case

The power of the Novikov-Thorne model is that the time-averaged radial structure of the accretion disk can be obtained by the conservation laws of rest-mass, energy, and angular momentum; that is, it is independent of the specific properties of the accretion fluid. In this subsection, we briefly discuss the master equations; the details of the calculations can be found in the original paper [40].

From the equation of the conservation of rest-mass, $\nabla_\mu (\rho u^\mu) = 0$, where ρ is the time-averaged rest-mass density and u^μ is the time-averaged 4-velocity of the fluid, we integrate over the 3-volume of the disk between r and $r + \Delta r$ and over the time Δt. By using Gauss's theorem, we obtain that the time-averaged mass accretion rate is independent of the radius r

$$\dot{M} = -2\pi \sqrt{-G} \tilde{\Sigma} u^r = \text{const.}, \tag{6.2}$$

where $G = -\alpha^2 g_{rr} g_{\phi\phi}$ is the determinant of the near equatorial plane metric (which means it is evaluated at $\theta = \pi/2$ and depends only on r) and $\alpha^2 = g_{t\phi}^2/g_{\phi\phi} - g_{tt}$ is the lapse function. $\tilde{\Sigma}$ is the time-averaged surface density

$$\tilde{\Sigma}(r) = \int_{-h}^{h} \rho\, dz, \tag{6.3}$$

where z is the usual z-coordinate of cylindrical-like coordinates.

From the conservation laws of energy, $\nabla_\mu T^{t\mu} = 0$, and angular momentum, $\nabla_\mu T^{\phi\mu} = 0$, we can obtain the time-averaged energy flux emitted from the surface of the disk $\mathcal{F}(r)$ (as measured in the rest-frame of the accretion fluid) and the time-averaged torque $W_\phi^r(r)$

$$\mathcal{F}(r) = \frac{\dot{M}}{4\pi M^2} F(r), \tag{6.4}$$

$$W_\phi^r(r) = \frac{\dot{M}}{2\pi M^2} \frac{(\Omega L_z - E)}{\partial_r \Omega} F(r). \tag{6.5}$$

Here E, L_z, and Ω are, respectively, the conserved specific energy, the conserved axial component of the specific angular momentum, and the angular velocity for equatorial circular geodesics introduced in Sect. 3.1. $F(r)$ is the dimensionless function

$$F(r) = -\frac{\partial_r \Omega}{(E - \Omega L_z)^2} \frac{M^2}{\sqrt{-G}} \int_{r_{\text{in}}}^{r} (E - \Omega L_z)(\partial_x L_z)\, dx, \tag{6.6}$$

where r_{in} is the inner edge of the disk and is assumed to be at the ISCO radius, namely $r_{\text{in}} = r_{\text{ISCO}}$. However, in the case of a disk truncated at a larger radius it is sufficient to set a different r_{in} without changing the form of this equation.

At first approximation, the total power of the accretion process L_{acc} is converted into radiation and kinetic energy of jets/outflows, so we can write the total efficiency η, defined as $L_{\text{acc}} = \eta \dot{M}$, as

$$\eta = \eta_r + \eta_k. \tag{6.7}$$

Here η_r is the radiative efficiency and can be measured from the bolometric lumi-
nosity[2] $L_{\rm bol}$ from the equation $L_{\rm bol} = \eta_r \dot{M}$ if the mass accretion rate is known. η_k is
the fraction of gravitational energy converted to kinetic energy of jets/outflows and
is usually assumed to vanish in the Novikov-Thorne model.

The accretion process in the Novikov-Thorne model can be summarized as follows. The particles of the accreting gas slowly fall onto the central black hole. When
they reach the ISCO radius, they quickly plunge onto the black hole without emitting
additional radiation. We usually assume that η_k is negligible and the Novikov-Thorne
radiative efficiency is

$$\eta_{\rm NT} = \eta_r = 1 - E_{\rm ISCO}\,, \qquad (6.8)$$

where $E_{\rm ISCO}$ is the specific energy of a test-particle at the ISCO radius.

Actually, not all the radiation emitted by the disk can escape to infinity. A part
of it leaves the disk, but it is then captured by the black hole. In this case, Eq. (6.8)
becomes

$$\eta_{\rm NT} = 1 - E_{\rm ISCO} - \zeta_E\,, \qquad (6.9)$$

where ζ_E takes into account the radiation captured by the black hole and can be
written as [52]

$$\zeta_E = \frac{1}{\dot{M}} \int_{r_{\rm ISCO}}^{\infty} \left[\int_0^{\pi/2} \int_0^{2\pi} C\,\Upsilon\,(-n_t)\cos\theta\,\sin\theta\,d\theta\,d\phi \right] \mathscr{F}(r)\,4r\,dr\,. \quad (6.10)$$

Here $C = 0$ (1) for the radiation that escapes to infinity (is captured by the black
hole), Υ takes into account possible angular dependence of the emission process
(for instance, $\Upsilon = 1$ for isotropic emission, and $\Upsilon \propto 1 + 2\cos\theta$ for limb-darkened
emission expected in an electron scattering atmosphere), and $n^\mu = k^\mu/k_{(t)}$ is the
normalized photon 4-momentum, where k^μ is the photon 4-momentum and $k_{(t)}$ is
the photon energy in the rest-frame of the emitter. See [52] and references therein
for more details.

Equation (6.9) receives an additional small correction if we include the returning
radiation; that is, some photons leave the disk and return to the disk due to the strong
light bending in the vicinity of the black hole [29].

6.1.3 Kerr Spacetime

In the special case of the Kerr metric, we can plug the expressions found in Sect. 3.2
into the equations of the time-averaged radial structure of the accretion disk. E, L_z,

---

[2]The bolometric luminosity is the total electromagnetic luminosity of an object, namely the electromagnetic luminosity integrated over all wavelengths.

and Ω are given, respectively, by Eqs. (3.20), (3.21), and (3.22). $\sqrt{-G} = r$ and $F(r)$ in Eq. (6.6) can be written as[3]

$$
F(x) = \frac{3}{2} \frac{1}{x^4 \left(x^3 - 3x + 2a_*\right)} \left[x - x_0 - \frac{3}{2} a_* \ln\left(\frac{x}{x_0}\right) \right.
$$

$$
- \frac{3 \left(x_1 - a_*\right)^2}{x_1 \left(x_1 - x_2\right) \left(x_1 - x_3\right)} \ln\left(\frac{x - x_1}{x_0 - x_1}\right) - \frac{3 \left(x_2 - a_*\right)^2}{x_2 \left(x_2 - x_1\right) \left(x_2 - x_3\right)} \ln\left(\frac{x - x_2}{x_0 - x_2}\right)
$$

$$
\left. - \frac{3 \left(x_3 - a_*\right)^2}{x_3 \left(x_3 - x_1\right) \left(x_3 - x_2\right)} \ln\left(\frac{x - x_3}{x_0 - x_3}\right) \right],
\tag{6.11}
$$

where $x = \sqrt{r/M}$, $x_0 = \sqrt{r_{\rm ISCO}/M}$, and x_1, x_2, and x_3 are the three roots of $x^3 - 3x + 2a_* = 0$, so

$$
x_1 = 2\cos\left(\frac{1}{3} \arccos a_* - \frac{\pi}{3}\right), \qquad x_2 = 2\cos\left(\frac{1}{3} \arccos a_* + \frac{\pi}{3}\right),
$$

$$
x_3 = -2\cos\left(\frac{1}{3} \arccos a_*\right).
\tag{6.12}
$$

Neglecting the radiation emitted by the disk and captured by the black hole or returning to the disk, the Novikov-Thorne radiative efficiency $\eta_{\rm NT} = 1 - E_{\rm ISCO}$ can be written with the help of Eq. (3.20) evaluated at the ISCO radius (3.27). $\eta_{\rm NT}$ only depends on the value of the black hole spin parameter a_* and is a monotonically increasing (decreasing) function of the spin for a corotating (counterrotating) disk, see Fig. 6.2. In particular, for the special cases of Schwarzschild and extremal Kerr black holes, we have

$$
\eta_{\rm NT}(a_* = 0) = 1 - \frac{2\sqrt{2}}{3} \approx 0.057,
$$

$$
\eta_{\rm NT}(a_* = 1) = 1 - \frac{1}{\sqrt{3}} \approx 0.423 \quad \text{(corotating disk)},
$$

$$
\eta_{\rm NT}(a_* = 1) = 1 - \frac{5}{\sqrt{27}} \approx 0.038 \quad \text{(counterrotating disk)}.
\tag{6.13}
$$

It is worth noting that the accretion process onto a black hole can be a very efficient mechanism to convert rest-mass into radiation. In the case of nuclear fusion inside stars, less than 1% of rest-mass is converted to energy. As we can see here, the accretion process onto a black hole is much more efficient and can reach $\sim$42% if the black hole has $a_* = 1$.

---

[3]The expression of F here is slightly different from that in [40] because our F has the factor $M^2/\sqrt{-G}$ in order to have this quantity dimensionless and independent of M.

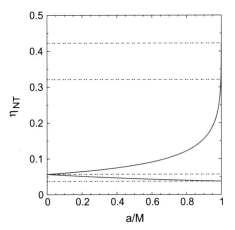

Fig. 6.2 Novikov-Thorne radiative efficiency $\eta_{NT} = 1 - E_{ISCO}$ as a function of the spin a in the Kerr metric. The *upper curve* is for corotating orbits, the *lower curve* for counterrotating orbits. The *dotted horizontal lines* correspond (from *top* to *bottom*) to the Novikov-Thorne radiative efficiency for $a/M = 1$ ($\eta_{NT} \approx 0.423$), $a/M = 0.998$ ($\eta_{NT} \approx 0.321$), $a/M = 0$ ($\eta_{NT} \approx 0.057$), and $a/M = 1$ in the case of a counterrotating disk ($\eta_{NT} \approx 0.038$)

6.2   Transfer Function for Thin Disks

Now we want to calculate the spectrum of the accretion disk of a black hole as seen by a distant observer. In the case of thin accretion disks, the calculations can be conveniently split into two parts with the approach of the *transfer function* [15]: (i) the calculation of the local spectrum of the radiation at the surface of the disk, and (ii) the calculation of the propagation of the radiation from the disk to the detector in the flat faraway region.

In some cases, the local spectrum of the radiation is only determined by the astrophysical model and is independent of the background metric. The local spectrum of the radiation on the surface of the disk is described by the specific intensity $I_e(\nu_e, r_e, \vartheta_e)$ as measured in the rest-frame of the accreting gas. I_e typically depends on the photon frequency ν_e, on the emission radius r_e, and possibly on the emission angle with respect to the normal to the disk ϑ_e. The frequency ν_e and the emission angle ϑ_e are given as measured in the rest-frame of the fluid. If the system is not axisymmetric, the local spectrum depends on the exact position on the disk, not just on the emission radius. In the same way, the spectrum may depend on two angles, not only ϑ_e.

The observed flux (for instance, in erg s^{-1} cm^{-2} Hz^{-1}) can be written as

$$F_{obs}(\nu_{obs}) = \int I_{obs}(\nu_{obs}, X, Y) d\tilde{\Omega} = \int g^3 I_e(\nu_e, r_e, \vartheta_e) d\tilde{\Omega}, \qquad (6.14)$$

where I_{obs} is the specific intensities of the radiation detected by the distant observer (for instance, in erg s^{-1} cm^{-2} str^{-1} Hz^{-1}), $d\tilde{\Omega} = dXdY/D^2$ is the element of the solid angle subtended by the image of the disk in the observer's sky, and D is the distance of the observer from the source. $I_{obs} = g^3 I_e$ follows from Liouville's theorem [31], where $g = \nu_{obs}/\nu_e$ is the redshift factor.

The transfer function $f(g^*, r_e, i)$ only depends on the metric of the spacetime and the position of the observer. It takes into account all the relativistic effects (gravitational redshift, Doppler boosting, light bending). For a specified metric, the transfer function depends on the emission radius r_e, the viewing angle of the distant observer i, and the relative redshift factor g^* defined as

$$g^* = \frac{g - g_{min}}{g_{max} - g_{min}}, \tag{6.15}$$

which ranges from 0 to 1. Here $g_{max} = g_{max}(r_e, i)$ and $g_{min} = g_{min}(r_e, i)$ are, respectively, the maximum and the minimum values of the redshift factor g for the photons emitted from the radial coordinate r_e and detected by a distant observer with polar coordinate i. Figure 6.3 remarks the difference between the emission angle ϑ_e and the viewing angle i. The former is the angle of propagation of a photon with the normal to the disk at its emission point (as measured in the rest-frame of the fluid). The latter is the inclination angle of the whole disk with respect to the distant observer.

Introducing the transfer function f, the observed flux can be rewritten as

$$F_{obs}(\nu_{obs}) = \frac{1}{D^2} \int_{r_{ISCO}}^{\infty} \int_0^1 \pi r_e \frac{g^2}{\sqrt{g^*(1-g^*)}} f(g^*, r_e, i) I_e(\nu_e, r_e, \vartheta_e) \, dg^* \, dr_e. \tag{6.16}$$

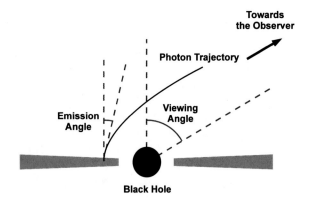

Fig. 6.3 The viewing angle i is the inclination angle of the disk, namely the angle between the axis normal to the disk and the line of sight of the distant observer; it is the same in Newtonian gravity and in general relativity. The emission angle ϑ_e is the angle, measured in the rest-frame of the gas and at a certain emission point, between the normal to the disk and the photon propagation direction. In Newtonian gravity, $i = \vartheta_e$, but in a curved spacetime the two angles are different in general

Fig. 6.4 Impact of the viewing angle i on the transfer function f. The spacetime is described by the Kerr metric with the spin parameter $a_* = 0.998$. The emission radius is $r_e = 4\,M$. The values of the viewing angle are indicated. From [7]

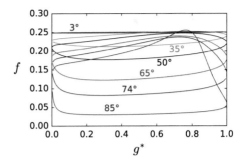

Fig. 6.5 Impact of the dimensionless spin parameter a_* on the transfer function f. The spacetime is described by the Kerr metric, the emission radius is $r_e = 7\,M$, and the viewing angle is $i = 30°$. The values of the spin parameter are indicated. From [7]

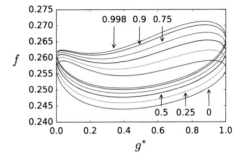

The expression of the transfer function f is

$$f(g^*, r_e, i) = \frac{1}{\pi r_e} g \sqrt{g^*(1 - g^*)} \left| \frac{\partial(X, Y)}{\partial(g^*, r_e)} \right|, \tag{6.17}$$

where $|\partial(X, Y)/\partial(g^*, r_e)|$ is the Jacobian. The transfer function thus acts as an integration kernel to calculate the spectrum detected by the distant observer starting from the local spectrum at any point of the disk. Let us note that in the specific intensity I_e, ν_e and ϑ_e must be written in terms of g^* and r_e.

In the Kerr metric, for a given emitting radius r_e and viewing angle i, the transfer function is a closed curve parameterized by g^*, see Figs. 6.4 and 6.5. This is true except in the special case $i = 0$. The points $g^* = 1$ and $g^* = 0$ are connected by two curves, so we have two branches of the transfer function, say $f_1(g^*, r_e, i)$ and $f_2(g^*, r_e, i)$. This is because two points of the same ring with radius r_e have the same redshift factor, but the points are different and therefore the transfer function is usually different. In non-Kerr metrics, it is possible to find some differences. For example, the transfer function may not be a closed curve if there is a region of the disk that cannot be seen by the distant observer. This can happen in some metrics for large values of the viewing angle, as a consequence of the different light bending.

In the case of isotropic emission (I_e independent of ϑ_e and of the emission azimuthal angle) in an axisymmetric system (e.g. no orbiting spots), Eq. (6.16) can be written as

$$F_{obs}(\nu_{obs}) = \frac{1}{D^2} \int_{r_{ISCO}}^{\infty} \int_0^1 \frac{\pi r_e\, g^2}{\sqrt{g^*(1-g^*)}} \left[f_1(g^*, r_e, i) + f_2(g^*, r_e, i) \right] I_e(\nu_e, r_e)\, dg^*\, dr_e\,.$$

$$(6.18)$$

If I_e does depend on ϑ_e, it is necessary to perform the integral twice, for the upper branch and for the lower one, so Eq. (6.16) becomes

$$F_{obs}(\nu_{obs}) = \frac{1}{D^2} \int_{r_{ISCO}}^{\infty} \int_0^1 \frac{\pi r_e\, g^2}{\sqrt{g^*(1-g^*)}} f_1(g^*, r_e, i) I_e(\nu_e, r_e, \vartheta_{e,1})\, dg^*\, dr_e$$

$$+ \frac{1}{D^2} \int_{r_{ISCO}}^{\infty} \int_0^1 \frac{\pi r_e\, g^2}{\sqrt{g^*(1-g^*)}} f_2(g^*, r_e, i) I_e(\nu_e, r_e, \vartheta_{e,2})\, dg^*\, dr_e\,, \quad (6.19)$$

where $\vartheta_{e,1}$ and $\vartheta_{e,2}$ indicate the emission angles with relative redshift factor g^*, respectively in the branches 1 and 2.

The accretion disk is optically thick, so it is opaque and cannot be crossed by photons. Moreover, higher order images produced by photons orbiting one or more times the compact object are usually neglected in the calculations.

6.3   Calculation of the Transfer Function

6.3.1   General Case

In this subsection, we discuss the calculation of the transfer function in a generic stationary and axisymmetric spacetime. We have to relate the position of the photon in the image plane of the distant observer, (X, Y), with the emission point in the disk, to know the emission radius r_e, the redshift factor g, and the emission angle ϑ_e. In the general case, it is necessary to consider a grid of photons and numerically compute their trajectories from the image plane to the disk by solving the geodesic equations. The photon initial conditions have been already presented in Sect. 3.4.

Solving the geodesic equations for a photon with position in the image plane (X, Y), we find the position of emission in the disk. The redshift factor g is, by definition, the ratio between the photon frequency measured by the distant observer ν_{obs} and that measured in the reference frame of the gas ν_e, and therefore

$$g = \frac{\nu_{obs}}{\nu_e} = \frac{-u_{obs}^{\mu} k_{\mu}}{-u_e^{\nu} k_{\nu}}\,, \quad (6.20)$$

where $u_{obs}^{\mu} = (1, 0, 0, 0)$ is the 4-velocity of the distant observer, k^{μ} is the 4-momentum of the photon, $u_e^{\nu} = u_e^t(1, 0, 0, \Omega)$ is the 4-velocity of the particles of the gas, and $\Omega = u_e^{\phi}/u_e^t$ is the angular velocity of the gas.[4] From the normalization

---

[4] In the notation of Chap. 3, $i = u_e^t$ and $\dot{\phi} = u_e^{\phi}$.

condition $g_{\mu\nu}u_e^\mu u_e^\nu = -1$, we can find the expression for u_e^t

$$u_e^t = \frac{1}{\sqrt{-g_{tt} - 2g_{t\phi}\Omega - g_{\phi\phi}\Omega^2}} \, . \tag{6.21}$$

Plugging Eq. (6.21) into Eq. (6.20), we obtain

$$g = \frac{\sqrt{-g_{tt} - 2g_{t\phi}\Omega - g_{\phi\phi}\Omega^2}}{1 - \lambda\Omega} \, , \tag{6.22}$$

where $\lambda = -k_\phi/k_t$ as in Eq. (3.73) ($k_t = -E$, where E is the photon energy) is a constant of motion along the photon trajectory (as a consequence of the fact that the spacetime is stationary and axisymmetric).

If the local spectrum depends on the emission angle ϑ_e, the latter must be rewritten in terms of the emission radius and redshift factor. The normal to the disk is

$$n^\mu = \left(0, 0, \sqrt{g^{\theta\theta}}, 0\right)\Big|_{r_e, \theta_e = \pi/2} \, , \tag{6.23}$$

and therefore the cosine of the emission angle ϑ_e is

$$\cos\vartheta_e = \pm \frac{n^\mu k_\mu}{u_e^\nu k_\nu}\Big|_e = \pm\sqrt{g^{\theta\theta}} \frac{\sqrt{-g_{tt} - 2g_{t\phi}\Omega - g_{\phi\phi}\Omega^2}}{1 - \lambda\Omega} \frac{k_\theta}{k_t} \, , \tag{6.24}$$

where k_θ is the θ-component of the 4-momentum of the photon at the point of emission in the disk and, in the general case, it is determined at the end of the geodesic integration. In Eq. (6.24) $\cos\vartheta_e > 0$ and therefore, for viewing angles $0 < i < \pi/2$, the sign is $+$ if we are considering photons moving from the distant observer to the disk and $-$ in the case of photons moving from the disk to the distant observer.

At the end of the integration of the photon trajectories we have $r_e = r_e(X, Y)$, $g = g(X, Y)$, and $\vartheta_e = \vartheta_e(X, Y)$. From the first two relations, it is possible to numerically compute the Jacobian in the transfer function

$$\left|\frac{\partial(X, Y)}{\partial(g^*, r_e)}\right| = (g_{max} - g_{min}) \left|\frac{\partial X}{\partial g}\frac{\partial Y}{\partial r_e} - \frac{\partial X}{\partial r_e}\frac{\partial Y}{\partial g}\right| \, . \tag{6.25}$$

This completes the calculation of the transfer function f for a specific background metric. If we know the local spectrum of the radiation, we can obtain the observed flux via Eq. (6.16).

The numerical calculation of the transfer function in a generic stationary, axisymmetric, and asymptotically flat spacetime is discussed in [7]. Since the calculation process takes time, the standard strategy is to tabulate the transfer function. The observed spectrum is calculated during the data analysis, by integrating Eq. (6.16) with the tabulated f and the specific intensity from the astrophysical model under consideration.

6.3.2   Kerr Spacetime

In the Kerr metric, it is possible to exploit the existence of the Carter constant and proceed in a different way which, after some more tedious analytical calculations, permits us to solve some elliptic integrals rather than the second order geodesic equations. The advantage of the elliptic integrals is that the numerical calculations become faster and more accurate.

First, since the photon trajectories are independent of the photon energy, as done at the very end of Sect. 3.4 it is convenient to use the two parameters $\lambda = -k_\phi / k_t$ and $q^2 = \mathcal{Q}/k_t^2$, which are constants of motion.

In the Kerr metric, the angular velocity is given by Eq. (3.22), and Eq. (6.21) becomes

$$u_e^t = \frac{r_e^{3/2} + aM^{1/2}}{r_e^{1/2}\sqrt{r_e^2 - 3Mr_e + 2aM^{1/2}r_e^{1/2}}} . \tag{6.26}$$

The redshift factor in Eq. (6.22) reduces to

$$g = \frac{r_e^{1/2}\sqrt{r_e^2 - 3Mr_e + 2aM^{1/2}r_e^{1/2}}}{r_e^{3/2} + aM^{1/2} - M^{1/2}\lambda} . \tag{6.27}$$

The normal to the disk at the point of emission becomes $n^\mu = (0, 0, 1/r_e, 0)$ and the emission angle ϑ_e assumes quite a compact form

$$\cos\vartheta_e = \frac{qg}{r_e} , \tag{6.28}$$

because $u_e^\nu k_\nu = k_t/g$ and $q = -k_\theta/k_t$.

As done in Sect. 3.4, the photon position in the image plane of the distant observer can be written in terms of the constants of motion λ and q

$$X = \frac{\lambda}{\sin i} , \qquad Y = \pm\sqrt{q^2 + a^2\cos^2 i - \lambda^2\cot^2 i} . \tag{6.29}$$

The Jacobian in the transfer function f can thus be rewritten as

$$\left| \frac{\partial(X,Y)}{\partial(g^*, r_e)} \right| = \frac{q\,(g_{max} - g_{min})}{Y\sin i} \left| \frac{\partial\lambda}{\partial g}\frac{\partial q}{\partial r_e} - \frac{\partial\lambda}{\partial r_e}\frac{\partial q}{\partial g} \right| . \tag{6.30}$$

From Eq. (6.27), we can write $\lambda = \lambda(g, r_e)$

$$\lambda = \frac{r_e^{3/2}}{M^{1/2}} + a - \frac{r_e^{1/2}}{gM^{1/2}}\sqrt{r_e^2 - 3Mr_e + 2aM^{1/2}r_e^{1/2}} . \tag{6.31}$$

In order to calculate $\partial q/\partial r_e$ and $\partial q/\partial g$ in Eq. (6.30), it is sufficient to solve the photon trajectories on the (r, θ) plane. If we define

$$\tilde{R} = \frac{R}{E^2} = r^4 + \left(a^2 - \lambda^2 - q^2\right) r^2 + 2M \left[q^2 + (\lambda - a)^2\right] r - a^2 q^2 \,, \quad (6.32)$$

$$\tilde{\Theta} = \frac{\Theta}{E^2} = q^2 + a^2 \cos^2 \theta - \lambda^2 \cot^2 \theta \,, \quad (6.33)$$

the equation to solve is

$$s_r \int_{r_e}^{D} \frac{dr'}{\sqrt{\tilde{R}}} = s_\theta \int_{\pi/2}^{i} \frac{d\theta'}{\sqrt{\tilde{\Theta}}} \,, \quad (6.34)$$

where $s_r = \pm 1$ and $s_\theta = \pm 1$, depending on the photon propagation direction.

The two integrals in Eq. (6.34) can be reduced to elliptic integrals, see Appendix B for more details. Roughly speaking, we have

$$\int_{r_e}^{D} \frac{dr'}{\sqrt{\tilde{R}}} = C_r F[\psi_r(r_e), \kappa_r] \,,$$

$$\int_{\pi/2}^{i} \frac{d\theta'}{\sqrt{\tilde{\Theta}}} = C_\theta F[\psi_\theta(\pi/2), \kappa_\theta] \,, \quad (6.35)$$

where F is the elliptic integral of the first kind with argument ψ_j and modulus κ_j $(j = r, \theta)$. C_r, C_θ, ψ_r, ψ_θ, κ_r, and κ_θ are functions of λ and q. From these equations, one can write r_e as a function of λ and q

$$r_e(\lambda, q) = \psi_r^{-1} \left\{ F^{-1} \left[\frac{C_\theta}{C_r} F[\psi_\theta(\pi/2), \kappa_\theta], \kappa_r \right] \right\} \,. \quad (6.36)$$

Special attention has to be paid for the direction of propagation, namely for s_r and s_θ. The latter change sign at the possible turning points r_t and θ_t, which occur when $\tilde{R} = 0$ and $\tilde{\Theta} = 0$, respectively [14].

If we plug Eq. (6.31) into Eq. (6.36), we can write $q = q(g, r_e)$. With $\lambda = \lambda(g, r_e)$ and $q = q(g, r_e)$, we can evaluate the Jacobian in the transfer function, Eq. (6.30). This is the approach employed in [49]. In [15], the calculations to arrive at the transfer function are slightly different. One obtains $g = g(\lambda, q)$ and $r_e = r_e(\lambda, q)$ to calculate the inverse Jacobian. However, since one typically wants to compute the transfer function for some specific values of g and r_e and does not know λ and q, the approach of [15] requires additional calculations.

It is worth noting that usually these calculations are not possible in non-trivial generalizations of the Kerr metric, because in the Kerr metric the function $\tilde{R}$ is already a polynomial of fourth order. In specific generalizations of the Kerr metric, the integrals (6.35) can be transformed to a hyper-elliptic integral [30]. In the more general case of a Petrov type D spacetime, this may not be possible and eventually there may not be a significant advantage to have the equations of motion separable and of the first order.

6.4   Evolution of the Spin Parameter

Accretion from a thin disk is an efficient mechanism to spin a black hole up. The accreting gas moves on nearly geodesic circular orbits in the equatorial plane. As the gas loses energy and angular momentum, it approaches the ISCO radius and then quickly plunges onto the central object without an appreciable emission of additional radiation. In this case, the mass and spin angular momentum of the black hole change as

$$M \to M + \delta M , \quad J \to J + \delta J , \tag{6.37}$$

where δM and δJ are, respectively, the mass and the angular momentum carried by the gas, namely

$$\delta M = E_{\text{ISCO}} \delta m , \quad \delta J = L_{\text{ISCO}} \delta m . \tag{6.38}$$

E_{ISCO} and L_{ISCO} are, respectively, the specific energy and the specific angular momentum of the gas at the ISCO radius, while δm is the gas rest-mass. The evolution of the spin parameter turns out to be governed by the following equation [10]

$$\frac{da_*}{d \ln M} = \frac{1}{M} \frac{L_{\text{ISCO}}}{E_{\text{ISCO}}} - 2a_* . \tag{6.39}$$

The black hole is spun up if the right hand side is positive, and spun down if the right hand side is negative. The equilibrium value of the spin parameter is reached when the right hand side vanishes and only depends on the background metric; that is, different types of black holes may have different equilibrium values of the spin parameter.

In the case of the Kerr metric, it is possible to integrate Eq. (6.39) and find an analytic expression for the spin parameter a_* as a function of the black hole mass M. The solution is [10]

$$a_* = \begin{cases} \sqrt{\frac{2}{3} \frac{M_0}{M}} \left[4 - \sqrt{18 \frac{M_0^2}{M^2} - 2} \right] & \text{if } M \leq \sqrt{6} \, M_0 , \\ 1 & \text{if } M > \sqrt{6} \, M_0 , \end{cases} \tag{6.40}$$

assuming an initially non-rotating black hole with mass M_0. The plot of the function $a_*(M)$ is shown in Fig. 6.6. From Eq. (6.40), the equilibrium value of the spin parameter is 1 and requires that the black hole increases its mass by the factor $\sqrt{6} \approx 2.4$. As shown in Fig. 6.6, at the beginning the spin increases quite quickly. The black

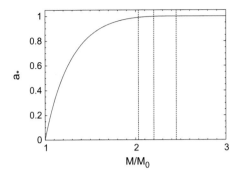

Fig. 6.6 Evolution of the spin parameter a_* of a Kerr black hole accreting from a thin disk according to the formula in Eq. (6.40), in which it is neglected the radiation emitted by the disk and captured by the *black hole* as well as the so-called returning radiation. The *black hole* has the initial mass M_0 and the initial spin parameter $a_* = 0$. The three *dashed vertical lines* indicate when the *black hole* spin reaches the values $a_* = 0.99\,(M/M_0 \approx 2.03), 0.998\,(M/M_0 \approx 2.20)$, and $1\,(M/M_0 = \sqrt{6} \approx 2.45)$

hole has roughly to double its mass to get the spin $a_* = 0.99$, but it still needs a non-negligible amount of gas to reach 1.

If we include the effect of the radiation emitted by the disk and captured by the black hole, Eq. (6.38) becomes

$$\delta M = (E_{\text{ISCO}} + \zeta_E)\,\delta m\,, \quad \delta J = (L_{\text{ISCO}} + \zeta_L)\,\delta m\,, \tag{6.41}$$

where ζ_E is given by Eq. (6.10) and ζ_L is [52]

$$\zeta_L = \frac{1}{\dot{M}} \int_{r_{\text{ISCO}}}^{\infty} \left[\int_0^{\pi/2} \int_0^{2\pi} C\,\Upsilon\,n_\phi\cos\theta\,\sin\theta d\theta d\phi \right] \mathscr{F}(r)\,4r dr\,. \tag{6.42}$$

As in Eq. (6.10), $C = 0$ (1) for the radiation that escapes to infinity (is captured by the black hole), while Υ takes into account the fact that the emission may not be isotropic. Equation (6.39) becomes

$$\frac{da_*}{d\ln M} = \frac{1}{M}\frac{L_{\text{ISCO}} + \zeta_L}{E_{\text{ISCO}} + \zeta_E} - 2a_*\,. \tag{6.43}$$

In the Kerr spacetime, from Eq. (6.43) the equilibrium value of the spin parameter is the well-known Thorne limit $a_*^{\text{Th}} \approx 0.998$ [52]. While this value is very close to 1, if $a_* = a_*^{\text{Th}}$ we find $\eta_{\text{NT}} \approx 0.321$, which is not very close to the maximum radiative efficiency $\eta_{\text{NT}}^{\text{max}} \approx 0.423$ for $a_* = 1$.

The fact that it is impossible to spin a black hole up to the extremal limit $a_* = 1$ is consistent with the third law of black hole mechanics [11]. In analogy with the third law of thermodynamics, in which it is impossible to reduce the temperature of a system to zero in a finite number of operations, the third law of black hole mechanics forbids to reach an extremal Kerr black hole from accretion of a finite amount of mass.

The Thorne limit receives corrections from the returning radiation, namely the radiation emitted by the disk and returning to the disk due to the strong light bending in the vicinity of the black hole. Taking this effect into account, the equilibrium value of the spin parameter becomes $a_* \approx 0.9983$ for isotropic emission and $a_* \approx 0.9986$ for limb-darkened emission [29].

Equation (6.39) and its extensions can be applied to non-Kerr black hole solutions in which a test-particle plunges from the ISCO radius onto the black hole (but exceptions in which this is not true are possible [6]). In this case, the equilibrium spin parameter may be either larger or smaller than 1, depending on the background geometry [3, 4].

In the case of a thick disk, it is possible to use an expression similar to Eq. (6.39), see Appendix E.2 and references therein. The accretion process from a thick disk may be, at least in principle, even more efficient to spin a black hole up [1, 44].

The presence of magnetic fields in the plunging region may reduce the equilibrium value of the spin parameter because magnetic fields can transport angular momentum outward. For instance, in the simulations in [19, 48], the equilibrium spin parameter is around 0.95.

6.4.1   Spins of Black Holes

Generally speaking, the value of the spin parameter of a black hole can be expected to be determined by the competition of three physical processes: the event creating the object, mergers, and gas accretion.

In the case of black hole binaries, it is usually thought that the spin of a black hole is mainly natal and that the effect of the accretion process is negligible [23]. However, see [17] for a different conclusion. The argument is that a stellar-mass black hole has a mass around 10 $M_\odot$. If the stellar companion is a few Solar masses, the black hole cannot significantly change its mass and spin angular momentum even swallowing the whole star. If the stellar companion is heavy, its lifetime is too short: even if the black hole accretes at the Eddington rate, there is not the time to transfer the necessary amount of matter to significantly change the black hole spin parameter. One may expect that a black hole cannot swallow more than a few $M_\odot$ from the companion star, and for a 10 $M_\odot$ object this is not enough to significantly changes a_* [23].

If the black hole spin were mainly natal, its value should be explained by studying the gravitational collapse of massive stars. While there are still uncertainties in the angular momentum transport mechanisms of the progenitors of stellar-mass black holes, it is widely accepted that the gravitational collapse of a massive star with Solar metallicity cannot create fast-rotating remnants [54, 55]. The birth spin of these black holes is expected to be low (see e.g. [17] and references therein). However, this is not what we observe. As it is discussed in Chaps. 7 and 8, black hole spin measurements show that some of these objects have a spin parameter close to 1. In the case of LMXBs, the black hole in GRS 1915+105 has $a_* > 0.98$ [35] and $M = 12.4 \pm 2.0 \ M_\odot$ [43], while the stellar companion's mass is $M = 0.52 \pm 0.41 \ M_\odot$. In the case of HMXBs, the black hole in Cygnus X-1 has $a_* > 0.98$ [20] and $M = 14.8 \pm 1.0 \ M_\odot$, while the stellar wind from the companion is not an efficient mechanism to transfer mass. Both the spin constraints are at 3-σ.

While black holes in LMXBs and HMXBs should form in different environments, in both cases the origin of so high spin values is not well understood. Fragos and McClintock [17] show indeed that the accretion process immediately after the formation of a black hole binary may be very important and eventually responsible for the high spins of black holes in LMXBs. A more speculative explanation is discussed in [5]. If the metric around astrophysical black holes had some particular deviations from the Kerr background, the accreting objects may look like very fast-rotating Kerr black holes after swallowing a few Solar masses from the companion star, and this would make observations consistent with the expected low value of the initial spin.

The case of supermassive black holes in galactic nuclei is different. The initial value of their spin parameter is likely completely irrelevant: their mass has increased by several orders of magnitude from its original value, and the spin parameter has evolved accordingly. On average, the capture of small bodies (minor merger) in randomly oriented orbits should spin the compact object down, since the magnitude of the orbital angular momentum for corotating orbits is always smaller than the one for counterrotating orbits [21]. The case of coalescence of two compact objects with comparable mass (major merger) can be rigorously computed only if we know the exact nature of these objects and the theory of gravity, as the background is not fixed and the emission of gravitational waves is important. In general relativity and in the case of random merger of two black holes, the most probable final product is a black hole with $a_* \approx 0.70$, while fast-rotating objects with $a_* > 0.9$ should be rare [13].

Accretion from a disk can potentially be a very efficient way to spin a compact object up [13]. If accretion proceeds via short episodes (chaotic accretion) [25], the net effect is not different from minor mergers in randomly oriented orbits and the compact object is spun down. On the contrary, prolonged disk accretion is a very efficient mechanism to spin the compact object up. In this case, Kerr black holes in AGN may have a spin parameter close to the Thorne limit [13]. Such a possibility seems to be supported by some observations [53]. Nearby supermassive black hole accreting from a more spherically symmetric gas distribution may instead have lost most of their spin angular momentum acquired in a previous stage and have lower spin, or they may be in the process of spinning down.

6.5 Deviations from the Kerr Metric

The Kerr metric should be able to well describe the gravitational field around an astrophysical black hole. In general relativity, initial deviations from the Kerr solution are indeed quickly radiated away through the emission of gravitational waves ("black

holes rapidly go bald") [42]. As shown in the next subsections, the gravitational field
of an accretion disk is normally completely negligible, and even the black hole
electric charge can be safely ignored in the metric of the spacetime.

6.5.1 Accretion Disk

The mass of the accretion disk inevitably introduces deviations from the Kerr background. However, on general grounds, it is clear that for typical accretion disks the
effect should be completely negligible. If M is the mass of the black hole and m is
the mass of the disk, in the case of a stellar-mass black hole in a binary the ratio is[5]
$m/M \sim 10^{-9}$–10^{-10}. This is a small parameter and it can be used as the expansion
parameter in the corrections of observational quantities from the Kerr case. The corrections will be of a similar order of magnitude and therefore negligible for present
and future observations.

For instance, the impact of the mass of the accretion disk in the measurement
of the black hole spin is discussed in [9] within a simple analytical model. In Weyl
coordinates $\{t, \rho, z, \phi\}$, the general form of a static and axisymmetric metric depends
only on the two functions $\lambda(\rho, z)$ and $\nu(\rho, z)$ in the line element

$$ds^2 = -e^{2\lambda}dt^2 + e^{2(\nu-\lambda)}\left(d\rho^2 + dz^2\right) + \rho^2 e^{-2\lambda}d\phi^2 . \tag{6.44}$$

Writing the Einstein equations for the vacuum, it is easy to see that, if λ_1 and λ_2 are
two solutions, then $\lambda_3 = \lambda_1 + \lambda_2$ is also a solution. The non-linearity of the Einstein
equations shows up in the function ν, which is given by

$$\nu_3 = \nu_1 + \nu_2 + 2\int \rho \left[\left(\frac{\partial\lambda_1}{\partial\rho}\frac{\partial\lambda_2}{\partial\rho} - \frac{\partial\lambda_1}{\partial z}\frac{\partial\lambda_2}{\partial z}\right)d\rho \right.$$
$$\left. + \left(\frac{\partial\lambda_1}{\partial\rho}\frac{\partial\lambda_2}{\partial z} + \frac{\partial\lambda_1}{\partial z}\frac{\partial\lambda_2}{\partial\rho}\right)dz\right], \tag{6.45}$$

where ν_1 and ν_2 are the solutions of the function ν associated, respectively, to the
solutions λ_1 and λ_2.

If we combine the Schwarzschild solution in Weyl coordinates, which is given
by the two functions (λ_{BH}, ν_{BH}), with the Lemos–Letelier solution (λ_D, ν_D), which
describes a thin disk, we obtain the metric of a spacetime with a central non-rotating
black hole and a thin accretion disk. The solution has three free parameters: the black
hole mass M, the disk mass m, and the inner radius of the disk b. The Lemos–Letelier

---

[5]A rough estimate of m/M can be obtained in the following way. For a black hole with $M = 10\,M_\odot$
accreting at $\sim$10% of the Eddington limit, the mass accretion rate is $\sim 10^{-8}\,M_\odot$/yr; see e.g. Eq. (G.2).
Considering that the disk may be essentially created and destroyed during an outburst, and that the
latter may last for about one month, we find that the mass of the disk is $m \sim 10^{-9}\,M_\odot$. In this case,
$m/M \sim 10^{-10}$.

disk is not the kind of thin disk expected around an astrophysical black hole, but it can work as a toy model to get a rough estimate of the impact of the mass of the disk in a spin measurement.

In the continuum-fitting and the iron line methods discussed in the next chapters, the key-quantity is the position of the ISCO radius. To be more precise, the radial coordinate of the ISCO has no physical meaning, being determined by the coordinate system, which is arbitrary. A good proxy for the measurement of the spin is the Novikov-Thorne radiative efficiency $\eta_{NT} = 1 - E_{ISCO}$ [26]. In the Kerr metric, the Novikov-Thorne radiative efficiency for small values of the spin is

$$\eta_{NT} = 1 - \frac{2\sqrt{2}}{3} \mp \frac{\sqrt{3}}{54} \left(\frac{a}{M}\right) + O\left(\frac{a^2}{M^2}\right). \tag{6.46}$$

Following Sect. 3.1, we can obtain the ISCO radius and E_{ISCO} for a generic stationary and axisymmetric spacetime. For $m \ll M$, $mb^2 \ll M^3$, and assuming the inner edge of the disk b at the ISCO radius, for the metric of a non-rotating black hole surrounded by a Lemos–Letelier disk, we find [9]

$$\eta_{NT} = 1 - \frac{2\sqrt{2}}{3} + \frac{49}{144\sqrt{3}} \left(\frac{m}{M}\right) + O\left(\frac{m^2}{M^2}\right). \tag{6.47}$$

Even in the most favorable conditions, the spin measurement may reach a precision at the level of some per cent. If $m/M \sim 10^{-9}$, the mass of the disk can be definitively neglected. It should be taken into account only if we could measure the spin with a similar precision, which is out of reach even in the future.

6.5.2 Electric Charge

In general, astrophysical bodies may acquire a non-vanishing electric charge because of the difference between the mass of protons and electrons [56]. For instance, in the case of an object surrounded by an atmosphere of positive ions and electrons in thermal equilibrium, when the object is uncharged the escape velocity can be reached easier by electrons, because their mass is smaller. As electrons leave the atmosphere, the electrostatic field around the object grows, and an equilibrium is eventually reached when the sums of all forces acting on ions and electrons are equal, which results in a non-vanishing electric charge of the object.

In the case of an accreting black hole, the gravitational force acting on protons and electrons is the same, but the electromagnetic radiation and the electrostatic force act differently on the two particle species. A rough estimate of the equilibrium electric charge of an accreting black hole can be obtained as follows. We consider spherically symmetric accretion of a plasma of protons and electrons onto a spherically symmetric object of mass M, ignoring relativistic effects. The equations of motion for the proton and electron fluids are

$$m_p \dot{v}_p = -\frac{m_p M}{r^2} + \frac{\sigma_{\gamma p} L}{4\pi r^2} + \frac{eQ}{r^2} , \tag{6.48}$$

$$m_e \dot{v}_e = -\frac{m_e M}{r^2} + \frac{\sigma_{\gamma e} L}{4\pi r^2} - \frac{eQ}{r^2} . \tag{6.49}$$

v_p and v_e are, respectively, the proton and electron fluid velocities, the dot $\dot{}$ is the derivative with respect to time, and L is the accretion luminosity. $\sigma_{\gamma p}$ and $\sigma_{\gamma e}$ are the scattering cross-sections of photons with, respectively, protons and electrons. Q is the electric charge of the black hole and e is the proton electric charge.

Considering that $m_p \gg m_e$ and $\sigma_{\gamma e} \gg \sigma_{\gamma p}$, accretion is possible ($\dot{v}_p, \dot{v}_e \le 0$) for a luminosity L below the Eddington luminosity

$$L_{\text{Edd}} = \frac{4\pi M m_p}{\sigma_{\gamma e}} = 1.257 \cdot 10^{38} \left(\frac{M}{M_\odot}\right) \text{ erg/s} . \tag{6.50}$$

The equilibrium electric charge is reached for $\dot{v}_p = \dot{v}_e$ and is

$$Q_{\text{eq}} = \frac{\sigma_{\gamma e} L}{4\pi e} = 0.962 \cdot 10^{21} \left(\frac{M}{M_\odot}\right) e , \tag{6.51}$$

where in the last passage it is assumed that the black hole is accreting at the Eddington luminosity. If this is not the case, the expression on the right hand side must be multiplied by the factor L/L_{Edd}.

Let us now write the Reissner–Nordström line element as

$$ds^2 = -\left(1 - \frac{r_{\text{Sch}}}{r} + \frac{r_Q^2}{r^2}\right) dt^2 + \left(1 - \frac{r_{\text{Sch}}}{r} + \frac{r_Q^2}{r^2}\right)^{-1} dr^2$$
$$+ r^2 d\theta^2 + r^2 \sin^2\theta d\phi^2 . \tag{6.52}$$

r_{Sch} and r_Q are defined as

$$r_{\text{Sch}} = \frac{2G_N M}{c^2} = 2.953 \cdot 10^3 \left(\frac{M}{M_\odot}\right) \text{ m} , \tag{6.53}$$

$$r_Q = \sqrt{\frac{G_N Q^2}{4\pi \varepsilon_0 c^4}} = 1.381 \cdot 10^{-36} \left(\frac{|Q|}{e}\right) \text{ m} , \tag{6.54}$$

where Newton's constant G_N and the speed of light c have been reintroduced for convenience. $1/(4\pi \varepsilon_0) = 8.988 \cdot 10^9 \text{ Nm}^2 \text{C}^{-2}$ is Coulomb's constant in vacuum. If we plug the equilibrium electric charge Q_{eq} in Eq. (6.54), we obtain

$$r_Q = 1.33 \cdot 10^{-15} \left(\frac{M}{M_\odot} \right) \text{ m} . \qquad (6.55)$$

$r_Q \ll r_{\text{Sch}}$ and deviations from the Schwarzschild metric are extremely small that can be neglected in any present and future observation. The electrostatic field around a black hole due to Q_{eq} may instead be important in the case of putative small black holes, with a mass much smaller than $M_\odot$ [8].

References

1. M.A. Abramowicz, J.P. Lasota, Acta Astron. **30**, 35 (1980)
2. N. Afshordi, B. Paczynski, Astrophys. J. **592**, 354 (2003), arXiv:astro-ph/0202409
3. C. Bambi, JCAP **1105**, 009 (2011), arXiv:1103.5135 [gr-qc]
4. C. Bambi, Phys. Lett. B **705**, 5 (2011), arXiv:1110.0687 [gr-qc]
5. C. Bambi, Eur. Phys. J. C **75**, 22 (2015), arXiv:1412.4987 [gr-qc]
6. C. Bambi, E. Barausse, Phys. Rev. D **84**, 084034 (2011), arXiv:1108.4740 [gr-qc]
7. C. Bambi, A. Cardenas-Avendano, T. Dauser, J.A. Garcia, S. Nampalliwar, arXiv:1607.00596 [gr-qc]
8. C. Bambi, A.D. Dolgov, A.A. Petrov, JCAP **0909**, 013 (2009), arXiv:0806.3440 [astro-ph]
9. C. Bambi, D. Malafarina, N. Tsukamoto, Phys. Rev. D **89**, 127302 (2014), arXiv:1406.2181 [gr-qc]
10. J.M. Bardeen, Nature **226**, 64 (1970)
11. J.M. Bardeen, B. Carter, S.W. Hawking, Commun. Math. Phys. **31**, 161 (1973)
12. J.M. Bardeen, J.A. Petterson, Astrophys. J. **195**, L65 (1975)
13. E. Berti, M. Volonteri, Astrophys. J. **684**, 822 (2008), arXiv:0802.0025 [astro-ph]
14. S. Chandrasekhar, *The Mathematical Theory of Black Holes* (Clarendon Press, Oxford, 1998)
15. C.T. Cunningham, Astrophys. J. **202**, 788 (1975)
16. P.C. Fragile, O.M. Blaes, P. Anninois, J.D. Salmonson, Astrophys. J. **668**, 417 (2007), arXiv:0706.4303 [astro-ph]
17. T. Fragos, J.E. McClintock, Astrophys. J. **800**, 17 (2015), arXiv:1408.2661 [astro-ph.HE]
18. T. Fragos, M. Tremmel, E. Rantsiou, K. Belczynski, Astrophys. J. **719**, L79 (2010), arXiv:1001.1107 [astro-ph.HE]
19. C.F. Gammie, S.L. Shapiro, J.C. McKinney, Astrophys. J. **602**, 312 (2004), arXiv:astro-ph/0310886
20. L. Gou et al., Astrophys. J. **790**, 29 (2014), arXiv:1308.4760 [astro-ph.HE]
21. S.A. Hughes, R.D. Blandford, Astrophys. J. **585**, L101 (2003), arXiv:astro-ph/0208484
22. A. Ingram, C. Done, P.C. Fragile, Mon. Not. Roy. Astron. Soc. **397**, L101 (2009)
23. A.R. King, U. Kolb, Mon. Not. Roy. Astron. Soc. **305**, 654 (1999), arXiv:astro-ph/9901296
24. A.R. King, S.H. Lubow, G.I. Ogilvie, J.E. Pringle, Mon. Not. Roy. Astron. Soc. **363**, 49 (2005), arXiv:astro-ph/0507098
25. A.R. King, J.E. Pringle, Mon. Not. Roy. Astron. Soc. **373**, L93 (2006), arXiv:astro-ph/0609598
26. L. Kong, Z. Li, C. Bambi, Astrophys. J. **797**, 78 (2014), arXiv:1405.1508 [gr-qc]
27. A.K. Kulkarni et al., Mon. Not. Roy. Astron. Soc. **414**, 1183 (2011), arXiv:1102.0010 [astro-ph.HE]
28. S. Kumar, J.E. Pringle, Mon. Not. Roy. Astron. Soc. **213**, 435 (1985)
29. L.X. Li, E.R. Zimmerman, R. Narayan, J.E. McClintock, Astrophys. J. Suppl. **157**, 335 (2005), arXiv:astro-ph/0411583
30. N. Lin, N. Tsukamoto, M. Ghasemi-Nodehi, C. Bambi, Eur. Phys. J. C **75**, 599 (2015), arXiv:1512.00724 [gr-qc]
31. R.W. Lindquist, Ann. Phys. **37**, 487 (1966)

32. G. Lodato, J.E. Pringle, Mon. Not. Roy. Astron. Soc. **368**, 1196 (2006), arXiv:astro-ph/0602306
33. R.G. Martin, J.E. Pringle, C.A. Tout, Mon. Not. Roy. Astron. Soc. **381**, 1617 (2007), arXiv:0708.2034 [astro-ph]
34. R.G. Martin, C.A. Tout, J.E. Pringle, Mon. Not. R. Astron. Soc. **387**, 188 (2008), arXiv:0802.3912 [astro-ph]
35. J.E. McClintock, R. Shafee, R. Narayan, R.A. Remillard, S.W. Davis, L.X. Li, Astrophys. J. **652**, 518 (2006), arXiv:astro-ph/0606076
36. S.C. Noble, J.H. Krolik, Astrophys. J. **703**, 964 (2009), arXiv:0907.1655 [astro-ph.HE]
37. S.C. Noble, J.H. Krolik, J.F. Hawley, Astrophys. J. **711**, 959 (2010), arXiv:1001.4809 [astro-ph.HE]
38. I.D. Novikov, K.S. Thorne, Astrophysics and black holes, in *Black Holes*, ed. by C. De Witt, B. De Witt (Gordon and Breach, New York, 1973)
39. B. Paczynski, arXiv:astro-ph/0004129
40. D.N. Page, K.S. Thorne, Astrophys. J. **191**, 499 (1974)
41. R.F. Penna, J.C. McKinney, R. Narayan, A. Tchekhovskoy, R. Shafee, J.E. McClintock, Mon. Not. R. Astron. Soc. **408**, 752 (2010), arXiv:1003.0966 [astro-ph.HE]
42. R.H. Price, Phys. Rev. D **5**, 2419 (1972)
43. M.J. Reid, J.E. McClintock, J.F. Steiner, D. Steeghs, R.A. Remillard, V. Dhawan, R. Narayan, Astrophys. J. **796**, 2 (2014), arXiv:1409.2453 [astro-ph.GA]
44. A. Sadowski, M. Bursa, M. Abramowicz, W. Kluzniak, J.P. Lasota, R. Moderski, M. Safarzadeh, Astron. Astrophys. **532**, A41 (2011), arXiv:1102.2456 [astro-ph.HE]
45. R. Shafee, J.C. McKinney, R. Narayan, A. Tchekhovskoy, C.F. Gammie, J.E. McClintock, Astrophys. J. **687**, L25 (2008), arXiv:0808.2860 [astro-ph]
46. R. Shafee, R. Narayan, J.E. McClintock, Astrophys. J. **676**, 549 (2008), arXiv:0705.2244 [astro-ph]
47. N.I. Shakura, R.A. Sunyaev, Astron. Astrophys. **24**, 337 (1973)
48. S.L. Shapiro, Astrophys. J. **620**, 59 (2005), arXiv:astro-ph/0411156
49. R. Speith, H. Riffert, H. Ruder, Comput. Phys. Commun. **88**, 109 (1995)
50. J.F. Steiner, J.E. McClintock, Astrophys. J. **745**, 136 (2012), arXiv:1110.6849 [astro-ph.HE]
51. J.F. Steiner, J.E. McClintock, R.A. Remillard, L. Gou, S. Yamada, R. Narayan, Astrophys. J. **718**, L117 (2010), arXiv:1006.5729 [astro-ph.HE]
52. K.S. Thorne, Astrophys. J. **191**, 507 (1974)
53. J.M. Wang, Y.M. Chen, L.C. Ho, R.J. McLure, Astrophys. J. **642**, L111 (2006), arXiv:astro-ph/0603813
54. S.E. Woosley, J.S. Bloom, Ann. Rev. Astron. Astrophys. **44**, 507 (2006), arXiv:astro-ph/0609142
55. S.C. Yoon, N. Langer, C. Norman, Astron. Astrophys. **460**, 199 (2006), arXiv:astro-ph/0606637
56. B. Ya Zeldovich, I.D. Novikov, *Relativistic Astrophysics. Volume 1: Stars and Relativity* (Chicago University Press, Chicago, 1971)
57. V.V. Zhuravlev, P.B. Ivanov, P.C. Fragile, D.M. Teixeira, Astrophys. J. **796**, 104 (2014), arXiv:1406.5515 [astro-ph.HE]

Chapter 7
Continuum-Fitting Method

Within the Novikov–Thorne model, it is relatively straightforward to compute the thermal spectrum of a thin accretion disk as detected by a distant observer in the flat faraway region. With the assumptions that the metric around the compact object is described by the Kerr solution and that the inner edge of the disk is at the ISCO radius, we can measure the black hole spin parameter a_*. This technique is called the *continuum-fitting method* and was proposed in [53]. Up to now, we have estimated the spin of about ten stellar-mass black holes with this technique. The continuum-fitting method can also be extended to non-Kerr spacetimes to test the metric around astrophysical black holes [1, 21].

The continuum-fitting method is probably the most robust technique available today [29, 30]. The physics behind is relatively well understood. However, it has also some weak points. First, corrections for non-blackbody effects are important and are taken into account by disk atmosphere models, but the validity of these models is sometimes criticized. Second, the measurement of the spin requires independent measurements of the mass M of the compact object, the distance D, and the inclination angle of the disk i. These measurements are usually obtained by optical observations, but systematics effects are not always under control and the uncertainty on these quantities is usually large (eventually these uncertainties provide the main contribution in the final uncertainty of the spin parameter). Third, the continuum-fitting method is normally applied to stellar-mass black holes only, because the thermal spectrum of a Novikov–Thorne disk is in the soft X-ray band ($\sim 1\,\text{keV}$) for $M \approx 10\,M_\odot$ and in the optical/UV band (~ 1–$10\,\text{eV}$) for a compact object of 10^5–$10^{10}\,M_\odot$. In the second case, extinction and dust absorption limit the ability of an accurate measurement. Attempts to apply the technique to supermassive black holes have been limited to very special cases [9, 12]. The spectrum is also relatively simple without specific features, so we can measure only one parameter of the near horizon geometry. If we assume the Kerr metric, this is enough to measure the spin parameter. If we want to test the Kerr metric, there is typically a degeneracy between the estimate of the spin and possible deviations from the Kerr solution.

© Springer Nature Singapore Pte Ltd. 2017
C. Bambi, *Black Holes: A Laboratory for Testing Strong Gravity*,
DOI 10.1007/978-981-10-4524-0_7

7.1 Calculation of the Spectrum

If we assume that the Novikov–Thorne disk is in local thermal equilibrium, its emission is blackbody-like and we can define an effective temperature T_{eff} from the time-averaged energy flux $\mathscr{F}$ in Eq. (6.4), namely

$$\mathscr{F} = \sigma T_{\text{eff}}^4 , \tag{7.1}$$

where $\sigma = 5.67 \cdot 10^{-5} \, \text{erg s}^{-1} \, \text{cm}^{-2} \, \text{K}^{-4}$ is the Stefan–Boltzmann constant. The resulting profile of the effective temperature in the Kerr spacetime is shown in Fig. 7.1 for different values of the spin parameter and assuming that the inner edge of the disk is at the ISCO radius.

Since a Novikov–Thorne disk is realized when the mass accretion rate is $\dot{M} \sim 0.1 \, \dot{M}_{\text{Edd}}$, from Eq. (6.4) we see that $\mathscr{F} \propto M^{-1}$ and therefore

$$T_{\text{eff}} \sim \left(\frac{0.1 \, \dot{M}_{\text{Edd}}}{4\pi \sigma M^2} \right)^{1/4} \sim \left(\frac{10 \, M_\odot}{M} \right)^{1/4} \text{keV} , \tag{7.2}$$

where $\dot{M}_{\text{Edd}}$ is given in Eq. (G.2). Equation (7.2) is a rough estimate of T_{eff}, but clearly shows why the temperature of thin disks is in the soft X-ray band in the case of black hole binaries and in the optical/UV band when M is 10^5–$10^{10} \, M_\odot$.

Since the temperature of the disk near the inner edge can be high, corrections for non-blackbody effects can be important. This is usually taken into account by introducing the color correction term (or hardening factor) f_{col}, which is largely due to electron scattering in the disk and in practice is obtained from disk atmosphere models [10, 11]. The color temperature is defined as $T_{\text{col}}(r) = f_{\text{col}} T_{\text{eff}}$. The local specific intensity of the radiation emitted by the disk is (reintroducing the speed of light c)

Fig. 7.1 Radial profile of the effective temperature T_{eff} of a Novikov–Thorne accretion disk in Kerr spacetime for different values of the spin parameter a_*. Here $M = 10 \, M_\odot$ and $\dot{M} = 10^{18} \, \text{g s}^{-1}$

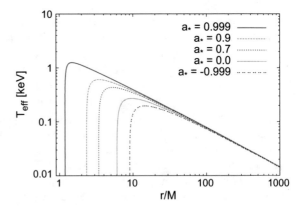

$$I_e(\nu_e) = \frac{2h\nu_e^3}{c^2} \frac{1}{f_{col}^4} \frac{\Upsilon}{\exp\left(\frac{h\nu_e}{k_B T_{col}}\right) - 1}, \tag{7.3}$$

where h is the Planck constant and k_B is the Boltzmann constant. Υ is a function of the angle ϑ_e between the propagation direction of the photon emitted by the disk and the normal of the disk surface [3]. The most common choices are

$$\Upsilon = \begin{cases} 1 & \text{(isotropic emission)}, \\ (3/7)(1 + 2\cos\vartheta_e) & \text{(limb-darkened emission)}, \end{cases} \tag{7.4}$$

The choice of the form of Υ affects the final measurement, but the impact is small. It is one of the uncertainties in the model.

The calculation of the thermal spectrum of thin accretion disks has been extensively discussed in the literature; see e.g. [1, 24] and references therein. The photon flux number density as measured by a distant observer is

$$N_{E_{obs}} = \frac{1}{E_{obs}} \int I_{obs}(\nu_{obs}) d\tilde{\Omega} = \frac{1}{E_{obs}} \int g^3 I_e(\nu_e) d\tilde{\Omega}$$
$$= A_1 \left(\frac{E_{obs}}{keV}\right)^2 \int \frac{1}{M^2} \frac{\Upsilon dXdY}{\exp\left[\frac{A_2}{gF^{1/4}}\left(\frac{E_{obs}}{keV}\right)\right] - 1}, \tag{7.5}$$

where $d\tilde{\Omega} = dXdY/D^2$ is, as in Sect. 6.2, the element of the solid angle subtended by the image of the disk in the observer's sky, F is the dimensionless function in Eq. (6.6), and A_1 and A_2 are two constants given by (reintroducing the constants G_N and c)

$$A_1 = \frac{2(keV)^2}{f_{col}^4} \left(\frac{G_N M}{c^3 h^{3/2} D}\right)^2$$
$$= \frac{0.07205}{f_{col}^4} \left(\frac{M}{M_\odot}\right)^2 \left(\frac{kpc}{D}\right)^2 \text{ photons keV}^{-1} \text{cm}^{-2} \text{s}^{-1}, \tag{7.6}$$

$$A_2 = \left(\frac{keV}{k_B f_{col}}\right) \left(\frac{G_N M}{c^3}\right)^{1/2} \left(\frac{4\pi\sigma}{\dot{M}}\right)^{1/4}$$
$$= \frac{0.1331}{f_{col}} \left(\frac{10^{18} \text{ g s}^{-1}}{\dot{M}}\right)^{1/4} \left(\frac{M}{M_\odot}\right)^{1/2}. \tag{7.7}$$

In Eq. (7.5), the redshift factor g and the function F depend on the emission point in the disk. The integral can be performed after ray-tracing calculations have related any point of emission in the disk to the point of detection in the plane of the distant observer. We have thus to find $g = g(X, Y)$ and $F = F(X, Y)$.

One can also employ the formalism of the transfer function discussed in the previous chapter. In such a case, it is necessary to write ν_e and ϑ_e in terms of g^* and

Fig. 7.2 Images of a Novikov–Thorne accretion disk around a Schwarzschild black hole (*top panel*) and a Kerr black hole with the spin parameter $a_* = 0.999$ (*bottom panel*). The viewing angle is $i = 80°$ and the colors are for gT_{col} (in keV), where g is the redshift factor and T_{col} is the color temperature. The other parameters are: $M = 10\,M_\odot$, $\dot{M} = 10^{18}\,\mathrm{g\,s}^{-1}$, and $f_{col} = 1.6$. The outer radius of the accretion disk is at $r_{out} = 25\,M$

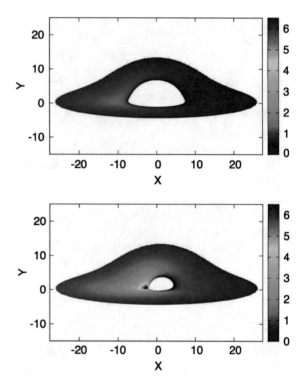

r_e, and perform the integral in Eq. (6.18) or in Eq. (6.19). In the case of the thermal spectrum of the disk, the local spectrum I_e does depend on the metric via Eq. (7.1). However, the time consuming block of the calculations is related to the ray-tracing part, and therefore it can be convenient to have the transfer function of the spacetime already tabulated.

Figure 7.2 shows the images of a Novikov–Thorne accretion disk around a Schwarzschild black hole (top panel) and a Kerr black hole with the spin parameter $a_* = 0.999$ (bottom panel). The color indicates the value of gT_{col} in keV, so the two images are essentially two plots of $gT_{col}(X, Y)$. The boundary of the hole at the centre of the image of the disk is the apparent image of the inner edge of the disk. The external boundary of the image corresponds instead to the apparent image of the outer edge of the disk, which is here assumed very small ($r_{out} = 25\,M$) for graphical reasons.

In the image of the accretion disk around a Schwarzschild black hole, the asymmetry with respect to the Y-axis is due to the Doppler shift. In the semi-plane $X < 0$, the gas is moving in the direction of the observer and the emitted radiation is blueshifted. In the semi-plane $X > 0$, the gas is moving in the opposite direction and the radiation is redshifted. In the image in the bottom panel referring to a rotating black hole, part of the asymmetry with respect to the Y-axis is also due to light bending, because photon trajectories depend on the photon angular momentum with respect to the black hole spin. This, in particular, causes the asymmetry with respect to the Y-axis in the apparent image of the inner edge of the disk. In both images, the asymmetry with respect to the X-axis is due to light bending.

In the Kerr metric, because of the presence of the Carter constant, in Boyer–Lindquist coordinates the photon equations of motion are separable and of first order. Since the spacetime is stationary and axisymmetric, eventually it is only necessary to solve the motion in the (r, θ) plane, which (in the specific case of the Kerr metric in four dimensions) can be reduced to the problem of solving some elliptic integrals [24]. In extensions of the Kerr metric, the presence of the Carter constant still results in separable equations of motion, but one has now to solve at least hyper-elliptic integrals. In the case of a general stationary and axisymmetric metric without Carter constant, it is necessary to solve the second order geodesic equations. Usually it is computationally more convenient to start from the point of detection in the plane of the distant observer and trace backward in time the photon trajectory to find the emission point in the disk. More details can be found, for instance, in [1].

In the Kerr metric, the thermal spectrum of a thin accretion disk has five "main" parameters: the black hole mass M, the mass accretion rate $\dot{M}$, the inclination angle of the disk with respect to the line of sight of the distant observer i, the distance of the source D, and the spin parameter a_*. The impact of these parameters on the spectrum of a thin disk are shown in Fig. 7.3. The hardening factor f_{col} may be computed from disk atmosphere models. For stellar-mass black holes accreting at 10% of the Eddington limit, one finds $f_{col} \approx 1.5$–1.7 [10, 11].

7.2 Spin Measurements

The analysis of the thermal spectrum of geometrically thin and optically thick accretion disks was suggested as a technique to measure the spin parameter of black holes in [53]. Assuming that the compact object is a Kerr black hole, the model depends on five parameters $(M, \dot{M}, i, D, a_*)$. However, it is not possible to infer all these parameters from the data of the spectrum of a thin disk, because there is a degeneracy. As shown in Fig. 7.3, the shape of the spectrum is quite simple and eventually the absence of peculiar features permits one to obtain the same spectrum with different combinations of $(M, \dot{M}, i, D, a_*)$.

However, if we can obtain independent measurement of M, D, and i,[1] one can fit the thermal component of the spectrum of a black hole and infer a_* and $\dot{M}$. This is the continuum-fitting method. The technique is relatively robust, because the Novikov–Thorne model is based on the conservation laws of rest-mass, energy, and angular momentum, while there are no assumptions concerning the properties of

---

[1]In the case of the viewing angle i, one usually measures the inclination angle of the orbital plane of the binary system (see Sect. 4.1.1) and then assumes that it is the same as the inclination angle of the accretion disk.

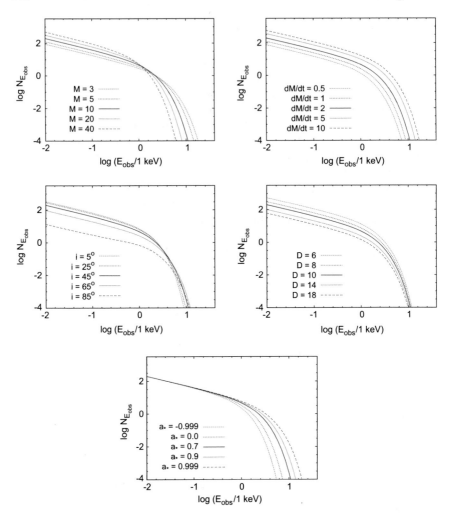

Fig. 7.3 Impact of the model parameters on the thermal spectrum of a thin disk: mass M (*top left panel*), mass accretion rate $\dot{M}$ (*top right panel*), viewing angle i (*central left panel*), distance D (*central right panel*), and spin parameter a_* (*bottom panel*). When not shown, the values of the parameters are: $M = 10\,M_\odot$, $\dot{M} = 2 \cdot 10^{18}\,\mathrm{g\,s^{-1}}$, $D = 10\,\mathrm{kpc}$, $i = 45°$, and $a_* = 0.7$. M in units of $M_\odot$, $\dot{M}$ in units of $10^{18}\,\mathrm{g\,s^{-1}}$, D in kpc, and flux density N_{E_obs} in photons $\mathrm{keV^{-1}\,cm^{-2}\,s^{-1}}$. From [2], reproduced by permission of IOP Publishing. All rights reserved

the accreting gas or the geometry of the system. Nevertheless, there are some weak points, as already discussed at the beginning of this chapter.

Current estimates with the continuum-fitting method of the spins of stellar-mass black holes under the assumption of the Kerr background are reported in the second column in Table 7.1. As we can see, some objects look like very fast-rotating black holes with a_* close to 1, some objects have an intermediate value of the spin

Table 7.1 Summary of the continuum-fitting and iron line measurements of the spin parameter of stellar-mass black holes under the assumption of the Kerr background. In some cases, only upper or lower bounds have been obtained. See the references in the last column for more details

BH Binary	a_* (Continuum)	a_* (Iron)	Principal references
GRS 1915-105	>0.98	0.98 ± 0.01	[28, 32]
Cyg X-1	>0.98	$0.97^{+0.014}_{-0.02}$	[14, 18, 19, 34, 48, 52]
GS 1354-645	—	>0.98	[13]
LMC X-1	0.92 ± 0.06	$0.97^{+0.02}_{-0.25}$	[16, 47]
GX 339-4	<0.9	0.95 ± 0.03	[15, 20, 35, 36]
MAXI J1836-194	—	0.88 ± 0.03	[38]
M33 X-7	0.84 ± 0.05	—	[27]
4U 1543-47	0.80 ± 0.10^a	—	[41]
IC10 X-1	$\gtrsim 0.7$	—	[43]
Swift J1753.5	—	$0.76^{+0.11}_{-0.15}$	[37]
XTE J1650-500	—	$0.84 \sim 0.98$	[51]
GRO J1655-40	0.70 ± 0.10^a	>0.9	[37, 41]
GS 1124-683	$0.63^{+0.16}_{-0.19}$	—	[4, 33]
XTE J1752-223	—	0.52 ± 0.11	[39]
XTE J1652-453	—	<0.5	[6]
XTE J1550-564	0.34 ± 0.28	$0.55^{+0.15}_{-0.22}$	[46]
LMC X-3	0.25 ± 0.15	—	[45]
H1743-322	0.2 ± 0.3	—	[44]
A0620-00	0.12 ± 0.19	—	[17]
XMMU J004243.6	<-0.2	—	[31]

[a] These sources were studied in an early work of the continuum-fitting method, within a more simple model, and therefore the published 1-σ error estimates are doubled following [29]

parameter, and other sources may be slow- or non-rotating black holes. There is (controversial) evidence for at least one negative value of the spin parameter; that is, this object would have a counterrotating disk. More details can be found in the references reported in the last column in Table 7.1.

The validity of the Novikov–Thorne model has been already discussed in Sect. 6.1. Here we just point out that the thermal spectrum of a thin disk of an astrophysical black hole should deviate from that calculated in the Novikov–Thorne model, but such deviations should not be important for the precision of current spin measurements.

The impact of deviations from the Novikov–Thorne model in the spin measurements have been discussed in [23]. Without entering the details, Kulkarni et al. [23] find that there are indeed deviations from the theoretical model: some radiation is emitted inside the ISCO and the emission peak seems to be at smaller radii with respect to the Novikov–Thorne prediction. Both effects would lead to overestimate the value of the spin. Figure 7.4 compares the disk luminosity from GRMHD simulations (solid lines) with the Novikov–Thorne model (dashed lines) for $a_* = 0, 0.7,$

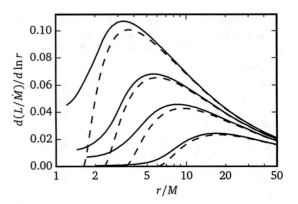

Fig. 7.4 Luminosity profile from GRMHD simulations (*solid lines*) and from the Novikov–Thorne model (*dashed lines*) for $a_* = 0$, 0.7, 0.9, and 0.98 (from *bottom* to *top*). See the text for more details. Figure 7.1 from [23], reproduced by permission of Oxford University Press

0.9, 0.98 (from bottom to top). The error in the spin estimate due to these deviations from the Novikov–Thorne model is smaller for low viewing angles and high spins, and larger for high viewing angles and low spins. The deviations decrease as the disk thickness h/r decreases. In the end, these effects do not seem to be important for current spin measurements, particularly because spin errors are dominated by the uncertainties in the measurements of M, D, and i, but also because any additional radiation appears to manifest distinctly as a non-thermal component [23, 54].

7.3 Polarization of the Disk's Spectrum

In the vacuum or in isotropic media, electromagnetic radiation propagates as a transverse wave in which the electric and magnetic fields are always orthogonal each other and oscillate in the plane perpendicular to the propagation direction. The polarization of an electromagnetic beam refers to the polarization of its electric field and is described by the polarization degree and the polarization vector. The *polarization degree*, δ, is the fraction of polarized radiation, and ranges from 0 (unpolarized radiation) to 1 (completely polarized radiation). The *polarization vector*, $\mathbf{f}$, indicates the oscillation direction of the electric field in the plane perpendicular to the propagation direction of the beam. Since only the oscillation direction matter, the polarization vector is usually normalized to 1, $\mathbf{f} \cdot \mathbf{f} = 1$. The 4-vector f^μ is space-like and has a gauge degree of freedom ($f^\mu \rightarrow f^\mu + \alpha k^\mu$, where α is a constant and k^μ is the photon 4-momentum). The polarization vector can also be replaced by the *polarization angle*, ψ, describing the angle of the electric field with respect to some reference axis in the plane perpendicular to the propagation direction.

Thermal radiation is unpolarized. However, because of Thomson scattering of photons off free electrons in the dense atmosphere of the disk, the thermal radiation from an accretion disk around a black hole becomes partially polarized. With reference to the rest frame of the gas, the degree of polarization depends on the emission angle ϑ_e, namely the angle between the normal to the disk surface and the direction

Fig. 7.5 Polarization degree as a function of the emission angle ϑ_e, i.e. the angle between the propagation direction of the photon and the normal to the disk surface. From [26] under the terms of the Creative Commons Attribution License

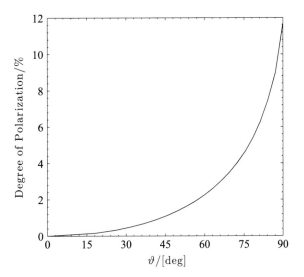

of propagation of the X-ray photon (see Fig. 6.3), and ranges from 0 ($\vartheta_e = 0°$, photon direction parallel to the normal to the disk) to about 12% ($\vartheta_e = 90°$, photon direction orthogonal to the normal to the disk) [3]. The polarization degree as a function of ϑ_e is shown in Fig. 7.5. The orientation of the polarization vector is instead parallel to the disk plane and orthogonal to the direction of propagation of the photon. The same scattering in the disk atmosphere causes a limb-darkened emission and the correct value for Υ can be found in the table in [3].

In Newtonian gravity, it would be straightforward to obtain the angle and the degree of polarization of an accretion disk around a black hole as measured by a distant observer. With reference to Fig. 7.2, the polarization vector would be parallel to the X-axis. The value of the polarization degree would be given by the curve plotted in Fig. 7.5 for $i = \vartheta_e$. In curved spacetime, there are some complications and it is necessary to perform some calculations. [7, 8, 42] were the first to study the polarization of the radiation of the electromagnetic spectrum of an accreting black hole. More recent studies have been reported in [25, 40] for the Kerr metric, and in [22, 26] for generic stationary, axisymmetric, and asymptotically flat spacetimes.

In a curved spacetime, it is first necessary to compute the polarization degree and the polarization angle at each point in the image of the distant observer, and then one can integrate over the image to get the spectra of the degree and of the angle of polarization. In terms of the Stokes parameters I, Q, U, and V [3], for each point in the image of the source we have

$$Q + iU = \delta I e^{2i\psi} , \qquad (7.8)$$

where $V = 0$ because the radiation is linearly polarized. The radiation field is decomposed into a completely polarized component $I^p = \delta I$ and an unpolarized one $I^u = (1 - \delta)I$. At the point of the detection, we have

$$
\begin{aligned}
\langle Q_{\text{obs}} \rangle + i \langle U_{\text{obs}} \rangle &= \frac{1}{\Delta\tilde{\Omega}} \int (Q_{\text{obs}} + i U_{\text{obs}}) \, d\tilde{\Omega} \\
&= \frac{1}{\Delta\tilde{\Omega}} \int g^3 \delta_e I_e e^{2i\psi_{\text{obs}}} d\tilde{\Omega} ,
\end{aligned}
\tag{7.9}
$$

where $\langle \cdot \rangle$ indicates the average over the image, $\Delta\tilde{\Omega}$ is the total solid angle subtended by the disk in the observer's sky, and the redshift factor $g = \nu_{\text{obs}}/\nu_e$ enters from the conservation of the quantity I/ν^3 along the photon path, namely $I_{\text{obs}}/\nu_{\text{obs}}^3 = I_e/\nu_e^3$.

In analogy to the calculations of the disk's thermal spectrum, Eq. (7.9) requires to ray-trace photons from the emission point in the disk to the detection point in the plane of the distant observer or, alternatively, backward in time, from the detection point in the plane of the distant observer to the emission point in the disk. In the case of polarization, both approaches are used, because the latter is not clearly more convenient than the former when it is necessary to take into account the effect of returning radiation. Equation (7.9) requires thus to solve the geodesic equations, but also to parallel transport the polarization vector along the photon trajectories. The polarization degree is a scalar, but the polarization vector is not, and therefore it is necessary to study the evolution of the polarization angle along the photon geodesic.

For every photon trajectory, one has to simultaneously solve both the geodesics equation and the equation of the parallel transport of the polarization vector, namely

$$
\frac{df^\mu}{d\tau} = -\Gamma^\mu_{\nu\rho} f^\nu \frac{dx^\rho}{d\tau} .
\tag{7.10}
$$

In the numerical integration of these two equations, it may be convenient to check at any step that the following relations are satisfied within a certain precision

$$
k_\mu k^\mu = 0 , \quad f^\mu f_\mu = 1 , \quad f^\mu k_\mu = 0 .
\tag{7.11}
$$

As for the integration of the geodesic equations, the cases of the Kerr metric (and, in general, of any Petrov type D spacetime) and of a generic black hole metric are different. The Walker–Penrose theorem asserts that, in any Petrov type D spacetime, for any 4-vector orthogonal to a null geodesic and parallel transported along such a null geodesic there is a complex-valued constant of motion (*Walker–Penrose constant*) along the photon path [3, 49, 50]. In the Kerr metric in Boyer–Lindquist coordinates, the Walker–Penrose constant reads

$$
\kappa = (\kappa_1 - i\kappa_2 \sin\theta)(r - ia\cos\theta) ,
\tag{7.12}
$$

where

$$\kappa_1 = k^t f^r - k^r f^t + a \sin^2 \theta \left(k^r f^\phi - k^\phi f^r \right) ,$$
$$\kappa_2 = \left(r^2 + a^2 \right) \left(k^\phi f^\theta - k^\theta f^\phi \right) - a \left(k^t f^\theta - k^\theta f^t \right) . \tag{7.13}$$

The existence of the Walker–Penrose constant is related to the existence of the Carter constant, and in the same way it is exploited in astrophysical codes for the Kerr metric to optimize the calculations. In the absence of the Walker–Penrose constant, which is the case for a generic stationary, axisymmetric, and asymptotically flat spacetime, one has to solve Eq. (7.10) as discussed in [22, 26].

We note that, in general, radiation that is initially completely polarized is detected in the observer's plane as partially polarized, because different points of the image have photons with different ψ_{obs}. The total intensity at the detection point is

$$\langle I_{\text{obs}} \rangle = \frac{1}{\Delta \tilde{\Omega}} \int g^3 I_e d\tilde{\Omega} = \langle I_{\text{obs}}^u \rangle + \langle I_{\text{obs}}^p \rangle . \tag{7.14}$$

The observed averaged polarization degree is [24]

$$\langle \delta_{\text{obs}} \rangle = \frac{\sqrt{\langle Q_{\text{obs}} \rangle^2 + \langle U_{\text{obs}} \rangle^2}}{\langle I_{\text{obs}} \rangle} , \tag{7.15}$$

and the observed averaged polarization angle is determined from the following two relations [24]

$$\sin \left(2\langle \psi_{\text{obs}} \rangle \right) = \frac{\langle U_{\text{obs}} \rangle}{\sqrt{\langle Q_{\text{obs}} \rangle^2 + \langle U_{\text{obs}} \rangle^2}} , \tag{7.16}$$

$$\cos \left(2\langle \psi_{\text{obs}} \rangle \right) = \frac{\langle Q_{\text{obs}} \rangle}{\sqrt{\langle Q_{\text{obs}} \rangle^2 + \langle U_{\text{obs}} \rangle^2}} . \tag{7.17}$$

With the above machinery, we can compute the spectrum of the polarization degree and of the polarization angle for a specific model. In the Kerr spacetime, there are five basic parameters, namely the mass of the object M, the distance of the source D, the inclination angle of the disk with respect to the line of sight of the observer i, the mass accretion rate $\dot{M}$, and the spin parameter a_*.

Figure 7.6 shows the spectrum of the polarization degree (left panel) and of the polarization angle (right panel) without including returning radiation, for the accretion disk around a Schwarzschild black hole (solid lines) and a Kerr black hole with $a_* = 0.9$ (dashed lines). The viewing angle is $i = 45°, 60°$, and $75°$ (from bottom to top in the left panel, from top to bottom in the right panel). The interpretation of these spectra is quite straightforward. For the polarization degree, low energy photons are mainly emitted at large radii, where the effect of light bending is small and therefore $\vartheta_e \approx i$ ($\vartheta_e = i$ in Newtonian gravity). High energy photons are mainly emitted from the inner part of the accretion disk, where light bending is stronger and therefore ϑ_e can be significantly smaller than i, reducing the polarization degree. For the polarization angle, the situation is similar. Low energy photons are not significantly

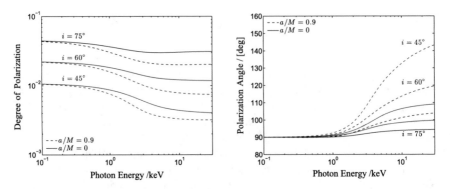

Fig. 7.6 Polarization degree (*left panel*) and polarization angle (*right panel*) of the thermal spectrum of a thin disk as a function of the photon energy and without including returning radiation. The central object is a Schwarzschild black hole (*solid lines*) and a Kerr black hole with the spin parameter $a_* = 0.9$ (*dashed lines*). The viewing angle is $i = 45°$, 60°, and 75°. From [26] under the terms of the Creative Commons Attribution License

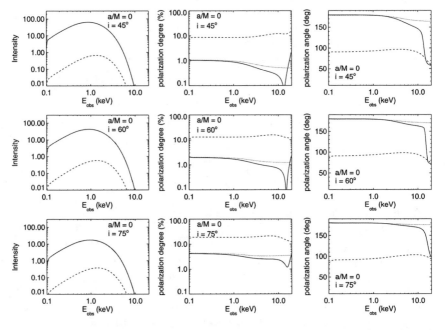

Fig. 7.7 Intensity (*left panels*), polarization degree (*central panels*), and polarization angle (*right panels*) of the thermal spectrum of a Novikov–Thorne thin disk around a Schwarzschild black hole as a function of the photon energy including direct and returning radiation. In each plot, the *dotted curve* represents the contribution from the direct radiation, the *dashed curve* the contribution from the returning radiation, and the *solid curve* is the total flux. The mass of the black hole is $M = 10 \, M_\odot$, the accretion luminosity is 10% of the Eddington limit, and the viewing angle is $i = 45°$ (*top panels*), 60° (*central panels*), and 75° (*bottom panels*). The polarization angle is defined in a different way with respect to that in Fig. 7.6. From [40]. © AAS. Reproduced with permission

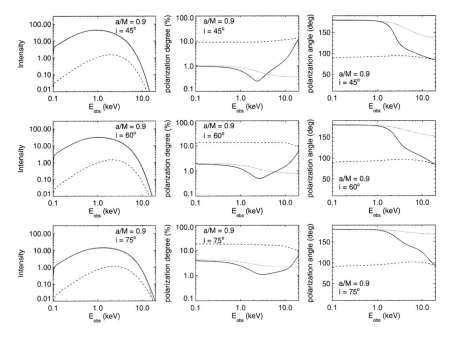

Fig. 7.8 As in Fig. 7.7 in the case of a Kerr black hole with $a_* = 0.9$. From [40]. © AAS. Reproduced with permission

affected by the curved spacetime near the compact object and therefore their angle reflects that at the emission point, where the polarization vector is parallel to the disk and orthogonal to the direction of propagation of the photon. High energy photons come from the strong gravity region and the orientation of their polarization vector has changed in a non-trivial way.

Figures 7.7 and 7.8 show the spectrum of the intensity (left panels), of the polarization degree (central panels), and of the polarization angle (right panels) including both direct and returning radiation (solid lines). The contribution of the direct radiation (dotted lines) is dominant at low energies, since these photons are mainly emitted at large radii where light bending is weak and there is not much returning radiation. High energy photons come from the inner part of the accretion disk, where gravity is strong and returning radiation is important. These plots show that the effect of returning radiation becomes non-negligible for energies above 1–2 keV.

The apparent image of an accretion disk around a Kerr black hole with the spin parameter $a_* = 0.99$ is shown in Fig. 7.9. In the top panel, only direct radiation is taken into account; in the bottom panel, the calculations include both direct and returning radiation. The contour map shows the relative intensity of the total (namely polarized and unpolarized) radiation, $I_{obs}/I_{obs,max}$ (logarithmic scale). The black segments show the polarization of the radiation: the length of the segment is proportional to the polarization degree, while its orientation corresponds to that of the polarization vector.

Fig. 7.9 Apparent images of an accretion disk around a Kerr black hole with the spin parameter $a_* = 0.99$, the mass $M = 10\,M\odot$, accreting at 10% of the Eddington limit, and observed from the viewing angle $i = 75°$. In the *top panel*, only direct radiation is taken into account; in the *bottom panel*, the calculations include both direct and returning radiation. The color map shows the total intensity of the radiation. The segments report the properties of the polarization of the radiation: the length of every segment is proportional to the polarization degree, while the orientation of every segment corresponds to the orientation of the polarization vector. From [40]. © AAS. Reproduced with permission

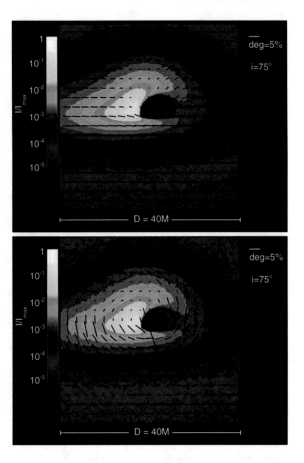

Since the measurement of the polarization can estimate the inclination angle of the accretion disk with respect to the line of sight of the distant observer, this technique will be hopefully able to test the assumption 3 of the Novikov–Thorne model in Sect. 6.1, namely whether the accretion disk is in the equatorial plane perpendicular to the black hole spin, as well as whether it is correct to assume that the inclination angle of the inner part of the disk coincides with the orbital plane of the binary (which is the standard assumption for the measurement of i necessary in the continuum-fitting method). In principle, future polarization measurements may be able to see whether a disk is warped at some radius as a result of the Bardeen–Petterson effect, due to an initial misalignment between the black hole spin and the inner edge of the disk [5].

References

1. C. Bambi, Astrophys. J. **761**, 174 (2012), arXiv:1210.5679 [gr-qc]
2. C. Bambi, J. Jiang, J.F. Steiner, Class. Quant. Grav. **33**, 064001 (2016), arXiv:1511.07587 [gr-qc]
3. S. Chandrasekhar, *The Mathematical Theory of Black Holes* (Clarendon Press, Oxford, 1998)
4. Z. Chen, L. Gou, J.E. McClintock, J.F. Steiner, J. Wu, W. Xu, J. Orosz, Y. Xiang, arXiv:1601.00615 [astro-ph.HE]
5. Y. Cheng, D. Liu, S. Nampalliwar, C. Bambi, Class. Quant. Grav. **33**, 125015 (2016), arXiv:1505.01562 [gr-qc]
6. C.Y. Chiang, R.C. Reis, D.J. Walton, A.C. Fabian, Mon. Not. R. Astron. Soc. **425**, 2436 (2012), arXiv:1207.0682 [astro-ph.HE]
7. P.A. Connors, R.F. Stark, Nature **269**, 128 (1977)
8. P.A. Connors, T. Piran, R.F. Stark, Astrophys. J. **235**, 224 (1980)
9. B. Czerny, K. Hryniewicz, M. Nikolajuk, A. Sadowski, Mon. Not. R. Astron. Soc. **415**, 2942 (2011), arXiv:1104.2734 [astro-ph.GA]
10. S.W. Davis, I. Hubeny, Astrophys. J. Suppl. **164**, 530 (2006), arXiv:astro-ph/0602499
11. S.W. Davis, O.M. Blaes, I. Hubeny, N.J. Turner, Astrophys. J. **621**, 372 (2005), arXiv:astro-ph/0408590
12. C. Done, C. Jin, M. Middleton, M. Ward, Mon. Not. R. Astron. Soc. **434**, 1955 (2013), arXiv:1306.4786 [astro-ph.HE]
13. A.M. El-Batal et al., Astrophys. J. **826**, L12 (2016), arXiv:1607.00343 [astro-ph.HE]
14. A.C. Fabian et al., Mon. Not. R. Astron. Soc. **424**, 217 (2012). arXiv:1204.5854 [astro-ph.HE]
15. J. Garcia et al., Astrophys. J. **813**, 84 (2015), arXiv:1505.03607 [astro-ph.HE]
16. L. Gou, J.E. McClintock, J. Liu, R. Narayan, J.F. Steiner, R.A. Remillard, J.A. Orosz, S.W. Davis, Astrophys. J. **701**, 1076 (2009), arXiv:0901.0920 [astro-ph.HE]
17. L. Gou, J.E. McClintock, J.F. Steiner, R. Narayan, A.G. Cantrell, C.D. Bailyn, J.A. Orosz, Astrophys. J. **718**, L122 (2010), arXiv:1002.2211 [astro-ph.HE]
18. L. Gou et al., Astrophys. J. **742**, 85 (2011), arXiv:1106.3690 [astro-ph.HE]
19. L. Gou et al., Astrophys. J. **790**, 29 (2014), arXiv:1308.4760 [astro-ph.HE]
20. M. Kolehmainen, C. Done, Mon. Not. R. Astron. Soc. **406**, 2206 (2010), arXiv:0911.3281 [astro-ph.HE]
21. L. Kong, Z. Li, C. Bambi, Astrophys. J. **797**, 78 (2014), arXiv:1405.1508 [gr-qc]
22. H. Krawczynski, Astrophys. J. **754**, 133 (2012), arXiv:1205.7063 [gr-qc]
23. A.K. Kulkarni et al., Mon. Not. R. Astron. Soc. **414**, 1183 (2011), arXiv:1102.0010 [astro-ph.HE]
24. L.X. Li, E.R. Zimmerman, R. Narayan, J.E. McClintock, Astrophys. J. Suppl. **157**, 335 (2005), arXiv:astro-ph/0411583
25. L.X. Li, R. Narayan, J.E. McClintock, Astrophys. J. **691**, 847 (2009), arXiv:0809.0866 [astro-ph]
26. D. Liu, Z. Li, Y. Cheng, C. Bambi, Eur. Phys. J. C **75**, 383 (2015), arXiv:1504.06788 [gr-qc]
27. J. Liu, J. McClintock, R. Narayan, S. Davis and J. Orosz, Astrophys. J. **679**, L37 (2008) [Erratum: Astrophys. J. **719**, L109 (2010)], arXiv:0803.1834 [astro-ph]
28. J.E. McClintock, R. Shafee, R. Narayan, R.A. Remillard, S.W. Davis, L.X. Li, Astrophys. J. **652**, 518 (2006), arXiv:astro-ph/0606076
29. J.E. McClintock, R. Narayan, J.F. Steiner, Space Sci. Rev. **183**, 295 (2014), arXiv:1303.1583 [astro-ph.HE]
30. J.E. McClintock et al., Class. Quant. Grav. **28**, 114009 (2011), arXiv:1101.0811 [astro-ph.HE]
31. M. Middleton, J. Miller-Jones, R. Fender, Mon. Not. R. Astron. Soc. **439**, 1740 (2014), arXiv:1401.1829 [astro-ph.HE]
32. J.M. Miller et al., Astrophys. J. **775**, L45 (2013), arXiv:1308.4669 [astro-ph.HE]
33. W.R. Morningstar, J.M. Miller, R.C. Reis, K. Ebisawa, Astrophys. J. **784**, L18 (2014), arXiv:1401.1794 [astro-ph.HE]

34. M.L. Parker et al., Astrophys. J. **808**, 9 (2015), arXiv:1506.00007 [astro-ph.HE]
35. M.L. Parker et al., Astrophys. J. **821**, L6 (2016), arXiv:1603.03777 [astro-ph.HE]
36. R.C. Reis, A.C. Fabian, R. Ross, G. Miniutti, J.M. Miller, C. Reynolds, Mon. Not. R. Astron. Soc. **387**, 1489 (2008), arXiv:0804.0238 [astro-ph]
37. R.C. Reis, A.C. Fabian, R.R. Ross, J.M. Miller, Mon. Not. R. Astron. Soc. **395**, 1257 (2009)
38. R.C. Reis, J.M. Miller, M.T. Reynolds, A.C. Fabian, D.J. Walton, Astrophys. J. **751**, 34 (2012), arXiv:1111.6665 [astro-ph.HE]
39. R.C. Reis et al., Mon. Not. R. Astron. Soc. **410**, 2497 (2011), arXiv:1009.1154 [astro-ph.HE]
40. J.D. Schnittman, J.H. Krolik, Astrophys. J. **701**, 1175 (2009), arXiv:0902.3982 [astro-ph.HE]
41. R. Shafee, J.E. McClintock, R. Narayan, S.W. Davis, L.X. Li, R.A. Remillard, Astrophys. J. **636**, L113 (2006), arXiv:astro-ph/0508302
42. R.F. Stark, P.A. Connors, Nature **266**, 429 (1977)
43. J.F. Steiner, D.J. Walton, J.A. Garcia, J.E. McClintock, S.G.T. Laycock, M.J. Middleton, R. Barnard, K.K. Madsen, arXiv:1512.03414 [astro-ph.HE]
44. J.F. Steiner, J.E. McClintock, M.J. Reid, Astrophys. J. **745**, L7 (2012), arXiv:1111.2388 [astro-ph.HE]
45. J.F. Steiner, J.E. McClintock, J.A. Orosz, R.A. Remillard, C.D. Bailyn, M. Kolehmainen, O. Straub, Astrophys. J. **793**, L29 (2014), arXiv:1402.0148 [astro-ph.HE]
46. J.F. Steiner et al., Mon. Not. R. Astron. Soc. **416**, 941 (2011), arXiv:1010.1013 [astro-ph.HE]
47. J.F. Steiner et al., Mon. Not. R. Astron. Soc. **427**, 2552 (2012), arXiv:1209.3269 [astro-ph.HE]
48. J.A. Tomsick et al., Astrophys. J. **780**, 78 (2014), arXiv:1310.3830 [astro-ph.HE]
49. R.M. Wald, *General Relativity* (University of Chicago Press, Chicago, 1984)
50. M. Walker, R. Penrose, Commun. Math. Phys. **18**, 265 (1970)
51. D.J. Walton, R.C. Reis, E.M. Cackett, A.C. Fabian, J.M. Miller, Mon. Not. R. Astron. Soc. **422**, 2510 (2012), arXiv:1202.5193 [astro-ph.HE]
52. D.J. Walton et al., Astrophys. J. **826**, 87 (2016), arXiv:1605.03966 [astro-ph.HE]
53. S.N. Zhang, W. Cui, W. Chen, Astrophys. J. **482**, L155 (1997), arXiv:astro-ph/9704072
54. Y. Zhu et al., Mon. Not. R. Astron. Soc. **424**, 2504 (2012), arXiv: 1202.1530 [astro-pi.HE]

Chapter 8
X-Ray Reflection Spectroscopy

Within the disk-corona model, the corona is a hot ($T \sim 100$ keV), usually optically thin, electron cloud, which enshrouds the central disk and acts as a source of X-rays due to inverse Compton scattering of thermal photons from the disk off free electrons in the corona. Multiple inverse Compton scattering produces a power-law spectrum for the corona. The geometry of the corona is currently unknown, and several configurations have been proposed in the literature, see Fig. 8.1. The corona illuminates also the disk, producing a reflected component with some emission lines. The most prominent line is usually the iron Kα one, which is at 6.4–6.97 keV, depending on the ionization degree of the disk surface.

The first detection of a broad iron line in the spectrum of a black hole was reported in [1], but its relativistic origin was not understood. X-ray reflection around black holes was first discussed in [15, 18]. In [7], it was pointed out that reflection of the inner part of the accretion disk produces emission lines that are broad and skewed, as a result of the strong Doppler boosting and gravitational redshift occurring in the strong gravitational field of the black hole. The first clear detection of a broad iron line was reported in [37], based on the ASCA data of the AGN MCG-6-30-15.

The analysis of the iron Kα line is a hot topic today, because it is potentially quite a powerful tool to test astrophysical black holes. The technique can be applied to both stellar-mass and supermassive black holes, because this line can be seen in both object classes and its shape does not directly depend on the mass of the compact object. While one has to fit the whole reflection spectrum of the source, information about the spacetime metric in the vicinity of the black hole is mainly extracted from the iron line. For this reason, the technique of fitting the reflection spectrum to determine the black hole spin is often called the *iron line method*. However, a simple model in which a broad iron line is added to a power-law component can only be useful in a preliminary study, and should not be employed to obtain the measurement of a black hole spin.

© Springer Nature Singapore Pte Ltd. 2017
C. Bambi, *Black Holes: A Laboratory for Testing Strong Gravity*,
DOI 10.1007/978-981-10-4524-0_8

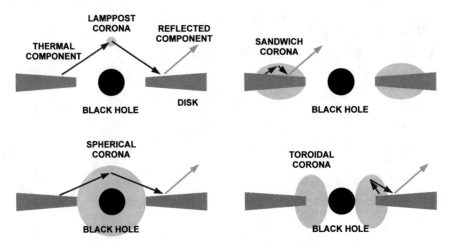

Fig. 8.1 Examples of possible corona geometries: lamppost geometry (*top left panel*), sandwich geometry (*top right panel*), spherical geometry (*bottom left panel*), and toroidal geometry (*bottom right panel*)

With respect to the continuum-fitting method discussed in the previous chapter, this technique has some advantages. It is independent of the mass and of the distance of the source. It does not require an independent measurement of the viewing angle of the observer, which can be inferred from the fit of the reflection spectrum. It is potentially much more powerful to probe the spacetime geometry, because the reflection spectrum has a more complicated structure than a multi-color blackbody spectrum and, in the presence of high quality data, it is possible to break possible parameter degeneracy. However, there are also some disadvantages. The validity of the Novikov-Thorne model is not as well studied as in the case of black holes in X-ray binaries in the high-soft state, for which the continuum-fitting method is applied. The geometry of the corona is not yet known. The emissivity profile is usually approximated by phenomenological models, which inevitably introduces systematics effects in the measurements of the parameters.

8.1   Reflection Process

The physics behind X-ray reflection and iron line fluorescence can be understood in terms of an X-ray power-law continuum illuminating a semi-infinite slab of cold gas, see Fig. 8.2. When an X-ray photon enters the slab, it interacts with the cold gas through Compton scattering off free/bound electrons, photoelectric absorption followed by fluorescent line emission, or photoelectric absorption followed by Auger de-excitation. Since the disk is optically thick, only the properties of its "skin" determine the reflection spectrum.

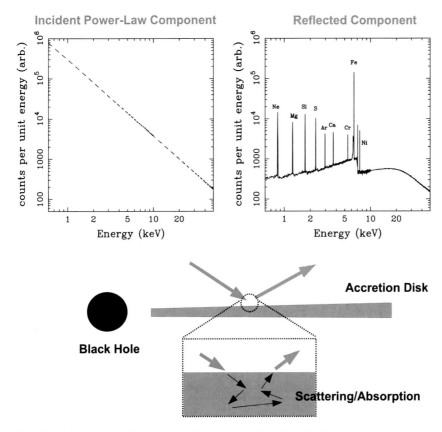

Fig. 8.2 Disk reflection. The accretion disk around a black hole is illuminated by the radiation from a hot corona. The spectrum of the incident radiation is described by a power-law $E^{-\Gamma}$, where Γ is the photon index. The reflection spectrum can be obtained from radiative transfer calculations and presents some emission lines. See the text for more details. Adapted from [29]

The reflection spectrum can be obtained by solving numerically radiative transfer equations describing the interaction of the X-ray photons with the gas on the surface of the accretion disk. The resulting reflection spectrum is characterized by *emission lines* in the 1–8 keV range and the so-called *Compton hump* around 20 keV (see top right panel in Fig. 8.2).

In the case of neutral iron, photoelectric absorption of an X-ray photon can eject one of the two electrons in the K-shell (principal quantum number $n = 1$). The absorption threshold is 7.1 keV. An L-shell electron ($n = 2$) moves to the K-shell and releases 6.4 keV of energy: 34% of the times this energy is released with the emission of a photon (fluorescent line emission) and 66% of the times this energy is transferred to another electron, which is ejected from the atom (Auger de-excitation). Let us note that the so-called neutral iron line at 6.4 keV is actually a combination of several transitions in the energy range 6.39–6.43 keV. The fluorescent line is called

Kα when an L-shell electron moves to the K-shell, Kβ when an M-shell electron ($n = 3$) moves to the K-shell, etc.

In the case of ionized iron, there are two differences. First, the absorption threshold and the de-excitation energy slightly increase. The fluorescent emission line can shift up to 6.97 keV for H-like iron ions. This is because electrons are more strongly bound to the nucleus. Second, the probabilities of fluorescent line emission and of Auger de-excitation change. From FeI (neutral iron) to FeXXIII (iron nucleus with four electrons), both processes are possible and the probability changes are modest. For FeXXIV, FeXXV, and FeXXVI (iron nucleus with, respectively, three, two, and one electron), the Auger de-excitation cannot take place because it requires at least two electrons in the L-shell. For FeXXV and FeXXVI, there are no L-shell electrons, so, strictly speaking, there is no fluorescence but only recombination. However, there is an effective fluorescence yield taking into account the probability of a Kα line emission in the recombination cascade.

The iron line is usually the most prominent feature in the X-ray reflection spectrum of black holes because the iron is more abundant than other heavy elements (the iron-26 nucleus is more tightly bound than lighter and heavier elements, so it is the final product of nuclear reactions) and the probability of fluorescent line emission is also high (for neutral matter, the fluorescence yield is proportional to Z^4, where Z is the atomic number). Moreover, around 6 keV the galactic absorption is negligible, the spectrum of the source is clean because there are just a few other ions emitting or absorbing radiation, and most X-ray detectors are quite sensitive around this energy.

The calculations of the reflection spectrum require a large number of atomic data. The model usually depends on four parameters: the photon power index of the incident radiation Γ, the ionization parameter ξ at the surface of the accretion disk, the iron abundance A_{Fe} (usually expressed in terms of the Solar iron abundance), and the emission angle of the radiation at the surface of the disk ϑ_{e}.

8.1.1   Photon Index of the Illuminating Radiation

The spectrum of the radiation illuminating the accretion disk has a power-law form, $E^{-\Gamma}$, where Γ is the photon index. This is indeed the kind of spectrum that can be expected by multiple inverse Compton scattering of the thermal photons from the disk off the free electrons in the corona. AGN have usually Γ around 2, but in particular sources Γ may either be close to 1 or larger than 3. The impact of the value of Γ on the spectrum of the reflected component is shown in Fig. 8.3 for $\xi = 10$, 100, 10^3, and 10^4 erg cm s^{-1} and $A_{\mathrm{Fe}} = 1$ (Solar iron abundance). Here Γ ranges from 1.2 to 3.4 (from yellow to violet according to the scale on the right).

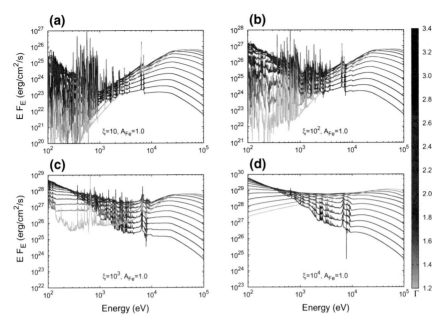

Fig. 8.3 Impact of the photon index Γ on the reflection spectrum. The panels (**a**), (**b**), (**c**), and (**d**) show, respectively, the spectra for $\xi = 10$, 100, 10^3, and 10^4. The photon index is $\Gamma = 1.2$, 1.4, 1.6, 1.8, 2.0, 2.2, 2.4, 2.6, 2.8, 3.0, 3.2, and 3.4 (from *yellow* to *violet* according to the scale on the *right*). No rescaling is applied. ξ in units of erg cm s^{-1}. The iron abundance is $A_{\mathrm{Fe}} = 1$. From [12]. ©AAS. Reproduced with permission

8.1.2 Ionization Parameter of the Disk Surface

The ionization parameter is traditionally defined as

$$\xi(r) = \frac{4\pi F_{\mathrm{X}}(r)}{n_{\mathrm{e}}(r)} , \qquad (8.1)$$

where $F_{\mathrm{X}}(r)$ and $n_{\mathrm{e}}(r)$ are, respectively, the X-ray flux and the comoving electron number density at the radius r. It is possible to define four regimes of ionization (for more details, see Sect. 3.5.1 in [12], which is the most recent study on the topic and correct some previous results present in the literature):

1. The disk is weakly ionized ($\xi \lesssim 100$ erg cm s^{-1}). The iron atoms/ions are in the form of FeI-FeXVII. Reflection produces an emission complex around 6.4 keV, which is the iron Kα line. The spectrum has also the iron Kβ line around 7.1 keV, which can indeed be produced up to FeXVII.
2. The disk is mildly ionized ($\xi \sim 200$–500 erg cm s^{-1}). The iron ions are in the form of FeXVIII-FeXXIV, which implies there is no Kβ emission. The spectrum shows a rich complex of emission lines at energies between 6.4 to 6.7 keV.

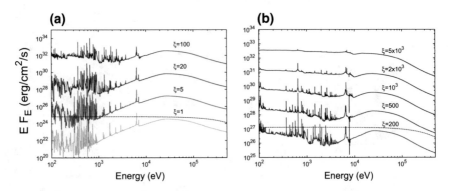

Fig. 8.4 Impact of the ionization parameter ξ on the reflection spectrum. The *left panel* shows the spectra for $\xi = 1, 5, 20$, and 100, multiplied, respectively, by 1, 10^2, 10^4, and 10^6 to avoid overlapping in the figure. The *dashed* curve is the incident power-law for the case $\xi = 1$. The *right panel* shows the spectra for $\xi = 200, 500, 1000, 2000$, and 5000, multiplied, respectively, by 1, 10, 10^2, 10^3, and 10^4. The *dashed* curve is the incident power-law for the case $\xi = 200$. ξ in units of erg cm s^{-1}. The photon index is $\Gamma = 2$ and the iron abundance is $A_{Fe} = 1$. From [12]. ©AAS. Reproduced with permission

3. The disk is highly ionized ($\xi \sim 1000$ erg cm s^{-1}). The emission is centered at about 6.7 and 6.9 keV, which implies that most of the iron ions are in the form of FeXXV (He-like ions) and FeXXVI (H-like ions). The line emission profile is thermally broadened into a symmetric profile via Compton scattering.
4. The disk is completely ionized ($\xi \gtrsim 5000$ erg cm s^{-1}). There is no iron line in the reflection spectrum, because the iron is completely ionized.

Let us note that the above values of ξ corresponds to the case of photon index $\Gamma = 2$. For a different value of Γ, the values of ξ slightly changes (see [12] for more details).

Figure 8.4 shows the impact of the ionization parameter ξ on the reflection spectrum. The left panel shows the spectra for $\xi = 1, 5, 20$, and 100 erg cm s^{-1}, multiplied, respectively, by 1, 10^2, 10^4, and 10^6 for graphical reasons. The right panel shows the spectra for $\xi = 200, 500, 1000, 2000$, and 5000 erg cm s^{-1}, multiplied, respectively, by 1, 10, 10^2, 10^3, and 10^4. The photon index is $\Gamma = 2$ and the iron abundance is $A_{Fe} = 1$.

8.1.3   Elemental Abundance of the Disk

The elemental abundance in the accretion disk affects the reflection spectrum, since it regulates the probability of interactions of the X-ray photons with the atoms/ions in the accretion disk. Figure 8.5 shows the impact of the iron abundance A_{Fe} on the reflection spectrum. A_{Fe} is given in terms of the Solar iron abundance. In the left panels, we can see the temperature profile of the accretion disk as a function of the Thomson optical depth. In the right panels, there is the corresponding spectrum. The

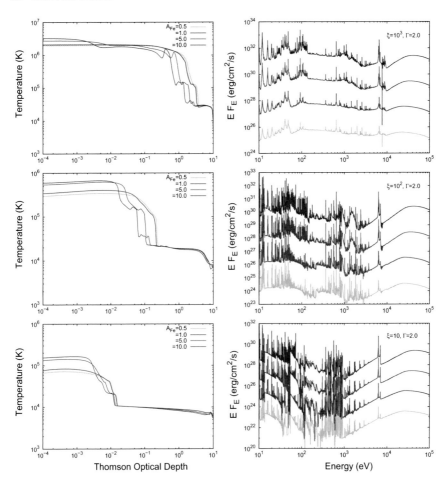

Fig. 8.5 Impact of the iron abundance A_{Fe} on the reflection spectrum. The *left panels* show the temperature profile of the accretion disk, while the *right panels* show the corresponding spectrum. The iron line abundance is $A_{Fe} = 0.5$, 1, 5, and 10 (from *yellow* to *violet*) and is expressed in terms of the Solar iron abundance. These models have been multiplied, respectively, by the factors 0.01, 1, 100, and 10^4 for graphical reasons. The ionization parameter is $\xi = 10$ (*top panels*), 100 (*central panels*), and 1000 erg cm s^{-1} (*bottom panels*), while the photon index is always assumed $\Gamma = 2$. From [12]. ©AAS. Reproduced with permission

ionization parameter is $\xi = 10$ (top panels), 100 (central panels), and 1000 erg cm s^{-1} (bottom panels). The photon index is $\Gamma = 2$ for all models. The spectra in every panel are rescaled to avoid overlapping.

8.1.4   Inclination Angle of the Reflected Radiation

The spectrum of the reflected radiation at the surface of the accretion disk does depend on the emission angle of the photon, ϑ_e. In general, the latter is not equivalent to the inclination angle of the disk with respect to the line of sight of the distant observer, i, due to light bending, and therefore the emission angle changes in different parts of the disk.

Figure 8.6 shows the reflection spectrum for different inclination angles of the reflected radiation at the disk surface. The ionization parameter is $\xi = 1, 10, 100$, and 1000 erg cm s^{-1} (from top to bottom). The spectrum for different emission angles is reported in the left panels. The central panels show the ratio between the spectra

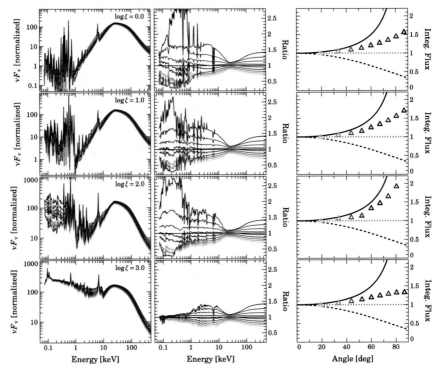

Fig. 8.6 Reflection spectrum for different inclination angles of the reflected radiation at the disk surface, ϑ_e. The *left panels* show the spectra. The *central panels* show the ratio between the spectra in the *left panels* and the angle-averaged spectrum. The *right panels* show the integrated flux as a function of the inclination angle. The *solid* and *dashed lines* in the *right panels* correspond, respectively, to the so-called limb-brightening and limb-darkening emissions, which are commonly used in relativistic blurring kernels. The inclination angle ranges from 20° (*yellow curves*) to 85° (*violet curves*). The ionization parameter is $\xi = 1, 10, 100$, and 1000 erg cm s^{-1} (from *top* to *bottom*). From [13]. ©AAS. Reproduced with permission

in the left panels and the angle-averaged spectrum. In the right panels, we see the integrated flux as a function of the emission angle, which ranges from 20° (yellow triangles) to 85° (violet triangles).

8.2 Iron Line Profile

The radiative transfer calculations briefly discussed in the previous section provide the shape of the reflection spectrum in the rest-frame of the accreting gas, say $S_e(\nu_e, \vartheta_e)$, where ν_e and ϑ_e are, respectively, the photon frequency and the polar angle of the emitted photon with respect to the normal of the disk. The normalization of the spectrum depends on the specific properties of the corona and on the position of the emission point in the disk. Assuming the disk is axisymmetric, the normalization can be described by some function of the emission radius only, say $N(r_e)$. The specific intensity of the reflected radiation can be written as

$$I_e(\nu_e, r_e, \vartheta_e) = N(r_e) S_e(\nu_e, \vartheta_e). \tag{8.2}$$

I_e takes into account the microphysics and the astrophysical model. If we know the transfer function f, which takes into account the photon relativistic effects, we can calculate the observed flux as discussed in Sect. 6.2

$$F_{obs}(\nu_{obs}) = \frac{1}{D^2} \int_{r_{ISCO}}^{\infty} \int_0^1 \pi r_e \frac{g^2}{\sqrt{g^*(1-g^*)}} f(g^*, r_e, i) I_e(\nu_e, r_e, \vartheta_e) \, dg^* \, dr_e. \tag{8.3}$$

With the formalism of the transfer function, we can thus separate the calculations involving the astrophysical model (reflection spectrum) from those involving the relativistic effects on the photons (Doppler boosting, gravitational redshift, light bending). The function $N(r_e)$ may depend on the background metric (see below the case of the lamppost set-up), but it is a minor job and the transfer function formalism is very convenient for the calculations.

While it is necessary to fit the full reflection spectrum to study a source and determine its parameters, in this section we will restrict the attention to the iron $K\alpha$ line only. This will permit to better understand the impact of the strong gravity region on the shape of the reflection spectrum.

The shape of the line is primarily determined by the background metric, the geometry of the emitting region, the disk emissivity, and the disk's inclination angle with respect to the line of sight of the distant observer. In the Kerr background, the relativistic emission line profile emitted by an accretion disk illuminated by an X-ray corona with arbitrary geometry is typically parametrized by the black hole spin a_*, the inner and the outer edge of the emission region r_{in} and r_{out}, and the viewing angle i. The intensity profile may be modeled as a power-law $I_e \propto r^{-q}$, where the emissivity index q is a free parameter to be determined by the fit. A more sophisticated choice is to assume a broken power-law, i.e.

$$I_e \propto \begin{cases} (r_{break}/r)^{q_{in}} & \text{if } r < r_{break}, \\ (r_{break}/r)^{q_{out}} & \text{if } r > r_{break}. \end{cases} \qquad (8.4)$$

The intensity profile in Eq. (8.4) has three free parameters, q_{in}, q_{out}, and r_{break}. A variant of Eq. (8.4) is to have two free parameters, q_{in} and r_{break}, and impose $q_{out} = 3$, which corresponds to the Newtonian limit in the lamppost geometry at large radii far from the X-ray source (see later Sect. 8.2.2). However, if we assume a specific corona geometry, it is possible to compute the intensity profile. For instance, [6] have computed the emissivity index $q = q(r)$ according to the height of the primary X-ray source h for a point-like hot corona located above the black hole on the axis of the accretion disk.

Galactic black hole binaries have line of sight velocities of just a few hundred km/s, which makes their relative motion negligible in the spectrum. In the case of AGN, the cosmological redshift of the source can instead be important for some objects and it can be an additional parameter of the model. However, it can be typically obtained from other observations and does not represent a free parameter in the fit of the reflection spectrum.

Figure 8.7 shows the contribution from different annuli of the accretion disk to the total iron line profile. The central object is a Kerr black hole with $a_* = 0.99$. The inclination angle of the disk with respect to the line of sight of the distant observer is $i = 45°$. The emissivity profile is modeled by a power-law with emissivity index $q = 3$. The inner edge of the disk is at the ISCO radius, $r_{ISCO} \approx 1.45\ M$ in Boyer-Lindquist coordinates for $a_* = 0.99$. The rest-frame energy of the line is $E_e = 6.4$ keV. The total photon flux is indicated by the black dotted line. The photons from the annulus $r_{ISCO} < r < 3\ M$ produce the extended low energy tail in the iron line profile. This is the innermost part of the accretion disk, where the gravitational redshift is stronger and thus some photons have energies significantly lower than 6.4 keV. As we move to larger radii, the annulus spectrum moves to higher energies. This is because the gravitational redshift becomes milder. Moreover, we clearly see that the spectrum of each annulus has two "horns", one at high and one at low energies. The former is produced by Doppler blueshift (photons emitted by gas moving in the direction of the observer), the latter by Doppler redshift (photons emitted by gas moving in the direction opposite to the observer). In the case of the spectrum of the innermost annulus, the region $r_{ISCO} < r < 3\ M$, the two horns are not evident because of the strong light bending, which makes the high energy horn (around 5 keV in Fig. 8.7) very broad. The photons with the highest energies come from the annulus $10\ M < r < 20\ M$, where the Doppler blueshift due to the motion of the gas is still high while the gravitational redshift is much weaker than in the inner part of the accretion disk. The range $10\ M < r < 20\ M$ is valid in the Kerr metric and for $i = 45°$, which is the case in Fig. 8.7. These radial coordinates change if we consider a different viewing angle i, while the exact value of the spin parameter a_* is not very important (at least for moderate values of the viewing angle i) because we

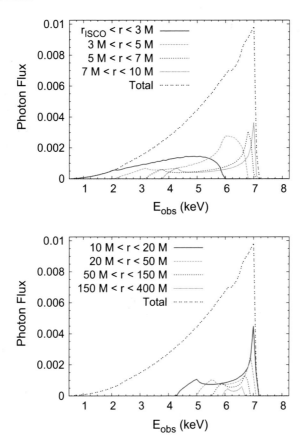

Fig. 8.7 Contribution from different annuli of the disk to the iron line profile. The spacetime is described by the Kerr metric with $a_* = 0.99$. The value of the other parameters is: viewing angle $i = 45°$, emissivity index $q = 3$, inner radius $r_{in} = r_{ISCO} \approx 1.45\ M$ (in Boyer-Lindquist coordinates), outer radius $r_{out} = 400\ M$. The radiation from the inner region $r_{ISCO} < r < 3\ M$ is strongly redshifted, so the spectrum is in the 0.5–6.0 keV range. At larger radii, the gravitational redshift is milder, and the spectrum of the annulus shifts to higher energies. The photons with the highest energies are emitted from the region $10\ M < r < 20\ M$, where the gravitational redshift is weaker but the Doppler blueshift is still important. For $r > 20\ M$, even the effect of the Doppler boosting gets weaker and the spectrum shrinks to its rest-frame energy of 6.4 keV. The characteristic horns in the spectrum of an annulus are due to the Doppler redshift and blueshift. Flux of the photon number in arbitrary units

are already relatively far from the black hole. If we move to larger radii, the energy range of the spectrum shrinks to the rest-frame energy 6.4 keV. This is the signal that even the Doppler boosting is becoming weaker and weaker.

8.2.1   Impact of the Model Parameters

The impact of the model parameters on the iron line profile is shown in Figs. 8.8, 8.9, 8.10, 8.11, 8.12, and 8.13. Figure 8.8 shows the effect of the inclination angle of the disk with respect to the line of sight of the distant observer, i. For small values of i (face-on disk), the Doppler boosting is weak. For high inclination angles (edge-on disk), the Doppler boosting is strong and moves the high energy peak of the profile to higher energies. The location of the high energy peak can indeed be helpful to infer the inclination angle i from the fit. However, there are some complications (the intensity profile and the ionization of the material) that make such a measurement not so straightforward as suggested by Fig. 8.8.

In Fig. 8.9 we consider different values of the emissivity index q, assuming that $I_e \propto r^{-q}$ at all radii. As the value of q increases, the contribution of the radiation emitted at small radii increases its weight on the total profile. For high values of

Fig. 8.8 Impact of the viewing angle i on the iron line profile. The emissivity index is $q = 3$, the spin parameter is $a_* = 0.7$, the inner radius is $r_{in} = r_{ISCO}$, the outer radius is $r_{out} = 400\,M$, and the rest-frame energy of the line is $E_e = 6.4$ keV. For $i = 10°$, the normalization of the line profile has been reduced for graphical reasons. Flux of the photon number in arbitrary units

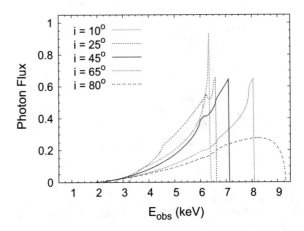

Fig. 8.9 Impact of the emissivity index q on the iron line profile. The viewing angle is $i = 45°$, the spin parameter is $a_* = 0.7$, the inner radius is $r_{in} = r_{ISCO}$, the outer radius is $r_{out} = 400\,M$, and the rest-frame energy of the line is $E_e = 6.4$ keV. Flux of the photon number in arbitrary units

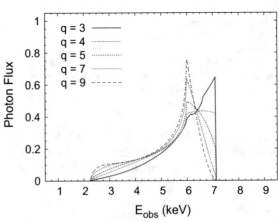

Fig. 8.10 Impact of the inner emissivity index q_{in} on the iron line profile assuming the intensity profile in Eq. (8.4) with $r_{break} = 6\,M$ and $q_{out} = 3$. The viewing angle is $i = 45°$, the spin parameter is $a_* = 0.7$, the inner radius is $r_{in} = r_{ISCO} \approx 3.393\,M$, the outer radius is $r_{out} = 400\,M$, and the rest-frame energy of the line is $E_e = 6.4$ keV. Flux of the photon number in arbitrary units

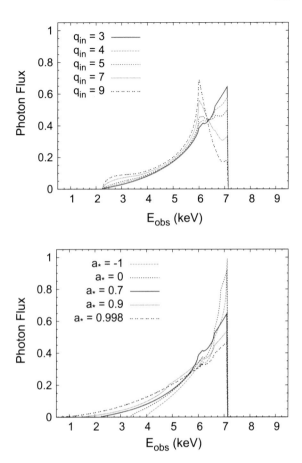

Fig. 8.11 Impact of the spin parameter a_* on the iron line profile. The viewing angle is $i = 45°$, the emissivity index is $q = 3$, the inner radius is $r_{in} = r_{ISCO}$, the outer radius is $r_{out} = 400\,M$, and the rest-frame energy of the line is $E_e = 6.4$ keV. For $a_* = -1$, the normalization of the line profile has been reduced for graphical reasons. Flux of the photon number in arbitrary units

q, the peak is not in the high energy part of the profile, but at lower energies. The Doppler blueshift is dominant at 10–20 gravitational radii from the central object for moderate values of the viewing angle. However, if q is very high, the emissivity is already low there. In Fig. 8.10 the emissivity profile is modeled as in Eq. (8.4) with $r_{break} = 6\,M$. The figure shows the iron line profiles for different values of q_{in} assuming $q_{out} = 3$. For larger values of r_{break}, the iron line profile is only weakly affected by the values of r_{break} and q_{out}.

The impact of the spin parameter a_* is illustrated in Fig. 8.11. Assuming that the inner edge of the disk is at the ISCO radius, the main effect of the spin is to set r_{in} and therefore the spin determines the extension of the low energy tail of the iron line profile.

If we relax the assumption that the inner edge of the disk is at the ISCO radius, r_{in} is another free parameter of the model. Figure 8.12 shows the iron line profile from a Kerr black hole with the spin parameter $a_* = 0.998$. To facilitate the comparison with Fig. 8.11, the inner edge of the disk is set at the same radial coordinate (in Boyer-

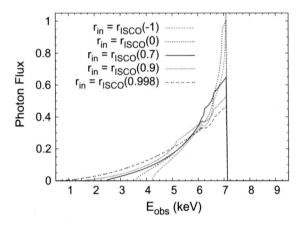

Fig. 8.12 Impact of the inner radius r_{in} on the iron line profile. To facilitate the comparison with Fig. 8.11, r_{in} is set to the ISCO radius corresponding to the spin parameter $a_* = -1$, 0, 0.7, 0.9, and 0.998. The viewing angle is $i = 45°$, the emissivity index is $q = 3$, the spin parameter is $a_* = 0.998$, the outer radius is $r_{out} = 400\,M$, and the rest-frame energy of the line is $E_e = 6.4$ keV. The normalization of the case with r_{in} set to the ISCO radius corresponding to the spin parameter $a_* = -1$ has been reduced for graphical reasons. Flux of the photon number in arbitrary units

Fig. 8.13 Impact of the outer radius r_{out} on the iron line profile. The viewing angle is $i = 45°$, the emissivity index is $q = 3$, the spin parameter is $a_* = 0.7$, the inner radius is $r_{in} = r_{ISCO}$, and the rest-frame energy of the line is $E_e = 6.4$ keV. Flux of the photon number in arbitrary units

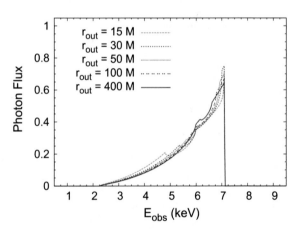

Lindquist coordinates) as the ISCO radius for a Kerr black hole with the values of a_* considered in Fig. 8.11. While the iron line profiles in Figs. 8.11 and 8.12 are not identical, the profiles with the same inner radius are definitively similar. This means that the main impact of the spin is the determination of the inner edge via the ISCO radius, while its effect on the iron line profile is much weaker when we relax the assumption $r_{in} = r_{ISCO}$.

Last, Fig. 8.13 shows how the outer radius of the disk r_{out} alters the iron line profile. For small values of r_{out}, the iron line profile has a cusp in its central part. Such a cusp is more or less pronounced, depending on the values of the other parameters. For instance, in Fig. 8.13, the cusp is at ~ 4.7 keV for $r_{out} = 15\,M$. This cusp results from the Doppler redshifted peak from the annulus at larger radii (see Fig. 8.7). Indeed, as the outer edge of the disk moves to larger radii, the position of the cusp moves to larger radii too. For very large values of r_{out}, we do not see any cusp. Iron line profiles in current data do not show any cusp, which can be explained either with the fact that the outer radius of the emitting region is large enough that the reflection flux is too low, or that the data are not good enough to identify such a small feature even if present.

8.2.2 Emissivity Profile from a Lamppost Corona

Modeling the emissivity profile with a power-law is a quite crude approximation. In the presence of a specific theoretical model, it would be possible to predict the emissivity profile as a function of the radical coordinate r. In the lamppost geometry, the corona is assumed to be a point-like source along the spin axis of the black hole. The set-up is sketched in Fig. 8.14. For the moment, the lamppost geometry is just one among other possible scenarios and we do not know if it is the set-up realized in Nature.

In Fig. 8.14, the height of the corona with respect the black hole is h. In Newtonian gravity, without light bending, a light ray path is like the red arrow in the figure. At the radius r in the accretion disk, the incident angle is ϑ_i. The intensity of the incident radiation is

$$I_i(r, h) \propto \frac{\cos \vartheta_i}{r^2 + h^2} = \frac{h}{\left(r^2 + h^2\right)^{3/2}} . \tag{8.5}$$

For $r \gg h$, we have $I_i \propto r^{-3}$. I_i (incident radiation) and I_e (reflected radiation) have the same radial profile, so the emissivity index $q = 3$ corresponds to the Newtonian limit at large radii for a lamppost corona.

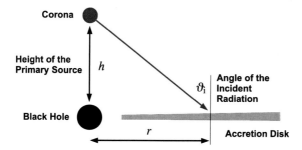

Fig. 8.14 Lamppost geometry. The primary X-ray source (corona) is located above the rotational axis of the black hole at the height h. The photons emitted by the primary X-ray source illuminate the accretion disk. Each photon hits the accretion disk at some radius r and with incident angle ϑ_i

In curved spacetime, light bending plays a non-negligible role, especially if the corona is just above the black hole. The calculations can be done numerically [6]. If we write

$$I_i(r, h) \propto r^{-\varepsilon} \,, \tag{8.6}$$

the results for the Kerr metric are shown in Fig. 8.15. ε is a non-trivial function of the radial coordinate and mainly depends on the height of the corona h. The exact value of the spin parameter of the spacetime is ignored here, because its impact on ε is weak. At large radii, we recover the Newtonian limit $\varepsilon = 3$. At small radii and with a corona just above the black hole, ε can be significantly larger than 3. At least some observations seem to require high, or even very high, values of q at small radii [41, 42]; see however [35].

It is clear that more realistic models of a lamppost corona may be more complicated. For instance, the corona may not be point-like but have a finite size, it may not be exactly along the spin axis, the emitting region may have a finite velocity, etc. See e.g. [42].

Fig. 8.15 The emissivity index ε in the lamppost corona geometry for different values of the height of the primary source h. $r_g = M$ is the gravitational radius. The *vertical dashed lines* mark the position of the ISCO radius for Kerr black holes with the spin $a_* = 0.99, 0$ (Schwarzschild), -1 (extremal Kerr black hole and counterrotating disk). Figure 3 from [6], reproduced by permission of Oxford University Press

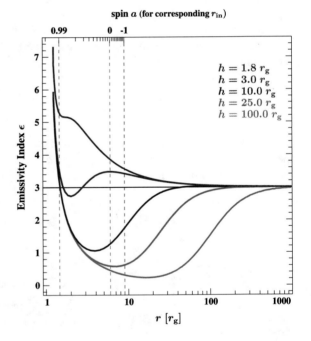

8.3 Spin Measurements

As shown in the previous section, the photons in the iron line are strongly affected by the relativistic effects occurring in the strong gravity region around the black hole. If the physics and astrophysics behind are properly understood, the analysis of the reflection spectrum can be a powerful tool to test the nature of the compact object. If we assume the Kerr metric, the spacetime metric affects the shape of the reflection spectrum only through the spin parameter a_*. The mass of the black hole M only sets the scale of the system and does not play any role in the shape of the reflection spectrum. While it is necessary to fit the whole reflection spectrum to measure the black hole spin, the iron $K\alpha$ line is the strongest feature sensitive to the gravitational field at a few gravitational radii from the compact object. This is the reason of the name "iron line method" to mean the study of the reflection spectrum to measure black hole spins. Two recent reviews on spin measurements with the iron line method are [2, 30].

With respect to the continuum-fitting method, the iron line one has some advantages. It is independents of the mass and the distant of the black hole, two quantities that are usually known only with large uncertainty. It does not require the viewing angle as input parameter, but i can be inferred from the iron line itself from the maximum Doppler blueshift. Since the iron line profile is independent of the mass of the compact object, this technique can be used to infer the spin of both stellar-mass and supermassive black holes. The iron line profile has also a more complicated structure than the thermal spectrum of a thin disk, so it can potentially measure more than one parameter of the strong gravity region around the black hole.

Current iron line spin measurements of stellar-mass black holes are summarized in the third column in Table 7.1 in Sect. 7.2. When the measurements from both the thermal spectrum and the iron line are available for the same object, the two measurements usually agree. This is a good indication of the validity of the two techniques. When the two measurements do not agree, it is likely that at least one of them is obtained from observations in which the Novikov-Thorne model is not valid (e.g. the inner edge of the disk is not at the ISCO radius) or there are systematics effects not under control.

Table 8.1 summarizes current spin measurements of supermassive black holes. Here the second column is for the spin measurement and the third column is for the luminosity of the source in Eddington unit. Several sources seem to have a spin parameter very close to 1. While it is possible that these objects are fast-rotating black holes, their spin measurement should be taken with some caution, because the validity of the Novikov-Thorne model is not guaranteed. For example, the accretion rate may be too high and the disk may not be geometrically thin. See Sect. 8.4.2 for more details. However, fast-rotating black holes in galactic nuclei can be expected in the case of prolonged disk accretion (see Sect. 6.4).

Figure 8.16 shows a sample of 17 supermassive black holes with a measurement of their mass M and of their spin parameter a_*. While the statistics is not high and the uncertainty in some measurements is quite large, we may see a few interesting

Table 8.1 Summary of the iron line measurements of the spin parameter of supermassive black holes under the assumption of the Kerr background. The third column shows the total luminosity in Eddington units (see the discussion in Sect. 8.4.2 for its relevance). See the references in the last column and [2] for more details

AGN	a_* (Iron)	L_{Bol}/L_{Edd}	Principal references
IRAS 13224-3809	>0.995	0.71	[40]
Mrk 110	>0.99	0.16 ± 0.04	[40]
NGC 4051	>0.99	0.03	[26]
MCG-6-30-15	>0.98	0.40 ± 0.13	[3, 23]
1H 0707-495	>0.98	~1	[28, 40, 45]
NGC 3783	>0.98	0.06 ± 0.01	[4, 26]
RBS 1124	>0.98	0.15	[40]
NGC 1365	$0.97^{+0.01}_{-0.04}$	$0.06^{+0.06}_{-0.04}$	[5, 32, 33]
Swift J0501.9-3239	>0.96	—	[40]
Ark 564	$0.96^{+0.01}_{-0.06}$	>0.11	[40]
3C 120	>0.95	0.31 ± 0.20	[20]
Ark 120	0.94 ± 0.01	0.04 ± 0.01	[25, 27, 40]
Ton S180	$0.91^{+0.02}_{-0.09}$	$2.1^{+3.2}_{-1.6}$	[40]
1H 0419-577	>0.88	1.3 ± 0.4	[40]
Mrk 509	$0.86^{+0.02}_{-0.01}$	—	[40]
IRAS 00521-7054	>0.84	—	[36]
3C 382	$0.75^{+0.07}_{-0.04}$	—	[40]
Mrk 335	$0.70^{+0.12}_{-0.01}$	0.25 ± 0.07	[27, 40]
Mrk 79	0.7 ± 0.1	0.05 ± 0.01	[10, 11]
Mrk 359	$0.7^{+0.3}_{-0.5}$	0.25	[40]
NGC 7469	0.69 ± 0.09	—	[27]
Swift J2127.4+5654	0.6 ± 0.2	0.18 ± 0.03	[24, 27]
Mrk 1018	$0.6^{+0.4}_{-0.7}$	0.01	[40]
Mrk 841	>0.56	0.44	[40]
Fairall 9	$0.52^{+0.19}_{-0.15}$	0.05 ± 0.01	[19, 27, 34, 40]

features. First, it seems that black holes with $M \approx 10^6$–$10^7 \, M_\odot$ have a spin parameter close to 1, while the spin of heavier black holes with $M \approx 10^7$–$10^9 \, M_\odot$ may be lower. Second, the objects in this sample are all radio-quiet AGN (see Appendix C). Their radio luminosity is low because they do not have powerful jets. Since their spin is instead high, this seems to exclude the idea that radio-quiet AGN would be slow-rotating black holes and radio-loud AGN would be fast-rotating black holes. If the jet of a black hole is powered by its spin, it is likely that, in analogy with stellar-mass black holes (see Sect. 4.5), jets appear only for relatively short times (see Appendix D).

Fig. 8.16 Sample of 17 supermassive black holes with a measurement of their mass M and of their spin parameter a_*. The masses have an error bar corresponding to 1-σ, the spins have an error bar corresponding to 90% confidence level. From [30]

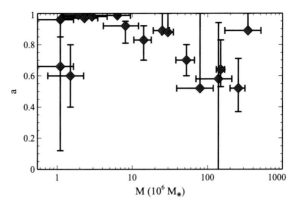

8.4 Validity of the Model

8.4.1 Relativistic Origin of Broad Iron Lines

The interpretation that broad iron lines are a feature of the reflection spectrum of the accretion disk is today well supported by observations and there is a common consensus on the fact that these iron lines are generated from the inner part of the disk. Most of the alternative proposals to explain broad iron lines are relatively easy to rule out [8]. In principle, warm absorbers, namely absorbing clouds that partially cover the power-law source, may somehow mimic the X-ray reflection spectrum and produce a broad iron line. The absorber model was proposed in [21, 22, 38], but is now ruled out by observations.

First, absorbing passing clouds may indeed be expected in the case of AGN, but they are definitively more unlikely in the case of black hole binaries. On the other hand, we observe broad and skewed iron lines in the spectrum of both supermassive and stellar-mass black holes, and the line shapes are very similar. It is thus natural to imagine that such a feature is more likely related to the background metric rather than to the astrophysical environment, which is different for AGN and black hole binaries.

Second, the absorber model can predict a broad iron line, but it does not predict any Compton hump at higher energies. X-ray missions like XMM-Newton or Chandra can only measure the X-ray spectrum up to about 10 keV (see Table 5.2), and therefore they cannot observe the Compton hump at 20–40 keV. After the launch of NuSTAR in 2012, it has been possible to measure the X-ray spectrum of black holes up to 70–80 keV and verify the presence of the Compton hump. Risaliti et al. [33] studied the supermassive black hole in the AGN NGC 1365 with both XMM-Newton and NuSTAR. The result is shown in Fig. 8.17, where we can see that both the absorber and the reflection models could explain the broad iron line at 3–7 keV, but the absorber model clearly fails to fit the spectrum above 10 keV. The NuSTAR data of NGC 1365 definitively support the reflection origin of the iron line and rule out the absorber model.

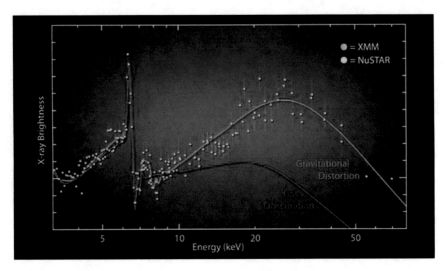

Fig. 8.17 Theoretical predictions of the absorber model (*red solid curve*) and of the reflection model (*green solid curve*), and observational data of XMM-Newton (*blue dots*) and NuSTAR (*yellow dots*) of the AGN NGC 1365. At low energies, the absorber and the reflection models predict a similar iron line, and the XMM-Newton data cannot distinguish the two scenarios. At higher energies, the two models predict different spectra. The reflection model has the Compton hump at 20–40 keV. The NuSTAR data support the reflection model and rule out the absorber model. See [33] and the text for more details. Image Credit: NASA/JPL-Caltech

Third, temporal variations of the corona luminosity inevitably cause temporal variations in the reflection spectrum. However, the speed of light is finite, and therefore the response of the reflection spectrum has some time delay with respect to the variability of the corona. Such a time delay, which is called the reverberation lag, can be inferred by studying the light curves made in different X-ray energy bands, exploiting the fact that the primary and the reflection spectra have different shapes. Zoghbi et al. [45] found a 30 s lag in the high frequency variations of the soft and iron-L line dominated bands relative to the harder, power-law dominated band in 1H 0707-495. This is exactly what one should expect from the reflection model, due to the additional light travel time of the reflection signal with respect to the direct power-law component from the corona.

Figure 8.18 shows the lag-energy spectra for five supermassive black holes. The amplitudes of the lags have been rescaled in order to have the relative lag between 3–4 and 6–7 keV equal for all the sources (the masses of these black holes are different). The iron line lag, namely the lag around the energy of the iron Kα line, is similar in all the sources, while the lag at lower energies is different. The interpretation is that the reflection spectrum cannot depend on the accretion flow, which is different in any source, but it must be related to metric of the spacetime, which is roughly the same for every object.

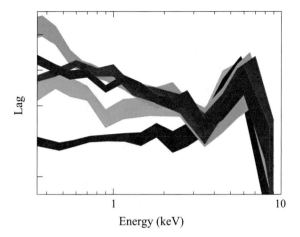

Fig. 8.18 Lag-energy spectra for 1H0707-495 (*blue*), IRAS 13224-380 (*red*), Ark 564 (*green*), Mrk 335 (*cyan*), and PG 1244+026 (*purple*). The shape of the iron line lag is similar in all the sources, while the lag at lower energies is different. This strongly supports the relativistic origin of the iron line and its importance as a tool to test strong gravity, because it suggests that the physics behind is quite independent of the properties of the accretion flow and depends on some fundamental property, namely the background geometry. Here the lags have been rescaled because the masses, and therefore the characteristic timescales, of these objects are different. From [39]

8.4.2 Spin Measurements

As the continuum-fitting method, even the analysis of the iron line relies on the validity of the Novikov-Thorne model. In particular, a crucial assumption is that the inner edge of the disk is at the ISCO radius. This is a subtle point and we should distinguish the cases of stellar-mass and supermassive black holes.

For stellar-mass black holes, the validity of the Novikov-Thorne model for the continuum-fitting method has already been discussed in Sect. 7.2. If the source is in the high-soft state with an accretion luminosity between 5 and 30% of its Eddington limit, deviations from the model are small. However, the reflection spectrum is dominant when the source is in the low-hard state. For such a spectral state, there has not yet been a systematic study on the validity of the Novikov-Thorne model as in the high-soft state. As discussed in Sect. 4.5, observations show that at the beginning of the low-hard state, when the mass accretion rate is low, the disk may be truncated at a radius larger than the ISCO radius. As the mass accretion rate increases, the inner edge of the disk approaches the central object. The measurement of the spin via the iron line approach may thus require that the source is at the end of the low-hard state or in the hard-intermediate state.

For example, in the case of the black hole binary GX 339-4, we have both observations in which the inner edge of the disk seems to be at the ISCO radius and whose data are thus suitable for spin measurements [14], as well as observations in which the disk is truncated and it would be wrong to assume $r_{in} = r_{ISCO}$ in the analysis of the

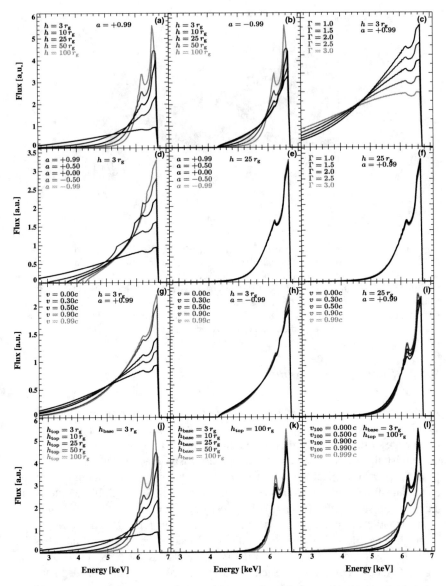

Fig. 8.19 Impact on the iron line profile of the model parameters studied in [6] within a lamppost geometry. Panels (**a**) and (**b**): impact of the height of the primary source h. Panels (**c**) and (**f**): impact of the photon index Γ of the spectrum of the primary source. Panels (**d**) and (**e**): impact of the spin parameter a_*. Panels (**g**), (**h**), and (**i**): irradiation by the base of a jet moving at the velocity v. Panels (**j**) and (**k**): irradiation by an elongated jet, where h_{top} is the top height of the jet and h_{base} is the height of the base of the jet. Panel (**l**): irradiation by an accelerated jet moving from the height $h = h_{\text{base}}$ to $h = h_{\text{top}}$, where v_{100} is the velocity at $h = 100\,r_{\text{g}}$. If not stated otherwise, the black hole spin is $a_* = 0.99$ and the height of the primary source is $h = 3\,r_{\text{g}}$. In all profiles, the viewing angle is $i = 30°$, the inner edge of the disk is at the ISCO radius, and the outer edge of the disk is at $400\,r_{\text{g}}$. $r_{\text{g}} = M$ is the gravitational radius. All profiles are normalized to have the same number of photons in the iron line. See the text for more details. Figure 9 from [6], reproduced by permission of Oxford University Press

reflection spectrum [9]. In the latter case, spin measurements could be still possible in principle, but with the current quality of the data we cannot break the degeneracy between a_* and r_{in} and therefore we cannot measure the black hole spin if we relax the assumption $r_{in} = r_{ISCO}$. In the presence of high quality data and with the correct astrophysical model, it may be possible to measure both a_* and r_{in} separately.

For supermassive black holes, the iron line method is used at any accretion luminosity, as shown by the third column in Table 8.1.[1] It is unclear if the ISCO assumption is correct. For low accretion luminosities, the disk may be thin but truncated at a radius larger than the one of the ISCO. The high values of the spin parameters reported in Table 8.1 are often in sources with a high accretion luminosity. In this case, it is possible that the disk becomes fat and the inner edge of the disk is inside the ISCO (see Appendix E). This would lead to an overestimate of the black hole spin. Such a high values of the black hole spin should thus be taken with some caution, because the uncertainty does not include possible systematic effects related to these issues.

Dauser et al. [6] have explored the actual capability of the iron line method to measure black hole spins within the lamppost set-up. They results are summarized in Fig. 8.19, which shows the iron line profile for different values of the parameters of the corona model. The conclusion of this work is that a spin measurement seems to be possible only in the case of a compact corona at not more than a few gravitational radii from the black hole. If the corona is not compact or it is not close to the black hole, the iron line is narrower and it is not possible to distinguish a fast-rotating black hole with an elongated corona from a slow-rotating black hole. In such a case, it would be only possible to infer a lower value for the spin parameter.

8.5 Reverberation Mapping

In the framework of the corona-disk model, "iron line reverberation" refers to the iron line signal as a function of time in response to a δ-function like pulse of radiation from the corona. The resulting line spectrum as a function of both time and across photon energy is called the 2D transfer function. As shown for the first time in [31, 43], the shape of the 2D transfer function is determined by the fundamental properties of the black hole and the system geometry. An accurate measurement of the 2D transfer function can better map the spacetime geometry in the strong gravity region than the corresponding time-integrated measurement [17].

Within the lamppost model, the set-up is that shown in Fig. 8.20. The corona illuminates the cold accretion disk. Because of the finite speed of light, photons emitted from different points in the accretion disk are detected at different times by

---

[1]It is worth noting that even the estimates of L_{Bol}/L_{Edd} in Table 8.1 have to be taken with some caution, because it is difficult to get reliable measurements of the mass and of the total luminosity for these objects. In the end, it is currently impossible to say if one of these sources is in the right spectral state to apply the Novikov-Thorne model and the iron line method.

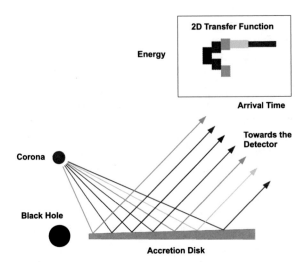

Fig. 8.20 Cartoon explaining the shape of the 2D transfer function of an iron line generated from a system with lamppost geometry. The corona above the rotational axis of the black hole illuminates the accretion disk and, because of the finite speed of light, photons emitted from different parts of the disk arrive at the detector at different times. The *purple* path is the shortest one, so the iron line photons from the point of emission in the disk associated to that path are the first to arrive. After these photons, we have those associated to the *blues* paths. The photons emitted at smaller (larger) radii than the photons from the *purple* path are more (less) affected by gravitational *redshifted*. The same is true for the photons of the *green* paths with respect to the photons of the *blue* paths. The 2D transfer function thus develops two branches, one with photons of higher energies and one with photons of lower energies emitted closer to the *black hole*. The low energy branch ends when we detect the last photons from the inner edge of the disk. The high energy branch, after having reached a maximum energy determined by the balance between gravitational *redshift* and Doppler *blueshift*, tends to the asymptotic value of the line (6.4 keV for neutral matter). The energy width of this branch at late time is determined by Doppler boosting

the distant observer. The 2D transfer function becomes nonzero as the first photons reach the observer (purple path in Fig. 8.20). After that, the observer detects photons with slightly longer path. These photons (blue paths in Fig. 8.20) are emitted at larger and smaller radii with respect to the first photons, and therefore they are, respectively, less and more affected by the gravitational redshift occurring near the black hole. The 2D transfer function thus develops two branches, one at higher and one at lower energies. Photons arriving later are even less and more affected by gravitational redshift (green paths in Fig. 8.20). The low energy branch of the 2D transfer function ends when the observer detects the photons emitted from the inner edge of the accretion disk. No photons are emitted from smaller radii. On the other hand, the high energy branch continues receiving photons, even if the signal is weaker and weaker because the intensity is lower and lower at larger radii. The photons of the high energy branch approach the rest-frame energy of the iron line (gravitational redshift becomes weaker and weaker), and the small energy width of the 2D transfer function is due to Doppler boosting for the gas orbital motion (red path in Fig. 8.20).

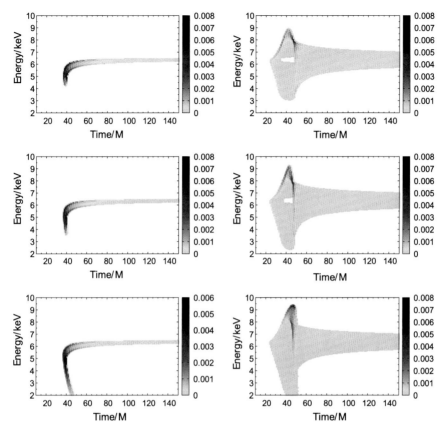

Fig. 8.21 2D transfer functions in the Kerr spacetime. The spin parameter is $a_* = 0$ (*top panels*), 0.5 (*central panels*), and 0.95 (*bottom panels*). The viewing angle is $i = 10°$ (*left panels*) and 80° (*right panels*). The height of the source is $h = 10\ M$ and the index of the intensity profile is $q = 3$. Flux of the photon number in arbitrary units. From [16], reproduced by permission of IOP Publishing. All rights reserved

Examples of 2D transfer functions in the Kerr metric are shown in Fig. 8.21. The black hole spin is $a_* = 0$ (top panels), 0.5 (central panels), and 0.95 (bottom panels). The inclination angle of the disk with respect to the line of sight of the distant observer is $i = 10°$ (left panels) and 80° (right panels).

The 2D transfer functions in the case of a low viewing angle are easier to interpret. Here the Doppler boosting is negligible, so the shape of the 2D transfer function is determined by the gravitational redshift. The extension of the low energy branch in these 2D transfer functions is determined by the position of the ISCO radius, which depends on the black hole spin. For a fast-rotating black hole, the low energy branch can reach very low energies. The radiation coming from larger radii and reaching the distant observer at later times is quite independent of the metric in the strong gravity region, and the three plots are very similar.

In the case of a large viewing angle (right panels in Fig. 8.21), the 2D transfer function has definitively a more complicated structure due to the combination of gravitational redshift and Doppler boosting. At later times, the 2D transfer function is wider than the low viewing angle case because the line is still broad due to Doppler boosting.

With the current X-ray facilities, iron line reverberation mapping is not possible because of the low photon count rate. For the time being, there are just measurements of reverberation lags in AGN between different X-ray bands. However, iron line reverberation mapping should not be out of reach for the next generation of X-ray satellites. Black hole binaries are brighter than AGN, but the timescale variability is much shorter. For the next generation of X-ray missions, like eXTP [44], black hole binaries are more promising sources for iron line reverberation measurements because the effective area of the detector is large enough to be able to count a sufficient number of photons in a short time.

References

1. P. Barr, N.E. White, G.C. Page, Mon. Not. R. Astron. Soc. **216**, 65 (1985)
2. L. Brenneman, *Measuring Supermassive Black Hole Spins in Active Galactic Nuclei* (Springer, New York, 2013), arXiv:1309.6334 [astro-ph.HE]
3. L.W. Brenneman, C.S. Reynolds, Astrophys. J. **652**, 1028 (2006), arXiv:astro-ph/0608502
4. L.W. Brenneman et al., Astrophys. J. **736**, 103 (2011), arXiv:1104.1172 [astro-ph.HE]
5. L.W. Brenneman, G. Risaliti, M. Elvis, E. Nardini, Mon. Not. R. Astron. Soc. **429**, 2662 (2013), arXiv:1212.0772 [astro-ph.HE]
6. T. Dauser, J. Garcia, J. Wilms, M. Bock, L.W. Brenneman, M. Falanga, K. Fukumura, C.S. Reynolds, Mon. Not. R. Astron. Soc. **430**, 1694 (2013), arXiv:1301.4922 [astro-ph.HE]
7. A.C. Fabian, M.J. Rees, L. Stella, N.E. White, Mon. Not. R. Astron. Soc. **238**, 729 (1989)
8. A.C. Fabian, K. Nandra, C.S. Reynolds, W.N. Brandt, C. Otani, Y. Tanaka, H. Inoue, K. Iwasawa, Mon. Not. R. Astron. Soc. **277**, L11 (1995), arXiv:astro-ph/9507061
9. F. Fuerst et al., arXiv:1604.08644 [astro-ph.HE]
10. L.C. Gallo, A.C. Fabian, T. Boller, W. Pietsch, Mon. Not. R. Astron. Soc. **363**, 64 (2005), arXiv:astro-ph/0508229
11. L.C. Gallo, G. Miniutti, J.M. Miller, L.W. Brenneman, A.C. Fabian, M. Guainazzi, C.S. Reynolds, Mon. Not. R. Astron. Soc. **411**, 607 (2011), arXiv:1009.2987 [astro-ph.HE]
12. J. Garcia, T. Dauser, C.S. Reynolds, T.R. Kallman, J.E. McClintock, J. Wilms, W. Eikmann, Astrophys. J. **768**, 146 (2013), arXiv:1303.2112 [astro-ph.HE]
13. J. Garcia et al., Astrophys. J. **782**, 76 (2014), arXiv:1312.3231 [astro-ph.HE]
14. J. Garcia et al., Astrophys. J. **813**, 84 (2015), arXiv:1505.03607 [astro-ph.HE]
15. P.W. Guilbert, M.J. Rees, Mon. Not. R. Astron. Soc. **233**, 475 (1988)
16. J. Jiang, C. Bambi, J.F. Steiner, JCAP **1505**, 025 (2015), arXiv:1406.5677 [gr-qc]
17. J. Jiang, C. Bambi, J.F. Steiner, Phys. Rev. D **93**, 123008 (2016), arXiv:1601.00838 [gr-qc]
18. A.P. Lightman, T.R. White, Astrophys. J. **233**, 57 (1988)
19. A.M. Lohfink, C.S. Reynolds, J.M. Miller, L.W. Brenneman, R.F. Mushotzky, M.A. Nowak, A.C. Fabian, Astrophys. J. **758**, 67 (2012), arXiv:1209.0468 [astro-ph.HE]
20. A.M. Lohfink et al., Astrophys. J. **772**, 83 (2013), arXiv:1305.4937 [astro-ph.HE]
21. L. Miller, T.J. Turner, J.N. Reeves, Astron. Astrophys. **483**, 437 (2008), arXiv:0803.2680 [astro-ph]
22. L. Miller, T.J. Turner, J.N. Reeves, Mon. Not. R. Astron. Soc. **399**, L69 (2009), arXiv:0907.3114 [astro-ph.HE]

23. G. Miniutti et al., Publ. Astron. Soc. Jpn. **59**, 315 (2007), arXiv:astro-ph/0609521
24. G. Miniutti, F. Panessa, A. De Rosa, A.C. Fabian, A. Malizia, M. Molina, J.M. Miller, S. Vaughan, Mon. Not. R. Astron. Soc. **398**, 255 (2009), arXiv:0905.2891 [astro-ph.HE]
25. E. Nardini, A.C. Fabian, R.C. Reis, D.J. Walton, Mon. Not. R. Astron. Soc. **410**, 1251 (2011), arXiv:1008.2157 [astro-ph.HE]
26. A.R. Patrick, J.N. Reeves, A.P. Lobban, D. Porquet, A.G. Markowitz, Mon. Not. R. Astron. Soc. **416**, 2725 (2011), arXiv:1106.2135 [astro-ph.HE]
27. A.R. Patrick, J.N. Reeves, D. Porquet, A.G. Markowitz, A.P. Lobban, Y. Terashima, Mon. Not. R. Astron. Soc. **411**, 2353 (2011), arXiv:1010.2080 [astro-ph.HE]
28. I. d. l. C. Perez et al., Astron. Astrophys. **524**, A50 (2010), arXiv:1007.4762 [astro-ph.HE]
29. C.S. Reynolds, Ph.D. thesis (University of Cambridge, UK, 1996)
30. C.S. Reynolds, Space Sci. Rev. **183**, 277 (2014), arXiv:1302.3260 [astro-ph.HE]
31. C.S. Reynolds, A.J. Young, M.C. Begelman, A.C. Fabian, Astrophys. J. **514**, 164 (1999), arXiv:astro-ph/9806327
32. G. Risaliti et al., Astrophys. J. **696**, 160 (2009), arXiv:0901.4809 [astro-ph.CO]
33. G. Risaliti et al., Nature **494**, 449 (2013), arXiv:1302.7002 [astro-ph.HE]
34. S. Schmoll et al., Astrophys. J. **703**, 2171 (2009), arXiv:0908.0013 [astro-ph.HE]
35. J. Svoboda, M. Dovciak, R.W. Goosmann, P. Jethwa, V. Karas, G. Miniutti, M. Guainazzi, Astron. Astrophys. **545**, A106 (2012), arXiv:1208.0360 [astro-ph.HE]
36. Y. Tan, J. Wang, X. Shu, Y. Zhou, Astrophys. J. **747**, L11 (2012), arXiv:1202.0400 [astro-ph.HE]
37. Y. Tanaka et al., Nature **375**, 659 (1995)
38. T.J. Turner, L. Miller, J.N. Reeves, S.B. Kraemer, Astron. Astrophys. **475**, 121 (2007), arXiv:0708.1338 [astro-ph]
39. P. Uttley, E.M. Cackett, A.C. Fabian, E. Kara, D.R. Wilkins, Astron. Astrophys. Rev. **22**, 72 (2014), arXiv:1405.6575 [astro-ph.HE]
40. D.J. Walton, E. Nardini, A.C. Fabian, L.C. Gallo, R.C. Reis, Mon. Not. R. Astron. Soc. **428**, 2901 (2013), arXiv:1210.4593 [astro-ph.HE]
41. D.R. Wilkins, A.C. Fabian, Mon. Not. R. Astron. Soc. **414**, 1269 (2011), arXiv:1102.0433 [astro-ph.HE]
42. D.R. Wilkins, A.C. Fabian, Mon. Not. R. Astron. Soc. **424**, 1284 (2012), arXiv:1205.3179 [astro-ph.HE]
43. A.J. Young, C.S. Reynolds, Astrophys. J. **529**, 101 (2000), arXiv:astro-ph/9910168
44. S.N. Zhang et al., [eXTP Collaboration], arXiv:1607.08823 [astro-ph.IM]
45. A. Zoghbi, A. Fabian, P. Uttley, G. Miniutti, L. Gallo, C. Reynolds, J. Miller, G. Ponti, Mon. Not. R. Astron. Soc. **401**, 2419 (2010), arXiv:0910.0367 [astro-ph.HE]

Chapter 9
Quasi-periodic Oscillations

Quasi-periodic oscillations (QPOs) are a common feature in the X-ray power density spectrum (PDS) of neutron stars and stellar-mass black holes [16, 26]. They are seen as relatively narrow peaks at characteristic frequencies. High-frequency QPOs in black hole binaries ($\sim$100 Hz) are particularly interesting because:

1. Their frequency is in the expected range for matter orbiting near the ISCO radius of the source.
2. They depend only very weakly on the observed X-ray flux, which suggests that the frequency is mainly determined by the metric of the spacetime and is not very sensitive to the properties of the accretion flow.
3. The centroid frequency can be measured with high precision.

Thanks to these properties, it is thought that, if properly understood, QPOs can be a powerful tool to test black holes and get precise measurements of the background metric. For the moment, the actual mechanism responsible for the production of QPOs is not known. There are several proposals in the literature, like the relativistic precession models [20–22], the resonance models [1, 2, 9], and the diskoseismology models [14, 19, 27].

Table 9.1 shows a summary of spin measurements of the stellar-mass black hole GRO J1655-40 from different QPO models. All these measurements assume that the metric around the compact object is described by the Kerr solution. This source has high-frequency QPOs at 300 and 450 Hz. The measurement with the relativistic precession model employs also a low-frequency QPO and does not need the mass of the object as an input parameter. The diskoseismology model requires the two high-frequency QPOs as input parameters, and provides an estimate of the black hole mass and spin. The other measurements require the black hole mass as an input parameter, and in the case of GRO J1655-40 the value of the mass is known from optical observations. As shown in this table, the correct interpretation of the two frequencies is crucial and different models provide completely different measurements of the black hole spin.

© Springer Nature Singapore Pte Ltd. 2017

C. Bambi, *Black Holes: A Laboratory for Testing Strong Gravity*,

DOI 10.1007/978-981-10-4524-0_9

Table 9.1 Summary of the spin measurements of GRO J1655-40 from different QPO models. All these measurements assume that the metric around the compact object is described by the Kerr solution. In the case of the measurements with the relativistic precession model and the diskoseismology model, there is no assumption about the mass of the object. In the other measurements, the value of the mass adopted to infer the black hole spin is reported in the third column

Model	a_*	Assumed mass ($M_\odot$)	References
Relativistic precession model	0.290 ± 0.003	–	[11]
Diskoseismology model	0.917 ± 0.024	–	[27]
Warp resonance model	0.9–0.99	5.1–5.7	[8]
3:2 Parametric resonance model	0.96–0.99	6.0–6.6	[25]
2:1 Forced resonance model	0.31–0.42	6.0–6.6	[25]
3:1 Forced resonance model	0.50–0.59	6.0–6.6	[25]
2:1 Keplerian resonance model	0.31–0.42	6.0–6.6	[25]
3:1 Keplerian resonance model	0.45–0.53	6.0–6.6	[25]

Table 9.2 As in Table 9.1 for GRS 1915+105. The lower value of the spin estimate from the forced and the Keplerian resonance models is negative, namely the disk would have an angular momentum antiparallel to the black hole spin

Model	a_*	Assumed mass ($M_\odot$)	References
Diskoseismology model	0.926 ± 0.020	–	[27]
Warp resonance model	0.93-0.99	14–16	[8]
3:2 Parametric resonance model	0.69–0.99	10.0–18.0	[25]
2:1 Forced resonance model	(−0.41)–0.44	10.0–18.0	[25]
3:1 Forced resonance model	(−0.15)–0.61	10.0–18.0	[25]
2:1 Keplerian resonance model	(−0.41)–0.44	10.0–18.0	[25]
3:1 Keplerian resonance model	(−0.13)–0.55	10.0–18.0	[25]

Table 9.2 shows the case of the black hole binary GRS 1915+105. This source has high-frequency QPOs at 113 and 168 Hz. Again, every model predicts a very different black hole spin.

On the basis that stellar-mass and supermassive black holes have a very similar behavior and that the scale of these systems is their mass M, high-frequency QPOs are expected in AGN as well. For objects with $M > 10^6 \, M_\odot$, high-frequency QPOs should be below 1 mHz. However, due to insufficient observation lengths and modeling problems, QPOs are more difficult to detect in AGN. The first robust detection of a QPO in an AGN was reported in [5] from the Seyfert galaxy RE J1034+396.

QPOs have been detected even in some ultra-luminous X-ray sources associated to intermediate-mass black hole candidates. For instance, Pasham et al. [13] have reported the detection of twin peak QPOs from M82 X-1 at 3.32 ± 0.06 Hz and 5.07 ± 0.06 Hz. If these frequencies are interpreted within the relativistic precession model, the mass of the black hole candidate would be $M = 415 \pm 63 \, M_\odot$ [13].

High-frequency QPOs have been also studied in the context of tests of the Kerr metric [3, 4, 7, 10]. In general, there is a strong correlation between the estimate of the black hole spin and possible deviations from the Kerr solution, because each frequency is just a number, and it is not difficult to obtain by simply adjusting the parameters of the background metric. However, if it is possible to break the parameter degeneracy, for instance by adding an independent constraint from other observations, and we suppose to know the correct astrophysical model, the constraints can be quite strong.

9.1 Observations

Detectors on board of X-ray missions can typically measure the time of arrival of every X-ray photon hitting the detector with an accuracy of microseconds. If we have the photon count rate $C(t)$, it can be Fourier transformed and squared to give a power density. If we use the Leahy normalization, the power density spectrum is

$$P(\nu) = \frac{2}{N} \left| \int_0^T C(t) e^{-2\pi i \nu t} dt \right|^2, \qquad (9.1)$$

where N is the total number of counts and T is the duration of the observation. QPOs are narrow features in the X-ray power density spectrum of a source.[1] In the case of black hole binaries, QPOs can be grouped into two classes: low-frequency QPOs (LFQPOs) and high-frequency QPOs (HFQPOs).

Low-frequency QPOs have a frequency in the range 0.1–30 Hz. They may either vary in frequency on very short timescales ($\sim$1 min) or be relatively stable. There are different kinds of LFQPOs and they are called, respectively, type-A, type-B, and type-C QPOs. For instance, type-A QPOs typically manifest when the source is in the high soft state, just after the intermediate state. The presence of type-B QPOs is used to define the soft intermediate state. Type-C QPOs are mainly detected when the source is in the bright end of the hard state or in the hard intermediate state.

High-frequency QPOs have a frequency in the range 40–450 Hz. In four systems, we have detections of pairs of HFQPOs (even if not always simultaneously) with the two frequencies that turn out to be in the ratio 3:2. These two QPOs are usually called, respectively, the upper and the lower HFQPOs. Table 9.3 shows the sources with detections of a pair of HFQPOs.

Figure 9.1 shows the power density spectrum obtained from an observation of the stellar-mass black hole XTE J11550-564. In the large panel, we see a type-B QPO at $\sim$5 Hz and a type-C QPO at 13 Hz. In the small panel, we have a HFQPO at 183 Hz.

---

[1] Broad structures in the X-ray power density spectrum are instead called "noise".

Table 9.3 Black hole binaries with a measurement of two high-frequency QPOs

BH binary	Lower frequency (Hz)	Upper frequency (Hz)	References
GRO J1655-40	300	450	[17, 18, 24]
XTE J1550-564	184	276	[18]
GRS 1915+105	113	168	[15, 16]
H1743-322	166	242	[6]

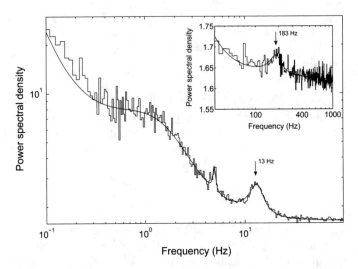

Fig. 9.1 Power density spectrum from an observation of XTE J1550-564. We see a type-B QPO at ~5 Hz, a type-C QPO at 13 Hz (marked by an *arrow*), and a HFQPO at 183 Hz in the inset (marked by an *arrow*). Figure 9.1 from [12], reproduced by permission of Oxford University Press

QPOs are usually modelled by a Lorentzian

$$P(\nu) \propto \frac{\lambda}{(\nu - \nu_0)^2 + (\lambda/2)^2}, \tag{9.2}$$

where ν_0 is the centroid frequency and λ is the full width at half maximum. A Lorentzian, strictly speaking, would correspond to the power spectrum of the signal of an exponentially damped sinusoid, namely

$$x(t) \propto \exp\left(-\frac{t}{\tau}\right) \cdot \cos\left(2\pi\nu_0 t\right). \tag{9.3}$$

The coherence time of the signal is $\tau = 1/(\pi\lambda)$ and its quality factor is $Q = \nu_0/\lambda$. The term QPO is usually reserved for a signal with $Q > 2$, while the signal is called peaked noise if $Q < 2$.

9.2 Fundamental Frequencies of a Test-Particle

Equatorial circular orbits of a test-particle are characterized by three fundamental frequencies:

1. *Orbital frequency* (or Keplerian frequency) ν_ϕ, which is the inverse of the orbital period.
2. *Radial epicyclic frequency* ν_r, which is the frequency of radial oscillations around the mean orbit.
3. *Vertical epicyclic frequency* ν_θ, which is the frequency of vertical oscillations around the mean orbit.

These three frequencies only depend on the metric of the spacetime and on the radius of the orbit. While they are defined as the characteristic frequencies for the orbital motion of a free particle, there is a direct relation between these frequencies and the ones of the oscillation modes of the fluid of an accretion flow [23].

In Newtonian gravity with the potential $V = -M/r$, the three fundamental frequencies have the same value:

$$\nu_\phi = \nu_r = \nu_\theta = \frac{1}{2\pi}\frac{M^{1/2}}{r^{3/2}}. \tag{9.4}$$

These frequencies are plotted as a function of the radial coordinate r in the top left panel in Fig. 9.2.

In the Schwarzschild metric, the main difference from Newtonian gravity is the existence of the ISCO radius and of the photon radius. Circular orbits with radii smaller than r_{ISCO} are radially unstable. The radial epicyclic frequency ν_r must thus reach a maximum at some radius $r_{\text{max}} > r_{\text{ISCO}}$ and then vanishes at the ISCO. The orbital and the vertical epicyclic frequencies are instead defined up to the innermost circular orbit, the so-called photon orbit (see Sect. 3.1). There are no circular orbits with radius smaller than the one of the photon orbit. The three fundamental frequencies as a function of the radial coordinate r in the Schwarzschild background are shown in the top right panel in Fig. 9.2. We always have $\nu_\phi = \nu_\theta > \nu_r$, see below Eq. (9.18).

In the Kerr metric, $\nu_\theta > \nu_r$ is still true and, for corotating orbits, $\nu_\phi \geq \nu_\theta$. The three fundamental frequencies as a function of the radial coordinate r in the Kerr spacetime with the spin parameter $a_* = 0.9$ and 0.998 and for corotating orbits are shown, respectively, in the bottom left and bottom right panels in Fig. 9.2.

In the case of a generic stationary and axisymmetric spacetime, the picture may be more complicated. The ISCO radius may be determined by the stability of the orbit along the vertical (instead of the radial) direction. In some metrics, there is also the possibility of the existence of one or more regions of stable circular orbits inside the ISCO and therefore at smaller radii. These regions of stable orbits are separated by a gap from the "traditional" ISCO; the details depend on the specific metric.

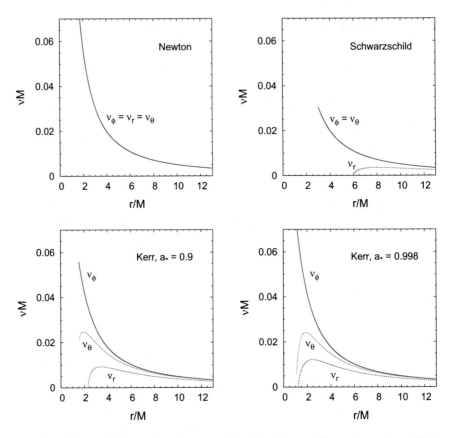

Fig. 9.2 Fundamental frequencies of a test-particle. *Top left panel* Newtonian gravity with the potential $V = -M/r$; the three fundamental frequencies have the same value. *Top right panel* Schwarzschild metric; the orbital and the vertical epicyclic frequencies have the same value, the radial epicyclic frequency vanishes at the ISCO radius. *Bottom left panel* Kerr metric with the spin parameter $a_* = 0.9$. *Bottom right panel* Kerr metric with the spin parameter $a_* = 0.998$

9.2.1 General Case

For a generic stationary and axisymmetric spacetime, we write the line element in the canonical form, namely

$$ds^2 = g_{tt}dt^2 + 2g_{t\phi}dtd\phi + g_{rr}dr^2 + g_{\theta\theta}d\theta^2 + g_{\phi\phi}d\phi^2. \qquad (9.5)$$

We can then proceed as in Sect. 3.1 and find the orbital angular velocity

$$\Omega_\phi = \frac{d\phi}{dt} = \frac{-\partial_r g_{t\phi} \pm \sqrt{\left(\partial_r g_{t\phi}\right)^2 - \left(\partial_r g_{tt}\right)\left(\partial_r g_{\phi\phi}\right)}}{\partial_r g_{\phi\phi}} \qquad (9.6)$$

This is the expression found in Eq. (3.11). The orbital frequency is $\nu_\phi = \Omega_\phi/2\pi$.

For the calculation of the radial and the vertical epicyclic frequencies, we can start from the conservation of the rest-mass $g_{\mu\nu}\dot{x}^\mu\dot{x}^\nu = -1$. As done in Sect. 3.1, we introduce the effective potential V_{eff} defined in Eq. (3.8) and we have

$$g_{rr}\dot{r}^2 + g_{\theta\theta}\dot{\theta}^2 = V_{\text{eff}}. \tag{9.7}$$

In the linear regime, we can consider separately small perturbations around circular equatorial orbits along, respectively, the radial and the vertical directions. For the radial direction, we assume $\dot{\theta} = 0$, we write $\dot{r} = \dot{t}(dr/dt)$, and Eq. (9.7) becomes

$$\left(\frac{dr}{dt}\right)^2 = \frac{1}{g_{rr}\dot{t}^2}V_{\text{eff}}. \tag{9.8}$$

We derive Eq. (9.8) with respect to the coordinate t and we obtain

$$\begin{aligned}
\frac{d^2r}{dt^2} &= \frac{1}{2}\frac{\partial}{\partial r}\left(\frac{1}{g_{rr}\dot{t}^2}V_{\text{eff}}\right) \\
&= \frac{V_{\text{eff}}}{2}\frac{\partial}{\partial r}\left(\frac{1}{g_{rr}\dot{t}^2}\right) + \frac{1}{2g_{rr}\dot{t}^2}\frac{\partial V_{\text{eff}}}{\partial r}.
\end{aligned} \tag{9.9}$$

If δ_r is a small displacement around the mean orbit, namely $r = r_0 + \delta_r$, we have

$$\frac{d^2r}{dt^2} = \frac{d^2\delta_r}{dt^2},$$

$$V_{\text{eff}}(r_0 + \delta_r) = V_{\text{eff}}(r_0) + \left(\frac{\partial V_{\text{eff}}}{\partial r}\right)_{r=r_0}\delta_r + O(\delta_r^2) = O(\delta_r^2),$$

$$\begin{aligned}
\left(\frac{\partial V_{\text{eff}}}{\partial r}\right)_{r=r_0+\delta_r} &= \left(\frac{\partial V_{\text{eff}}}{\partial r}\right)_{r=r_0} + \left(\frac{\partial^2 V_{\text{eff}}}{\partial r^2}\right)_{r=r_0}\delta_r + O(\delta_r^2) \\
&= \left(\frac{\partial^2 V_{\text{eff}}}{\partial r^2}\right)_{r=r_0}\delta_r + O(\delta_r^2).
\end{aligned} \tag{9.10}$$

A similar expression can be derived for the coordinate θ to find the vertical epicyclic frequency, introducing a small displacement around the mean orbit δ_θ, i.e. $\theta = \pi/2 + \delta_\theta$. Neglecting, respectively, terms $O(\delta_r^2)$ and $O(\delta_\theta^2)$, we find the following differential equations

$$\frac{d^2\delta_r}{dt^2} + \Omega_r^2\delta_r = 0, \tag{9.11}$$

$$\frac{d^2\delta_\theta}{dt^2} + \Omega_\theta^2\delta_\theta = 0, \tag{9.12}$$

where

$$\Omega_r^2 = -\frac{1}{2g_{rr}\dot{t}^2}\frac{\partial^2 V_{\text{eff}}}{\partial r^2}, \tag{9.13}$$

$$\Omega_\theta^2 = -\frac{1}{2g_{\theta\theta}\dot{t}^2}\frac{\partial^2 V_{\text{eff}}}{\partial \theta^2}. \tag{9.14}$$

The radial epicyclic frequency is $\nu_r = \Omega_r/2\pi$ and the vertical one is $\nu_\theta = \Omega_\theta/2\pi$.

9.2.2 Kerr Spacetime

In the Kerr metric, the three fundamental frequencies can be written in an analytic and compact form:

$$\nu_\phi = \pm\frac{1}{2\pi}\frac{M^{1/2}}{r^{3/2} \pm aM^{1/2}}, \tag{9.15}$$

$$\nu_r = \nu_\phi\sqrt{1 - \frac{6M}{r} \pm \frac{8aM^{1/2}}{r^{3/2}} - \frac{3a^2}{r^2}}, \tag{9.16}$$

$$\nu_\theta = \nu_\phi\sqrt{1 \mp \frac{4aM^{1/2}}{r^{3/2}} + \frac{3a^2}{r^2}}. \tag{9.17}$$

The Schwarzschild limit follows by imposing $a = 0$ and we obtain that the orbital and the vertical epicyclic frequencies coincide

$$\nu_\phi = \nu_\theta = \frac{1}{2\pi}\frac{M^{1/2}}{r^{3/2}}, \quad \nu_r = \nu_\phi\sqrt{1 - \frac{6M}{r}}. \tag{9.18}$$

The Newtonian limit can be quickly recovered from the Schwarzschild case by considering only the leading order term in M/r and the result is given in Eq. (9.4). In the Kerr spacetime $\nu_\theta \geq \nu_r$, but it may not be true in general. If the ISCO is marginally vertically stable, $\nu_\theta = 0$ at the ISCO, and therefore we have $\nu_r > \nu_\theta$.

It may be useful to have an estimate of the order of magnitude of these frequencies. For a Schwarzschild black hole, the orbital frequency is

$$\nu_\phi(a_* = 0) = 220 \left(\frac{10 M_\odot}{M}\right)\left(\frac{6M}{r}\right)^{3/2} \text{Hz}. \tag{9.19}$$

HFQPOs at 40–450 Hz are thus of the right magnitude to be associated to the orbital frequencies near the ISCO radius of stellar-mass black holes.

9.3   Relativistic Precession Models

The relativistic precession model was proposed in [20, 21] to explain the frequencies
of the QPOs in the power density spectrum of neutron stars, for which there are more
data and of higher quality, and was later extended to the QPOs of stellar-mass black
holes in [22]. The model does not provide a mechanism responsible for the production
of the observed QPOs, but it simply relates the observed frequencies with those of
the orbital motion of a test-particle.

Motta et al. [11] have proposed a variant of the original version of this model,
which seems to be supported by some observations of the black hole binary in
GRO J1655-40. See also [12], where the authors discuss the case of XTE J1550-564.
From the three fundamental frequencies discussed in the previous section, one finds
the *periastron precession frequency* ν_p and the *nodal precession frequency* ν_n

$$\nu_p = \nu_\phi - \nu_r,$$
$$\nu_n = \nu_\phi - \nu_\theta. \tag{9.20}$$

In the Kerr metric, all these frequencies depend on three parameters, which are the
black hole mass, the black hole spin, and the radial coordinate r. In the case of a
simultaneous detection of three QPOs, one may argue that they are associated to the
same event, and therefore the three frequencies would have the same radial coordinate
r. The detection of three frequencies in the same observation would thus permit one
to solve the system of equations and get the values of M, a_*, and r.

GRO J1655-40 is the only source for which we currently have the detection of
three QPOs, with two of them HFQPOs, at the same time. Motta et al. [11] have
proposed the following interpretation. The low-frequency type-C QPO ν_C would
correspond to the nodal precession frequency ν_n, while the lower HFQPO ν_L and
the upper HFQPO ν_U would be associated, respectively, to the periastron precession
frequency ν_p and to the orbital frequency ν_ϕ:

$$\nu_C = \nu_n, \quad \nu_L = \nu_p, \quad \nu_U = \nu_\phi. \tag{9.21}$$

Such an interpretation well fits the data of GRO J1655-40 and provides the
following measurements of the mass and the spin of the black hole

$$M = (5.31 \pm 0.07)\, M_\odot, \quad a_* = 0.290 \pm 0.003. \tag{9.22}$$

The measurement of the mass is consistent with some dynamical estimates from
optical observations (but not with others, as there are several measurements reported
in the literature and they are not in agreement). The spin estimate is not consistent
with those from the continuum-fitting and the iron line methods, see Table 7.1. For the
moment, it is not clear which measurement, if any, is correct. However, it is evident
that QPOs have the capability of providing quite precise estimates of the metric. The
uncertainty on the spin parameter in Eq. (9.22) is ~1%. This is definitively smaller

than the spin uncertainties from the continuum-fitting and iron line approaches in Table 7.1, which can be easily explained by noticing that it is indeed easier to measure the centroid frequency of QPOs than to fit the spectrum of the disk.

9.4 Resonance Models

It is remarkable that in the four sources for which we have two HFQPOs the ratio of the two frequencies is 3:2, see Table 9.3. This cannot be accidental and the correct theoretical model should be able to explain this feature. The resonance models proposed by Abramowicz and Kluzniak start from this point to explain the origin of HFQPOs [1, 2, 9].

In Eqs. (9.11) and (9.12), the radial and the vertical modes are decoupled. However, it is natural to expect that in a more realistic description there are non-linear effects coupling the two epicyclic modes. In this case, the equations can be written as

$$\frac{d^2\delta_r}{dt^2} + \Omega_r^2\delta_r = \Omega_r^2 F_r\left(\delta_r, \delta_\theta, \frac{d\delta_r}{dt}, \frac{d\delta_\theta}{dt}\right), \tag{9.23}$$

$$\frac{d^2\delta_\theta}{dt^2} + \Omega_\theta^2\delta_\theta = \Omega_\theta^2 F_\theta\left(\delta_r, \delta_\theta, \frac{d\delta_r}{dt}, \frac{d\delta_\theta}{dt}\right), \tag{9.24}$$

where F_r and F_θ are some functions that depend on the specific properties of the accretion flow. If we knew the details of the physical mechanisms of the accretion process, we could write the explicit form of these two functions and solve the system. Since this is not the case, we can guess possible properties and consequences of these equations and see if they can be fitted in a plausible physical scenario.

9.4.1 Parametric Resonances

A simple but physically interesting scenario is to imagine that vertical oscillations are governed by the Mathieu equation [1, 25]:

$$\frac{d^2\delta_\theta}{dt^2} + \Omega_\theta^2\delta_\theta = -\Omega_\theta^2 h\cos(\Omega_r t)\delta_\theta. \tag{9.25}$$

This is the case with $F_r = 0$ and $F_\theta = -\delta_r\delta_\theta$. The solution of Eq. (9.23) is simply $\delta_r = h\cos(\Omega_r t)$ and so we obtain Eq. (9.25). The Mathieu equation describes a parametric resonance with

$$\frac{\Omega_r}{\Omega_\theta} = \frac{2}{n}, \qquad n = 1, 2, 3, \ldots \tag{9.26}$$

The resonance is stronger for smaller values of n. In the Kerr background, $\nu_\theta > \nu_r$, and therefore the resonance $n = 3$ can naturally explain the observed 3:2 ratio if the upper frequency ν_U is associated with ν_θ and the lower frequency ν_L with ν_r. In more general spacetimes, the phenomenology may be richer [3].

9.4.2 Forced Resonances

The equation for vertical oscillations may also include a forced resonance, in which the force frequency is equal to the one of the radial oscillations [25]. In this case the equation is

$$\frac{d^2\delta_\theta}{dt^2} + \Omega_\theta^2 \delta_\theta + [\text{non linear terms in } \delta_\theta] = h(r)\cos(\Omega_r t). \tag{9.27}$$

The non-linear terms allow resonant solutions for δ_θ, with frequencies like

$$\Omega_- = \Omega_\theta - \Omega_r, \tag{9.28}$$
$$\Omega_+ = \Omega_\theta + \Omega_r. \tag{9.29}$$

In the Kerr metric, the observed 3:2 ratio may be explained with $\nu_\theta : \nu_r = 3 : 1$ ($\nu_U = \nu_\theta$ and $\nu_L = \nu_-$) or with $\nu_\theta : \nu_r = 2 : 1$ ($\nu_U = \nu_+$ and $\nu_L = \nu_\theta$). Again, in non-Kerr backgrounds in which the ISCO is marginally vertically stable, we may have other possibilities [3].

9.4.3 Keplerian Resonances

The possibility of a coupling between the Keplerian and the radial epicyclic frequencies might exist, even if it seems to be less theoretically motivated than the one in which the coupling is between the two epicyclic oscillations. In the Kerr metric, the simplest combinations are

$$\begin{aligned}
\nu_\phi : \nu_r &= 3 : 2 \quad (\nu_U = \nu_\phi \ \text{and} \ \nu_L = \nu_r), \\
\nu_\phi : \nu_r &= 3 : 1 \quad (\nu_U = \nu_\phi \ \text{and} \ \nu_L = 2\nu_r), \\
\nu_\phi : \nu_r &= 2 : 1 \quad (\nu_U = 3\nu_r \ \text{and} \ \nu_L = \nu_\phi).
\end{aligned} \tag{9.30}$$

References

1. M.A. Abramowicz, V. Karas, W. Kluzniak, W.H. Lee, P. Rebusco, Publ. Astron. Soc. Jap. **55**, 467 (2003)
2. M.A. Abramowicz, W. Kluzniak, Astron. Astrophys. **374**, L19 (2001), arXiv:astro-ph/0105077
3. C. Bambi, JCAP **1209**, 014 (2012), arXiv:1205.6348 [gr-qc]
4. C. Bambi, Eur. Phys. J. C **75**, 162 (2015), arXiv:1312.2228 [gr-qc]
5. M. Gierlinski, M. Middleton, M. Ward, C. Done, Nature **455**, 369 (2008), arXiv:0807.1899 [astro-ph]
6. J. Homan, J.M. Miller, R. Wijnands, M. van der Klis, T. Belloni, D. Steeghs, W.H.G. Lewin, Astrophys. J. **623**, 383 (2005), arXiv:astro-ph/0406334
7. T. Johannsen, D. Psaltis, Astrophys. J. **726**, 11 (2011), arXiv:1010.1000 [astro-ph.HE]
8. S. Kato, Publ. Astron. Soc. Jap. **64**, 139 (2012), arXiv:1207.5882 [astro-ph.HE]
9. W. Kluzniak, M.A. Abramowicz, arXiv:astro-ph/0105057
10. A. Maselli, L. Gualtieri, P. Pani, L. Stella, V. Ferrari, Astrophys. J. **801**, 115 (2015), arXiv:1412.3473 [astro-ph.HE]
11. S.E. Motta, T.M. Belloni, L. Stella, T. Muoz-Darias, R. Fender, Mon. Not. R. Astron. Soc. **437**, 2554 (2014), arXiv:1309.3652 [astro-ph.HE]
12. S.E. Motta, T. Muoz-Darias, A. Sanna, R. Fender, T. Belloni, L. Stella, Mon. Not. R. Astron. Soc. **439**, 65 (2014), arXiv:1312.3114 [astro-ph.HE]
13. D.R. Pasham, T.E. Strohmayer, R.F. Mushotzky, Nature **513**, 74 (2014), arXiv:1501.03180 [astro-ph.HE]
14. C.A. Perez, A.S. Silbergleit, R.V. Wagoner, D.E. Lehr, Astrophys. J. **476**, 589 (1997), arXiv:astro-ph/9601146
15. R.A. Remillard, AIP Conf. Proc. **714**, 13 (2004)
16. R.A. Remillard, J.E. McClintock, Ann. Rev. Astron. Astrophys. **44**, 49 (2006), arXiv:astro-ph/0606352
17. R.A. Remillard, E.H. Morgan, J.E. McClintock, C.D. Bailyn, J.A. Orosz, Astrophys. J. **522**, 397 (1999), arXiv:astro-ph/9806049
18. R.A. Remillard, M.P. Muno, J.E. McClintock, J.A. Orosz, Astrophys. J. **580**, 1030 (2002), arXiv:astro-ph/0202305
19. A.S. Silbergleit, R.V. Wagoner, M. Ortega-Rodriguez, Astrophys. J. **548**, 335 (2001), arXiv:astro-ph/0004114
20. L. Stella, M. Vietri, Astrophys. J. **492**, L59 (1998), arXiv:astro-ph/9709085
21. L. Stella, M. Vietri, Phys. Rev. Lett. **82**, 17 (1999), arXiv:astro-ph/9812124
22. L. Stella, M. Vietri, S. Morsink, Astrophys. J. **524**, L63 (1999), arXiv:astro-ph/9907346
23. O. Straub, E. Sramkova, Class. Quant. Grav. **26**, 055011 (2009), arXiv:0901.1635 [astro-ph.SR]
24. T.E. Strohmayer, Astrophys. J. **552**, L49 (2001), arXiv:astro-ph/0104487
25. G. Torok, M.A. Abramowicz, W. Kluzniak, Z. Stuchlik, Astron. Astrophys. **436**, 1 (2005)
26. M. van der Klis, arXiv:astro-ph/0410551
27. R.V. Wagoner, A.S. Silbergleit, M. Ortega-Rodriguez, Astrophys. J. **559**, L25 (2001), arXiv:astro-ph/0107168

Chapter 10
Imaging Black Holes

Under certain conditions, a precise observation of the direct image of the accretion flow around a black hole may provide information about the spacetime metric around the compact object. In particular, if the black hole is surrounded either by an optically thin emitting medium or by a Novikov–Thorne accretion disk, the image will have a dark area over a brighter background. Such a dark area is commonly called the *shadow* of the black hole [6], but the name may be misleading because it is not produced as common shadows, in which a body blocks light. In the case of a black hole surrounded by an optically thin emitting medium, the boundary of the shadow corresponds to the apparent photon capture sphere. For a black hole surrounded by a Novikov–Thorne accretion disk, the boundary of the shadow corresponds to the apparent image of the ISCO. In other situations, e.g. a black hole surrounded by an optically thick emitting medium, there may not be any shadow.

The current interest in the possibility of imaging astrophysical black holes is mainly motivated by the special situation of SgrA*, the supermassive black hole at the center of the Galaxy. SgrA* is the black hole with the largest angular size in the sky. Its mass is about $4 \cdot 10^6 \, M_\odot$ and its distance from us is around 8 kpc. The angular size of its gravitational radius $r_g = M$ in the sky is (ignoring light bending)

$$\frac{M}{D} = 2.4 \cdot 10^{-11} \left(\frac{M}{4 \cdot 10^6 \, M_\odot} \right) \left(\frac{8 \, \text{kpc}}{D} \right)$$

$$= 5 \left(\frac{M}{4 \cdot 10^6 \, M_\odot} \right) \left(\frac{8 \, \text{kpc}}{D} \right) \mu \text{as}, \tag{10.1}$$

where μas stands for micro arc second. It is widely thought that sub-mm very-long baseline interferometric (VLBI) facilities will be soon able to image SgrA* and detect its shadow [5]. This can be achieved thanks to three particular conditions:

1. The angular resolution θ_{res} of VLBI facilities scales as λ/d, where λ is the electromagnetic radiation wavelength and d is the distance among different stations. For $\lambda < 1$ mm and with stations located in different countries ($d > 10^3$ km), it

© Springer Nature Singapore Pte Ltd. 2017
C. Bambi, *Black Holes: A Laboratory for Testing Strong Gravity*,
DOI 10.1007/978-981-10-4524-0_10

is possible to reach an angular resolution of the order of the angular size of the apparent image of SgrA* ($\lambda/d \sim 10~\mu$as).

2. The emitting medium around SgrA* is optically thick at wavelengths $\lambda > 1$ mm, but becomes optically thin for $\lambda < 1$ mm.
3. The interstellar scattering dominates over intrinsic source structures at wavelengths $\lambda > 1$ mm, but becomes subdominant at sub-mm wavelengths.

Another potential candidate is the supermassive black hole at the center of the giant elliptical galaxy M87: its distance from us is about 10^3 times the distance between us and SgrA*, but its mass is about $10^9~M_\odot$. Its apparent image in the sky should be somewhat smaller than the one of SgrA*, but maybe only by a factor 2–3. However, the supermassive black hole in M87 may not be surrounded by an optically thin medium for sub-mm wavelengths.

In the case of stellar-mass black holes in our Galaxy, the distance is 1–10 kpc, but their mass is about 10 $M_\odot$. The result is that the angular size of their gravitational radius in the sky is about $10^{-5}~\mu$as. A similar high angular resolution is out of reach for present and near future observational facilities, but it could be possible in the case of X-ray interferometry techniques. Observational facilities with similar properties are already under discussion, but they will be unlikely realized in the next 20 years.

In this chapter, we will discuss the calculations of the apparent image of the photon capture sphere and of the ISCO. A different issue is the comparison of these predictions with future observations. For example, VLBI observations sample the Fourier space conjugate to the image plane at a small number of points. An analysis with the images discussed here could only be performed either after image reconstruction has taken place, or if these images are taken into the Fourier domain and sampled there, so that inference takes place at the level of the Fourier plane. Moreover, at the moment it is not completely clear whether such observations can reach the necessary precision to test the spacetime metric around black holes. However, it is not strictly necessary to get very precise measurements of the boundary of the shadow to probe the metric around a black hole. In the presence of the correct astrophysical model, the direct image of the accretion flow may already provide the necessary information, see, for instance, [2] or [10].

10.1 Imaging the Photon Capture Sphere

Let us consider an ideal experiment, in which a distant observer can shoot photons at a black hole. This is the set-up already introduced to compute the thermal and the reflection spectrum of a thin accretion disk, but now we do not consider any disk. We have just the observer and the black hole. We can then define a region in the image plane of the distant observer such that, if a photon with 3-momentum orthogonal to the image plane is fired inside this region, it will be captured by the black hole, while the photon will be scattered back to infinity is it is outside such a region.

Neglecting the possibility of exotic situations like the ones studied in [4], such a region in the image plane of the distant observer will have a simple shape and its boundary will correspond to the photon capture sphere (or photon ring) as seen by the distant observer. The photon capture sphere is the surface (in general, it is not a sphere) separating scattered and captured photons. More specifically, if a photon coming from infinity crosses the photon capture sphere, it is captured by the black hole and then crosses the event horizon. If this is not the case, the photon is scattered and comes back to infinity. In general, the photon capture sphere is different for photons with angular momentum parallel and antiparallel to the black hole spin. Typically the photon capture sphere is the photon orbit, but its image on the observer's sky is larger due to light bending.

In a real observation, we cannot fire photons to a black hole and perform a similar scattering experiment. However, if the black hole is completely surrounded by an optically thin emitting medium, its image is a dark area over a bright background. The dark area is the so-called black hole shadow and its boundary is the image of the photon capture sphere as seen by the distant observer. Neglecting the interstellar scattering, the boundary of the shadow is very sharp and only depends on the space-time metric in the strong gravity region. See the left panel in Fig. 10.1. The reason of this result will be explained in Sect. 10.1.4. The intensity map of the image is instead determined by the properties of the accretion flow and the emission mechanisms.

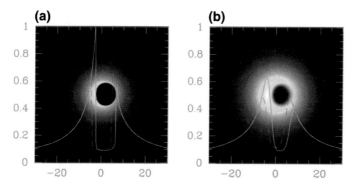

Fig. 10.1 Direct image of a *black hole* surrounded by an optically thin emitting medium with the characteristics of that of SgrA*. The *black hole* spin parameter is $a_* = 0.998$ and the viewing angle is $i = 45°$. The emitting gas is in free fall and has an emissivity $\propto 1/r^2$. *Left panel* image from ray-tracing calculations. *Right panel* image from a simulated observation of an idealized VLBI experiment at 0.6 mm wavelength taking interstellar scattering into account. The *solid green curve* and the *dashed purple curve* show, respectively, the intensity variations of the image along the x-axis and the y-axis. From [6]. © AAS. Reproduced with permission

10.1.1   Spherically Symmetric Spacetimes

In the case of a spherically symmetric black hole spacetime, the apparent photon capture sphere is a circle, for obvious symmetry reasons. The calculation of the radius of this circle in the sky of the observer is straightforward. Without loss of generality, the line element of a static, spherically symmetric, and asymptotically flat spacetime can be written as

$$ds^2 = -A(r)dt^2 + B(r)dr^2 + r^2 d\theta^2 + r^2 \sin^2 \theta d\phi^2, \tag{10.2}$$

where $A(r)$ and $B(r)$ must reduce to 1 for $r \rightarrow \infty$. Now we want to determine the apparent size of the central object as seen by a distant observer; that is, the photon impact parameter separating photons captured by the black hole and photons scattered back to infinity.

As the spacetime is spherically symmetric, we can do the calculations in the equatorial plane $\theta = \pi/2$. The metric coefficients in Eq. (10.2) do not depend on the t and ϕ coordinates, and therefore there are two constants of motion, the energy E and the angular momentum L_z. From the Euler-Lagrange equations, we find

$$\dot{t} = \frac{E}{A(r)}, \quad \dot{\phi} = \frac{L_z}{r^2}. \tag{10.3}$$

We plug the above expressions for $\dot{t}$ and $\dot{\phi}$ into $g_{\mu\nu}\dot{x}^\mu \dot{x}^\nu = 0$ (null geodesics), and we find

$$\dot{r}^2 = \frac{1}{B(r)} \left[\frac{E^2}{A(r)} - \frac{L_z^2}{r^2} \right]. \tag{10.4}$$

If $\dot{r}^2$ vanishes before hitting the object, the photon reaches a turning point and then comes back to infinity. In the opposite case, the photon falls onto the object. The impact parameter is $\lambda = L_z/E$ and the critical one separating captured and scattered photon orbits is given by the minimum value of λ for which the equation $\dot{r}^2 = 0$ has a solution; that is,

$$\lambda_c = \frac{r}{\sqrt{A(r)}}, \quad A(r) - \frac{1}{2}rA'(r) = 0. \tag{10.5}$$

λ_c corresponds to the value of the radius of the shadow as seen by a distant observer. Let us note that (i) λ_c only depends on $A(r)$, not on $B(r)$, and (ii) we are assuming the existence of a photon capture sphere outside the compact object. The assumption (ii) may not be true in some spacetimes: there may be no photon capture sphere or there may be more than one.

A simple example is the Schwarzschild spacetime, where $A(r) = 1 - 2M/r$. In this case, the solution of the system (10.5) is

$$\lambda_{\text{Sch}} = 3\sqrt{3}\,M \approx 5.196\,M. \tag{10.6}$$

10.1.2 *Kerr Metric*

Starting from [1], the apparent image of the photon capture sphere of the Kerr space-time has been studied by many authors. The master equations are Eq. (3.52) for null orbits, i.e.

$$\Sigma^2 \dot{r}^2 = R, \tag{10.7}$$

where[1]

$$R = r^4 + \left(a^2 - \lambda^2 - q^2\right) r^2 + 2M\left[(a - \lambda)^2 + q^2\right] r - a^2 q^2, \tag{10.8}$$

and Eqs. (3.71) and (3.72) for the photon position in the image plane of the distant observer, namely

$$X = \frac{\lambda}{\sin i}, \quad Y = \pm\sqrt{q^2 + a^2 \cos^2 i - \lambda^2 \cot^2 i}. \tag{10.9}$$

Motion is only possible when $R \geq 0$, and therefore the analysis of the position of the roots of the function R can be used to distinguish the capture from the scattering orbits. The three kinds of photon orbits are:

1. *Capture orbits*: R has no roots for $r \geq r_+$, where r_+ is the radial coordinate of the black hole event horizon. In this case, photons come from infinity and then cross the horizon.
2. *Scattering orbits*: R has real roots for $r \geq r_+$, which correspond to the photon turning points. If the photons come from infinity, they reach a minimum distance from the black hole, and then go back to infinity.
3. *Unstable orbits of constant radius*: these orbits separate the capture and the scattering orbits and are determined by

$$R(r_*) = \frac{\partial R}{\partial r}(r_*) = 0, \quad \text{and} \quad \frac{\partial^2 R}{\partial r^2}(r_*) \geq 0, \tag{10.10}$$

where r_* is the larger real root of R.

The apparent image of the photon capture sphere of a Kerr black hole can be determined by finding the unstable orbits of constant radius and employing Eq. (10.9) to obtain the image in the observer's sky.

The apparent image of the photon capture sphere is represented by a closed curve determined by the set of unstable circular orbits (λ_c, q_c) in the plane of the distant observer. The equations $R = 0$ and $\partial R/\partial r = 0$ are

---

[1]In this chapter, the function R indicates the function R of Chap. 3 divided by E^2, where E is the photon energy. Since we are considering photon orbits, the geodesics are independent of E.

$$R = r^4 + (a^2 - \lambda_c^2 - q_c^2)r^2 + 2M[q_c^2 + (\lambda_c - a)^2]r - a^2 q_c^2 = 0,$$

$$\frac{\partial R}{\partial r} = 4r^3 + 2(a^2 - \lambda_c^2 - q_c^2)r + 2M[q_c^2 + (\lambda_c - a)^2] = 0. \tag{10.11}$$

In the Schwarzschild background ($a = 0$), the shadow of the black hole is a circle of radius

$$r = 3\sqrt{3}\,M \approx 5.196\,M, \tag{10.12}$$

and we recover the result in Eq. (10.6). For $a \neq 0$, one finds

$$\lambda_c = \frac{M(r_*^2 - a^2) - r_*(r_*^2 - 2Mr_* + a^2)}{a(r_* - M)},$$

$$q_c = \frac{r_*^{3/2}\sqrt{4a^2 M - r_*(r_* - 3M)^2}}{a(r_* - M)}, \tag{10.13}$$

where r_* is the radius of the unstable orbit. After some manipulation of these equations, we can find the image of the photon capture sphere in the plane of the distant observer [1, 3]

$$X = \frac{M}{a_* \sin i}\left[t^2 + a_*^2 - 3 - \frac{2\left(1 - a_*^2\right)}{t}\right], \tag{10.14}$$

$$Y = \pm\left\{\frac{(1+t)^3}{a_*^2}\left[3 - t - \frac{4\left(1 - a_*^2\right)}{t^2}\right] + \left(a_*^2 - \frac{X^2}{M^2}\right)\cos^2 i\right\}^{1/2}M, \tag{10.15}$$

where here t is an auxiliary parameter whose range is determined by the fact that the expression inside the square root in Eq. (10.15) must be positive and depends on the specific values of the spin parameter a_* and the viewing angle i.

Figure 10.2 shows the impact of the spin parameter a_* on the shape of the apparent photon capture sphere of a Kerr black hole. The inclination angle is $i = 85°$, to maximize the relativistic effects. For $a_* = 0$, the shadow is a circle with radius $r = 3\sqrt{3}\,M$, which is larger than the radius of the event horizon $r_+ = 2M$ because of light bending and because actually the radial coordinate of the event horizon has no physical meaning, as the choice of the coordinate system is arbitrary in general relativity. As the spin parameter increases, we see that both the left and the right parts of the boundary of the shadow move to the right in the image plane of the distant observer. The reason is that photon orbits are different if the photon angular momentum is parallel/antiparallel to the black hole spin. As seen in Sect. 3.2, and in Fig. 3.4, the value of the photon radius decreases (increases) if the the photon angular momentum is parallel (antiparallel) to the black hole spin. For this reason, the image of the black hole that we can obtain firing photons from the distant observer to the black hole is like the image obtained considering photons moving from the black hole to the distant observer with $X \to -X$.

Fig. 10.2 Impact of the spin parameter a_* on the shape of the apparent photon capture sphere of a Kerr *black hole*. The viewing angle is $i = 85°$. The spin parameter is $a_* = 0$ (*magenta dotted curve*), 0.5 (*blue dotted curve*), 0.9 (*green dashed curve*), and 0.998 (*red solid curve*). See the text for more details

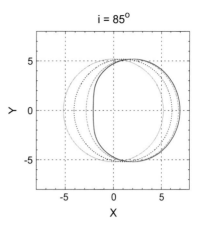

Fig. 10.3 Impact of the viewing angle i on the shape of the apparent photon capture sphere of a Kerr *black hole*. The spin parameter is $a_* = 0.998$. The viewing angle is $i = 5°$ (*magenta dotted curve*), 30° (*blue dotted curve*), 60° (*green dashed curve*), and 85° (*red solid curve*). See the text for more details

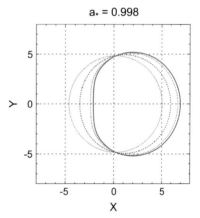

With a viewing angle smaller than $i = 85°$ employed in Fig. 10.2, the difference between a non-rotating and a fast-rotating black hole would be smaller. For $a_* = 0$, the shadow is independent of the viewing angle i, because of the spherical symmetry of the system. For $i = 0°$, the apparent image of the photon capture sphere is a circle for any a_*, but the radius is slight different, with the larger one for the Schwarzschild case and the smallest radius for $a_* = 1$.

Figure 10.3 shows the apparent image of the photon capture sphere for a Kerr black hole with $a_* = 0.998$ and different values of the inclination angle. For a small inclination angle, the shadow is almost a circle, while the shape becomes more and more distorted as the viewing angle increases.

10.1.3   General Case

If the line element of the spacetime can be written as

$$ds^2 = -\left(1 - \frac{2mr}{\Sigma}\right)dt^2 - \frac{4amr\sin^2\theta}{\Sigma}dtd\phi + \frac{\Sigma}{\Delta}dr^2 + \Sigma d\theta^2$$
$$+ \left(r^2 + a^2 + \frac{2a^2mr\sin^2\theta}{\Sigma}\right)\sin^2\theta d\phi^2, \tag{10.16}$$

where $\Sigma = r^2 + a^2\cos^2\theta$, $\Delta = r^2 - 2mr + a^2$, and $m = m(r)$ is a function of r only, we can adopt the approach discussed in the previous subsection. Equation (10.16) includes as trivial case the Kerr metric when $m = M$, but also the Kerr-Newman metric for $m = M - Q^2/(2r)$ and other black hole spacetimes for proper choices of the function $m(r)$.

It is straightforward to check that this family of metrics has the nice properties of the Kerr solution: we have a Carter-like constant and the equations of motion are separable. The equation for the radial coordinate of null orbits is still Eq. (10.7), but now R has m in the place of M. The photon position in the image plane of the distant observer is still given by Eq. (10.9). The system in Eq. (10.11) is replaced by [12]

$$R = r^4 + (a^2 - \lambda_c^2 - q_c^2)r^2 + 2m[q_c^2 + (\lambda_c - a)^2]r - a^2q_c^2 = 0,$$
$$\frac{\partial R}{\partial r} = 4r^3 + 2(a^2 - \lambda_c^2 - q_c^2)r + 2m[q_c^2 + (\lambda_c - a)^2]f = 0, \tag{10.17}$$

where f is

$$f = 1 + \frac{r}{m}\frac{dm}{dr}. \tag{10.18}$$

Equation (10.13) becomes

$$\lambda_c = \frac{m_*[(2 - f_*)r_*^2 - f_*a^2] - r_*(r_*^2 - 2m_*r_* + a^2)}{a(r_* - f_*m_*)},$$
$$q_c = \frac{r_*^{3/2}\sqrt{4(2 - f_*)a^2m_* - r_*[r_* - (4 - f_*)m_*]^2}}{a(r_* - f_*m_*)}, \tag{10.19}$$

where $m_* = m(r_*)$ and $f_* = f(r_*)$.

In more general cases, when the line element of the spacetime cannot be written in the form (10.16) with $m = m(r)$, it may be necessary to solve the geodesic equations of the photons from the image plane of the distant observer to the black hole to check when the photons are captured by the object or scattered back to infinity. Eventually the shadow is the set of photons in the image plane of the distant observer that are captured by the black hole, and the boundary of the shadow is the closed curve on the observer's sky separating captured and scattered photons.

10.1.4 Intensity Map

The calculation of the intensity map of the image of the emitting region requires some assumptions about the accretion process and the emission mechanisms. The observed specific intensity at the observed photon frequency ν_{obs} at the point (X, Y) in the observer's sky (for instance, in erg s^{-1} cm^{-2} str^{-1} Hz^{-1}) can be found integrating the specific emissivity along the photon path [9]

$$I_{obs}(\nu_{obs}, X, Y) = \int_\gamma g^3 j(\nu_e)d\ell, \tag{10.20}$$

where $g = \nu_{obs}/\nu_e$ is the redshift factor, ν_e is the photon frequency in the rest-frame of the emitter, $j(\nu_e)$ is the emissivity per unit volume in the rest-frame of the emitter, and $d\ell$ is the infinitesimal proper length as measured in the rest-frame of the emitter.

The redshift factor can be evaluated from

$$g = \frac{k_\mu u_{obs}^\mu}{k_\nu u_e^\nu}, \tag{10.21}$$

where k^μ is the 4-momentum of the photon, $u_{obs}^\mu = (1, 0, 0, 0)$ is the 4-velocity of the distant observer, while u_e^μ is the 4-velocity of the accreting gas emitting the radiation.

For instance, let us consider the simple case of a static and spherically symmetric spacetime, whose line element is given by Eq. (10.2), and gas in free fall starting with vanishing velocity at infinity. We have

$$u_e^t = \frac{1}{A(r)}, \quad u_e^r = -\sqrt{\frac{1 - A(r)}{A(r)B(r)}}, \quad u_e^\theta = u_e^\phi = 0. \tag{10.22}$$

Since the spacetime is static and spherically symmetric, without loss of generality we can restrict the calculations to the equatorial plane $\theta = \pi/2$, so $k^\theta = 0$. The intensity map in the observer's sky will be circularly symmetric, and therefore it is sufficient to compute the intensity map along a line from the center of symmetry. Since k_t and k_ϕ are constants of motion, k_r can be inferred from $k_\mu k^\mu = 0$, that is:

$$k_r = \pm k_t \sqrt{B(r) \left[\frac{1}{A(r)} - \frac{b^2}{r^2} \right]}, \tag{10.23}$$

where b is the impact parameter and the sign $+$ $(-)$ is when the photon approaches (leaves) the massive object (because $k_t = -E$). We see here that the intensity map of a static and spherically symmetric black hole spacetime depends on both $A(r)$ and $B(r)$, while the value of λ_c was only determined by the function $A(r)$

The infinitesimal proper length is $d\ell = k_\mu u_e^\mu d\tau$, where τ is the affine parameter of the photon trajectory. For the simple static and spherically symmetric spacetime with gas in free fall, it reduces to

$$d\ell = \frac{k_t}{g|k_r|} dr. \tag{10.24}$$

Integrating Eq. (10.20) over all the observed frequencies, we get the observed photon flux $\Phi_{obs}(X, Y)$.

The left panel in Fig. 10.1 shows the result of the calculation of the direct image of a Kerr black hole surrounded by an optically thin emitting medium with the characteristics of that of SgrA*. The black hole spin parameter is $a_* = 0.998$ and the viewing angle is $i = 45°$. The emitting gas is in free fall and has an emissivity $\propto 1/r^2$. The right panel is instead a simulated observation of an idealized VLBI experiment at 0.6 mm wavelength, in which the interstellar scattering is taken into account.

10.2 Imaging Thin Accretion Disks

If the black hole has a geometrically thin and optically thick accretion disk, the boundary of the shadow describes the image of the inner edge of the disk as seen by the distant observer. Because of light bending, the shape of the inner edge of the disk is not an ellipse. Only in the special case $i = 0°$, one sees a circle.

The first calculations of the image of a thin accretion disk were presented in [11] and later in [7]. These calculations have already been discussed in Chap. 7, see Fig. 7.2. In the case of a Novikov–Thorne accretion disk, in which the disk is orthogonal to the black hole spin and the inner edge of the disk is at the ISCO radius, we obtains the images shown in Figs. 10.4 and 10.5. Figure 10.4 shows the impact of the spin parameter a_* on the boundary of the shadow of a Kerr black hole, assuming that the viewing angle is $i = 60°$. In Fig. 10.5, the spin parameter is $a_* = 0.9$ and we vary the value of the viewing angle i.

10.3 Description of the Boundary of the Shadow

Let us now assume that we have (somehow) a precise detection of the shadow of a black hole. We want a simple formalism to describe the boundary of this shadow and compare it with a theoretical model, for instance to determine the black hole spin or to test the Kerr metric. The detection method may be important and actually favors

Fig. 10.4 Impact of the spin parameter a_* on the boundary of the shadow of a Kerr black hole accreting from a Novikov–Thorne disk. The viewing angle is $i = 60°$. From [13]

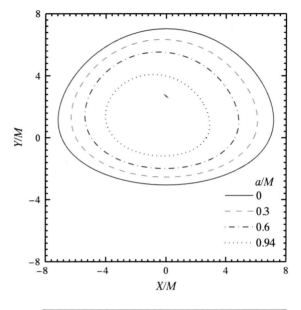

Fig. 10.5 Impact of the viewing angle i on the boundary of the shadow of a Kerr *black hole* accreting from a Novikov–Thorne disk. The spin parameter is $a_* = 0.9$. From [13]

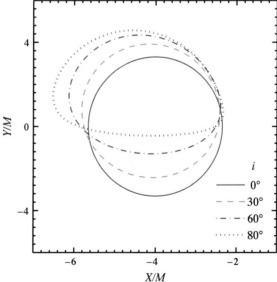

a different approach to describe and fit the shadow, but here we want to introduce a simple way to describe its shape assuming that we can have a precise detection of the boundary.

We have the image of the boundary of the shadow as shown in Fig. 10.6. This image is symmetric with respect to the X-axis, which is typically the case for the apparent image of the photon capture sphere and is usually not the case for the image

Fig. 10.6 The function $R(\phi)$ is defined as the distance between the center C and the boundary of the shadow at the angle ϕ as shown in this picture. See the text for more details

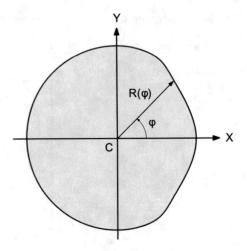

of the inner edge of the disk. We will briefly discuss the most general case without any symmetry at the end of this section.

We follow the description method presented in [8]. First, we find the "center" C of the shadow. Its Cartesian coordinates in the image plane of the observer are

$$X_C = \frac{\int \rho(X,Y)X\,dX\,dY}{\int \rho(X,Y)\,dX\,dY},$$

$$Y_C = \frac{\int \rho(X,Y)Y\,dX\,dY}{\int \rho(X,Y)\,dX\,dY}, \qquad (10.25)$$

where $\rho(X,Y) = 1$ inside the shadow and $\rho(X,Y) = 0$ outside. Such a definition reminds that of center of mass.

Since this shadow is symmetric with respect to the X-axis, we define $R(0)$ as the shorter segment between C and the shadow boundary along the X-axis. Defining the angle ϕ as shown in Fig. 10.6, $R(\phi)$ is the distance between the point C and the boundary at the angle ϕ. The function $R(\phi)/R(0)$ completely characterizes the shape of the black hole shadow. We prefer $R(\phi)/R(0)$ instead of $R(\phi)$ because the latter also depends on the mass and the distance of the black hole, two quantities that may not be known with high precision or, in any case, that we want to ignore.

Once we have sampled the shadow in this way, we have the set $\{R(\phi_i)/R(0)\}$, where $\{\phi_i\}$ is a set of angle for which we have the measurement of $R(\phi)/R(0)$. At this point, it is possible, for example, to compare two shadows or to fit an observed shadow with a theoretical model. The theoretical model may depend on the black hole spin a_*, the viewing angle i, and a number of extra parameters p_1, p_2, etc. If now we indicate with $R^{\text{obs}}(\phi)/R^{\text{obs}}(0)$ the observed shadow, we can infer the value of the parameters of the theoretical model by employing some goodness-of-fit statistical test. For example, we may introduce the function S defined as

$$S(a_*, i, p_1, p_2, \ldots) = \sum_k \left(\frac{R(a_*, i, p_1, p_2, \ldots; \phi_k)}{R(a_*, i, p_1, p_2, \ldots; 0)} - \frac{R^{\mathrm{obs}}(\phi_k)}{R^{\mathrm{obs}}(0)} \right)^2, \quad (10.26)$$

and find the best fit by minimizing S.

In the case of thin accretion disks, the shadow is not symmetric with respect to the X-axis. Nevertheless, it is easy to extend the previous formalism to any shadow without special symmetries. We can simply define as $R(0)$ the minimum or the maximum value of the function R and set the $\phi = 0$ angle there.

References

1. J.M. Bardeen, Timelike and null geodesics in the Kerr metric, in *Black Holes*, ed. by C. De Witt, B. De Witt (Gordon and Breach, New York, 1973)
2. A.E. Broderick et al., Perimeter Institute for Theoretical Physics Collaboration. Astrophys. J. **784**, 7 (2014), arXiv:1311.5564 [astro-ph.HE]
3. S. Chandrasekhar, *The Mathematical Theory of Black Holes* (Clarendon Press, Oxford, 1998)
4. P.V.P. Cunha, C.A.R. Herdeiro, E. Radu, H.F. Runarsson, Phys. Rev. Lett. **115**, 211102 (2015), arXiv:1509.00021 [gr-qc]
5. S. Doeleman et al., Nature **455**, 78 (2008), arXiv:0809.2442 [astro-ph]
6. H. Falcke, F. Melia, E. Agol, Astrophys. J. **528**, L13 (2000), arXiv:astro-ph/9912263
7. J. Fukue, Publ. Astron. Soc. Jap. **55**, 155 (2003)
8. M. Ghasemi-Nodehi, Z. Li, C. Bambi, Eur. Phys. J. C **75**, 315 (2015), arXiv:1506.02627 [gr-qc]
9. M. Jaroszynski, A. Kurpiewski, Astron. Astrophys. **326**, 419 (1997), arXiv:astro-ph/9705044
10. T. Johannsen, C. Wang, A.E. Broderick, S.S. Doeleman, V.L. Fish, A. Loeb, D. Psaltis, Phys. Rev. Lett. **117**, 091101 (2016), arXiv:1608.03593 [astro-ph.HE]
11. J.P. Luminet, Astron. Astrophys. **75**, 228 (1979)
12. N. Tsukamoto, Z. Li, C. Bambi, JCAP **1406**, 043 (2014), arXiv:1403.0371 [gr-qc]
13. L. Yang, Z. Li, Int. J. Mod. Phys. D **25**, 1650026 (2015), arXiv:1511.00086 [astro-ph.HE]

Chapter 11
Gravitational Waves

There are some important differences between the use of electromagnetic radiation and gravitational waves to study astrophysical black holes. Electromagnetic radiation is emitted by the particles in the accretion disk. Its properties are determined by the motion of the particles around the black hole and by the propagation of the photons from the point of emission to the point of detection. Assuming geodesic motion, electromagnetic observations can test the metric around astrophysical black holes and check whether it corresponds to the Kerr solution. We can also consider a specific framework with deviations from geodesic motion, but still we can only test the motion of test-particles. Gravitational waves are generated by dynamical systems, in which the metric changes with time. The properties of the gravitational waves are determined by both the spacetime metric and the field equations of the gravity theory. They can thus test both the Kerr metric and the Einstein equations.

The aim of this chapter is to provide a short review on the possible approaches to study astrophysical black holes using gravitational waves. A detailed discussion of the topic would require a whole book, and it is thus well beyond the scope of the present work. There are already a large number of papers in the literature studying how present and future gravitational wave detectors can probe the strong gravity region around astrophysical black holes and test general relativity and alternative theories of gravity. It is likely that such a line of research will grow much more in the near future.

The most promising sources to test the nature of astrophysical black holes are inspiral binaries, extreme-mass ratio inspirals (systems in which a stellar-mass compact object slowly falls onto a supermassive black hole), and perturbed black holes (e.g. after merger). They are very clean astrophysical systems only governed by gravity, which is a significant advantage with respect to the electromagnetic observations plagued by uncertainties associated to the astrophysical model. On the other hand, the gravitational wave approach is limited by the low signal-to-noise ratio and is affected by the systematics associated with the modeling of gravitational waves.

© Springer Nature Singapore Pte Ltd. 2017
C. Bambi, *Black Holes: A Laboratory for Testing Strong Gravity*,
DOI 10.1007/978-981-10-4524-0_11

11.1   Emission of Gravitational Waves

For the next sections, it is useful to start with a short review on the theory of the emission of gravitational waves. More details can be found in standard textbooks, like [48] or [45].

Let us define the *gravitational-field amplitude* as

$$h^{\mu\nu} = \sqrt{-g}\,g^{\mu\nu} - \eta^{\mu\nu}\,, \tag{11.1}$$

where $\eta^{\mu\nu} = \mathrm{diag}(-1, 1, 1, 1)$ is an auxiliary Minkowski metric. We then choose the coordinate system $\{x^{\mu}\}$ such that

$$\partial_{\mu} h^{\mu\nu} = 0\,. \tag{11.2}$$

The condition (11.2) is called the *harmonic gauge* (or the Hilbert gauge or the de Donder gauge). The choice of the harmonic gauge is similar to the choice of the Lorentz gauge in Maxwell's theory, $\partial_{\mu} A^{\mu} = 0$, where A_{μ} is the vector potential of the electromagnetic field.

In harmonic coordinates, the Einstein equations read

$$\Box_{\eta} h^{\mu\nu} = 16\pi\,\mathscr{T}^{\mu\nu}\,, \tag{11.3}$$

where $\Box_{\eta} = \eta^{\rho\sigma}\partial_{\rho}\partial_{\sigma}$ is the d'Alembertian operator of flat spacetime, $\mathscr{T}^{\mu\nu}$ is

$$\mathscr{T}^{\mu\nu} = (-g)\,T^{\mu\nu} + \frac{1}{16\pi}t^{\mu\nu}\,, \tag{11.4}$$

$T^{\mu\nu}$ is the matter energy-momentum tensor, and $t^{\mu\nu}$ is the pseudo-tensor (i.e. Lorentz-covariant tensor) of the gravitational field

$$
\begin{aligned}
t^{\mu\nu} =\ & -h^{\alpha\beta}\partial_{\alpha}\partial_{\beta}h^{\mu\nu} + \left(\partial_{\alpha}h^{\mu\beta}\right)\left(\partial_{\beta}h^{\nu\alpha}\right) + \frac{1}{2}g^{\mu\nu}g_{\alpha\beta}\left(\partial_{\gamma}h^{\alpha\delta}\right)\left(\partial_{\delta}h^{\beta\gamma}\right) \\
& + \frac{1}{8}\left(2g^{\mu\alpha}g^{\nu\beta} - g^{\mu\nu}g^{\alpha\beta}\right)\left(2g_{\gamma\delta}g_{\theta\varepsilon} - g_{\delta\theta}g_{\gamma\varepsilon}\right)\left(\partial_{\alpha}h^{\gamma\varepsilon}\right)\left(\partial_{\beta}h^{\delta\theta}\right) \\
& - g^{\mu\alpha}g_{\beta\gamma}\left(\partial_{\delta}h^{\nu\gamma}\right)\left(\partial_{\alpha}h^{\beta\delta}\right) - g^{\nu\alpha}g_{\beta\gamma}\left(\partial_{\delta}h^{\mu\gamma}\right)\left(\partial_{\alpha}h^{\beta\delta}\right) \\
& + g_{\alpha\beta}g^{\gamma\delta}\left(\partial_{\gamma}h^{\mu\alpha}\right)\left(\partial_{\delta}h^{\nu\beta}\right)\,.
\end{aligned}
\tag{11.5}
$$

The formal solution of Eq. (11.3) is

$$h^{\mu\nu}\,(t, \mathbf{x}) = 4\int d^{3}\mathbf{x}'\,\frac{\mathscr{T}^{\mu\nu}\left(t - |\mathbf{x} - \mathbf{x}'|, \mathbf{x}'\right)}{|\mathbf{x} - \mathbf{x}'|}\,, \tag{11.6}$$

where the integral is performed in the flat 3-dimensional space and $|\mathbf{x} - \mathbf{x}'|$ is the Euclidean distance between the point $\mathbf{x}$ and the point $\mathbf{x}'$. In Cartesian coordinates $\{x, y, z\}$, we have

$$|\mathbf{x} - \mathbf{x}'| = \sqrt{(x - x')^2 + (y - y')^2 + (z - z')^2} \, . \tag{11.7}$$

If the metric $g_{\mu\nu}$ slightly deviates from the Minkowski metric and we neglect terms of second and higher order in $h^{\mu\nu}$, Eq. (11.1) can be written as[1]

$$g_{\mu\nu} = \eta_{\mu\nu} - h_{\mu\nu} \, , \quad |h_{\mu\nu}| \ll 1 \, . \tag{11.8}$$

The Einstein equations (11.3) reduce to

$$\Box_\eta h^{\mu\nu} = 16\pi T^{\mu\nu} \, , \tag{11.9}$$

which is a simple wave equation describing "ripples" of the spacetime propagating with the speed of light. These ripples are the gravitational waves.

Let us now consider small changes in the coordinates that leave $\eta_{\mu\nu}$ unchanged but induce small changes in $h_{\mu\nu}$. We employ the coordinate transformation

$$x^\mu \rightarrow x'^\mu = x^\mu + \xi^\mu \, , \tag{11.10}$$

where ξ^μ are four arbitrary functions of x^μ of the same order as $h^{\mu\nu}$. We find that

$$h_{\mu\nu} \rightarrow h'_{\mu\nu} = h_{\mu\nu} - \partial_\mu \xi_\nu - \partial_\nu \xi_\mu \, . \tag{11.11}$$

If $\Box_\eta \xi^\mu = 0$, such a transformation preserves the harmonic gauge. This is the analogous of the transformation $A_\mu \rightarrow A_\mu + \partial_\mu \Lambda$ in Maxwell's theory, where the Lorentz gauge is preserved if $\Box_\eta \Lambda = 0$.

$h_{\mu\nu}$ is symmetric, and therefore has ten independent components. The harmonic gauge in Eq. (11.2) gives four conditions and reduces the number of independent components to six. However, $h_{\mu\nu}$ still depends on four arbitrary functions ξ_μ satisfying the equation $\Box_\eta \xi^\mu = 0$. We can thus further simplify $h_{\mu\nu}$ by imposing four new conditions. We can choose ξ^0 such that the trace of $h_{\mu\nu}$, $h = h^\mu_\mu$, vanishes, and the three functions ξ^i such that $h^{0i} = 0$. The harmonic gauge implies that h_{00} is independent of time and therefore corresponds to the Newtonian potential of the source. Restricting the attention to gravitational waves (i.e. the time-dependent part of h_{00}), we can set $h_{00} = 0$. Eventually we have

$$h^{0\mu} = 0 \, , \quad h^i_i = 0 \, , \quad \partial_i h^{ij} = 0 \, , \tag{11.12}$$

---

[1] With these assumptions, the indices of $h_{\mu\nu}$ are raised and lowered by the Minkowski metric $\eta_{\mu\nu}$, e.g. $h^{\mu\nu} = \eta^{\mu\rho} \eta^{\nu\sigma} h_{\rho\sigma}$. Let us note the sign in front of $h_{\mu\nu}$ and $h^{\mu\nu}$: $g_{\mu\nu} = \eta_{\mu\nu} - h_{\mu\nu}$ and $g^{\mu\nu} = \eta^{\mu\nu} + h^{\mu\nu}$.

which defines the *transverse-traceless gauge* (TT gauge). Quantities in the TT gauge are often indicated with TT, e.g. $h_{\mu\nu}^{TT}$. For a gravitational wave propagating along the z direction in vacuum, we have

$$h_{\mu\nu}^{TT} = \begin{pmatrix} 0 & 0 & 0 & 0 \\ 0 & h_+ & h_\times & 0 \\ 0 & h_\times & h_+ & 0 \\ 0 & 0 & 0 & 0 \end{pmatrix} \cos\left[\omega\left(t - z\right)\right], \qquad (11.13)$$

where ω is the angular frequency of the gravitational wave, while h_+ and $h_\times$ are the amplitudes of the gravitational wave in the two polarizations. If we consider the effect of the passage of a similar gravitational wave on a ring of particles in the (x, y) plane, we find the situation sketched in the panels (a) and (b) in Fig. 5.8.

11.1.1 Quadrupole Formula

In Maxwell's theory, the emission of electromagnetic radiation by slowly varying charge distributions can be decomposed into a series of multipoles ($l = 0, 1, 2,\ldots$). The monopole moment ($l = 0$) is proportional to the total electric charge, which is conserved and therefore does not change with time, and there is no monopole emission. The series thus consists of $l \geq 1$ multipoles, and the electric dipole radiation $l = 1$ is usually the leading term. In Einstein's gravity, we can proceed with a similar expansion to describe the emission of gravitational waves of slowly varying mass distributions. Now there is no monopole emission, because the mass is a conserved quantity. The mass dipole moment and the current dipole moment are proportional, respectively, to the momentum and the angular momentum of the system, which are both conserved quantities and therefore there is no dipole emission of gravitational waves. The leading order term is the quadrupole moment ($l = 2$).

The formal solution of Eq. (11.9) is

$$h^{\mu\nu}\left(t, \mathbf{x}\right) = 4 \int d^3\mathbf{x}' \frac{T^{\mu\nu}\left(t - |\mathbf{x} - \mathbf{x}'|, \mathbf{x}'\right)}{|\mathbf{x} - \mathbf{x}'|}. \qquad (11.14)$$

If we can assume that the region where the source is confined ($T^{\mu\nu}$ is non-vanishing) is much smaller than the wavelength of the emitted radiation

$$h^{\mu\nu}\left(t, \mathbf{x}\right) = 4 \int d^3\mathbf{x}' \frac{T^{\mu\nu}\left(t - |\mathbf{x}|, \mathbf{x}'\right)}{|\mathbf{x}|}. \qquad (11.15)$$

In the linearized theory, $h_{\mu\nu}$ and $T_{\mu\nu}$ are of the same order and therefore $\partial_\mu T^{\mu\nu} = 0$ with the ordinary derivative. We write $\partial_\mu T^{\mu i} = 0$ as $\partial_0 T^{0i} = -\partial_k T^{ki}$. We multiply both sides by x^j and we integrate over the volume V. We obtain

$$\frac{\partial}{\partial t} \int_V T^{i0} x^j \, d^3\mathbf{x} = -\int_V \frac{\partial T^{ik}}{\partial x^k} x^j \, d^3\mathbf{x}$$

$$= -\int_V \frac{\partial}{\partial x^k} \left(T^{ik} x^j \right) d^3\mathbf{x} + \int_V T^{ik} \frac{\partial x^j}{\partial x^k} d^3\mathbf{x}$$

$$= -\int_\Sigma T^{ik} x^j \, d\Sigma_k + \int_V T^{ij} d^3\mathbf{x}$$

$$= \int_V T^{ij} d^3\mathbf{x}, \tag{11.16}$$

where Σ is the surface of the volume V and $T^{\mu\nu} = 0$ on Σ. Since $T^{\mu\nu}$ is a symmetric tensor, we can also write Eq. (11.16) by exchanging i and j and

$$\frac{\partial}{\partial t} \int_V \left(T^{i0} x^j + T^{j0} x^i \right) d^3\mathbf{x} = 2 \int_V T^{ij} d^3\mathbf{x}. \tag{11.17}$$

Let us now write $\partial_\mu T^{\mu 0} = 0$ as $\partial_0 T^{00} = -\partial_k T^{k0}$. This time we multiply both sides by $x^i x^j$. We integrate over the volume V and we find

$$\frac{\partial}{\partial t} \int_V T^{00} x^i x^j \, d^3\mathbf{x} = -\int_V \frac{\partial T^{k0}}{\partial x^k} x^i x^j \, d^3\mathbf{x}$$

$$= -\int_V \frac{\partial}{\partial x^k} \left(T^{k0} x^i x^j \right) d^3\mathbf{x} + \int_V \left(T^{k0} \frac{\partial x^i}{\partial x^k} x^j + T^{k0} x^i \frac{\partial x^j}{\partial x^k} \right) d^3\mathbf{x}$$

$$= -\int_\Sigma T^{k0} x^i x^j \, d\Sigma_k + \int_V \left(T^{i0} x^j + T^{j0} x^i \right) d^3\mathbf{x}$$

$$= \int_V \left(T^{i0} x^j + T^{j0} x^i \right) d^3\mathbf{x}. \tag{11.18}$$

We derive with respect to the time t and we use Eq. (11.17)

$$\frac{\partial^2}{\partial^2 t} \int_V T^{00} x^i x^j \, d^3\mathbf{x} = \frac{\partial}{\partial t} \int_V \left(T^{0i} x^j + T^{0j} x^i \right) d^3\mathbf{x}$$

$$= 2 \int_V T^{ij} d^3\mathbf{x}. \tag{11.19}$$

We define the *quadrupole moment* of the source as

$$Q^{ij}(t) = \int_V T^{00}(t, \mathbf{x}) \, x^i x^j \, d^3\mathbf{x}. \tag{11.20}$$

Equation (11.15) becomes (reintroducing c and G_N)

$$h_{\mu 0} = 0,$$

$$h_{ij} = \frac{2}{r} \frac{G_N}{c^4} \ddot{Q}_{ij}(t - r), \tag{11.21}$$

where the double dot stands for the double derivative with respect to t. It is worth noting that a spherical or axisymmetric distribution of matter has a constant quadrupole moment, even if the body is rotating. This implies, for instance, that there is no emission of gravitational waves in a perfectly spherically symmetric collapse, in a perfectly axisymmetric rotating body, etc. Gravitational waves are emitted when there is a certain "degree of asymmetry", e.g. the coalescence of two objects, non-radial pulsation of a body, etc.

If we want Eq. (11.21) in the TT gauge, we have to apply to both sides a projector to get transverse waves and traceless tensors. The unit vector normal to the wavefront is $\mathbf{n} = \mathbf{x}/r$. The operator to project a vector onto the plane orthogonal to the direction of $\mathbf{n}$ is

$$P_{ij} = \delta_{ij} - n_i n_j \, . \tag{11.22}$$

The transverse-traceless projector is

$$P_{ijkl} = P_{ik} P_{jl} - \frac{1}{2} P_{ij} P_{kl} \, . \tag{11.23}$$

The result is

$$h_{\mu 0}^{\mathrm{TT}} = 0 \, ,$$
$$h_{ij}^{\mathrm{TT}} = \frac{2}{r} \frac{G_{\mathrm{N}}}{c^4} \ddot{Q}_{ij}^{\mathrm{TT}} (t - r) \, , \tag{11.24}$$

where

$$Q_{ij}^{\mathrm{TT}} = P_{ijkl} Q_{kl} \, . \tag{11.25}$$

In order to compute the energy emitted in the form of gravitational waves, we can proceed as follows. We expand the Einstein tensor in terms of $h_{\mu\nu}$, namely

$$G_{\mu\nu} = G_{\mu\nu}^{(1)} + G_{\mu\nu}^{(2)} + \cdots \tag{11.26}$$

where $G_{\mu\nu}^{(1)}$ is linear in $h_{\mu\nu}$, $G_{\mu\nu}^{(2)}$ is quadratic in $h_{\mu\nu}$, and we neglect higher orders terms in $h_{\mu\nu}$. We rewrite the Einstein equations as in Eqs. (11.3) and (11.4)

$$G_{\mu\nu}^{(1)} = 8\pi \left(T_{\mu\nu} + t_{\mu\nu}^{\mathrm{GW}} \right) \, , \tag{11.27}$$

where $t_{\mu\nu}^{\mathrm{GW}}$ describes (within our approximation) the energy-momentum tensor of the gravitational field itself

$$t_{\mu\nu}^{\mathrm{GW}} = \frac{1}{8\pi} G_{\mu\nu}^{(2)} \, . \tag{11.28}$$

In the TT gauge, it becomes

$$t_{\mu\nu}^{GW} = \frac{1}{32\pi} \langle (\partial_\mu h_{ij}^{TT}) (\partial_\nu h_{ij}^{TT}) \rangle , \tag{11.29}$$

where the angular brackets $\langle ... \rangle$ are introduced to indicate that we are averaging over several wavelengths. Indeed, the energy-momentum tensor associated to the gravitational waves cannot be localized within a wavelength. We can instead say that a certain amount of energy and momentum is contained in a region of space extending over several wavelengths.

In the case of the gravitational wave propagating along the z direction, Eq. (11.13), $t_{\mu\nu}^{GW}$ has only four non-vanishing components, namely

$$t_{00}^{GW} = -t_{0z}^{GW} = -t_{z0}^{GW} = t_{zz}^{GW} = \frac{1}{32\pi} \omega^2 \left(h_+^2 + h_\times^2 \right) . \tag{11.30}$$

The 00 component describes the energy density of the gravitational wave, the $0z$ and $z0$ components describe the energy flux along the z direction, while the zz component is the momentum flux.

From Eqs. (11.29) and (11.30), we find the quadrupole formula for the energy emission of gravitational waves

$$L_{GW} = \frac{1}{5} \frac{G_N}{c^5} \langle \dddot{Q}_{ij}^{TT} \cdot \dddot{Q}_{ij}^{TT} \rangle . \tag{11.31}$$

11.2 Response of Interferometer Detectors

As discussed in Sect. 5.2, the passage of a gravitational wave in an interferometer detector may cause a change in the difference in the photon travel time along the two arms. Let us consider the six possible polarization modes permitted in metric theories of gravity [69]. The space-space components of the metric perturbation $h^{\mu\nu}$ at the detection point can be written as a sum over the six polarization modes ($p = +, \times, x, y, b,$ and l) and each mode can be written as an expansion of plane waves, namely

$$\begin{aligned} h_{ij}(t, \mathbf{x}) &= \sum_p h_{ij}^p(t, \mathbf{x}) \\ &= \sum_p \int_{-\infty}^{+\infty} dv \int_{S^2} d\hat{\mathbf{k}} \, e^{2\pi i v \left(t - \hat{\mathbf{k}} \cdot \mathbf{x} \right)} \tilde{h}^p \left(v, \hat{\mathbf{k}} \right) \varepsilon_{ij}^p \left(\hat{\mathbf{k}} \right) \\ &= \sum_p \int_{S^2} d\hat{\mathbf{k}} \, h^p \left(t - \hat{\mathbf{k}} \cdot \mathbf{x} \right) \varepsilon_{ij}^p \left(\hat{\mathbf{k}} \right) , \end{aligned} \tag{11.32}$$

where v is the gravitational wave frequency, $\hat{\mathbf{k}}$ is the unit 3-vector pointing in the propagation direction of the gravitational wave, ε_{ij}^p is the p-mode polarization tensor, and we have introduced the strain h^p of the polarization mode p caused by the possible gravitational wave with propagation direction $\hat{\mathbf{k}}$

$$h^p\left(t - \hat{\mathbf{k}} \cdot \mathbf{x}\right) = \int_{-\infty}^{+\infty} dv\, e^{2\pi i v\left(t - \hat{\mathbf{k}} \cdot \mathbf{x}\right)} \tilde{h}^p\left(v, \hat{\mathbf{k}}\right) . \qquad (11.33)$$

For a gravitational wave propagating along the z direction, the polarization tensors are given by

$$||\varepsilon_{ij}^+ \left(\hat{\mathbf{z}}\right)|| = \begin{pmatrix} 1 & 0 & 0 \\ 0 & -1 & 0 \\ 0 & 0 & 0 \end{pmatrix} \quad \text{(plus mode)} ,$$

$$||\varepsilon_{ij}^\times \left(\hat{\mathbf{z}}\right)|| = \begin{pmatrix} 0 & 1 & 0 \\ 1 & 0 & 0 \\ 0 & 0 & 0 \end{pmatrix} \quad \text{(cross mode)} ,$$

$$||\varepsilon_{ij}^x \left(\hat{\mathbf{z}}\right)|| = \begin{pmatrix} 0 & 0 & 1 \\ 0 & 0 & 0 \\ 1 & 0 & 0 \end{pmatrix} \quad \text{(vector-x mode)} ,$$

$$||\varepsilon_{ij}^y \left(\hat{\mathbf{z}}\right)|| = \begin{pmatrix} 0 & 0 & 0 \\ 0 & 0 & 1 \\ 0 & 1 & 0 \end{pmatrix} \quad \text{(vector-y mode)} ,$$

$$||\varepsilon_{ij}^b \left(\hat{\mathbf{z}}\right)|| = \begin{pmatrix} 1 & 0 & 0 \\ 0 & 1 & 0 \\ 0 & 0 & 0 \end{pmatrix} \quad \text{(breathing mode)} ,$$

$$||\varepsilon_{ij}^l \left(\hat{\mathbf{z}}\right)|| = \begin{pmatrix} 0 & 0 & 0 \\ 0 & 0 & 0 \\ 0 & 0 & 1 \end{pmatrix} \quad \text{(longitudinal mode)} . \qquad (11.34)$$

See Fig. 5.8 for the physical interpretation of each polarization mode.

Now we want to derive the response of an interferometer detector due to the passage of gravitational waves [51]. Let us consider the coordinate systems of the detector and of a gravitational wave as in Fig. 11.1. The three unit 3-vectors of the detector coordinate system are

$$\hat{\mathbf{x}} = \begin{pmatrix} 1 \\ 0 \\ 0 \end{pmatrix} , \quad \hat{\mathbf{y}} = \begin{pmatrix} 0 \\ 1 \\ 0 \end{pmatrix} , \quad \hat{\mathbf{z}} = \begin{pmatrix} 0 \\ 0 \\ 1 \end{pmatrix} . \qquad (11.35)$$

The gravitational wave coordinate system is rotated by the angles (θ, ϕ, ψ) with respect to the detector coordinate system (see Fig. 11.1). If we rotate the detector coordinate system by the angles (θ, ϕ), we find

Fig. 11.1 Sketch of the
detector coordinate system
$(\hat{\mathbf{x}}, \hat{\mathbf{y}}, \hat{\mathbf{z}})$ and of the
gravitational wave
coordinate system $(\hat{\mathbf{l}}, \hat{\mathbf{m}}, \hat{\mathbf{k}})$.
The figure also shows the
angles θ, ϕ, and ψ

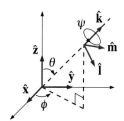

$$\hat{\mathbf{x}}' = \begin{pmatrix} \cos\theta\cos\phi \\ \cos\theta\sin\phi \\ -\sin\theta \end{pmatrix}, \quad \hat{\mathbf{y}}' = \begin{pmatrix} -\sin\phi \\ \cos\phi \\ 0 \end{pmatrix}, \quad \hat{\mathbf{z}}' = \begin{pmatrix} \sin\theta\cos\phi \\ \sin\theta\sin\phi \\ \cos\theta \end{pmatrix}. \quad (11.36)$$

With the additional rotation about the propagation direction of the gravitational wave, $\hat{\mathbf{k}}$, we recover the gravitational wave coordinate system $(\hat{\mathbf{l}}, \hat{\mathbf{m}}, \hat{\mathbf{k}})$

$$\hat{\mathbf{l}} = \hat{\mathbf{x}}'\cos\psi + \hat{\mathbf{y}}'\sin\psi,$$
$$\hat{\mathbf{m}} = -\hat{\mathbf{x}}'\sin\psi + \hat{\mathbf{y}}'\cos\psi,$$
$$\hat{\mathbf{k}} = \hat{\mathbf{z}}'. \quad (11.37)$$

In terms of the unit 3-vectors of the gravitational wave coordinate system, the six polarization modes are

$$||\varepsilon_{ij}^{+}|| = \hat{\mathbf{l}} \otimes \hat{\mathbf{l}} - \hat{\mathbf{m}} \otimes \hat{\mathbf{m}},$$
$$||\varepsilon_{ij}^{\times}|| = \hat{\mathbf{l}} \otimes \hat{\mathbf{m}} + \hat{\mathbf{m}} \otimes \hat{\mathbf{l}},$$
$$||\varepsilon_{ij}^{x}|| = \hat{\mathbf{l}} \otimes \hat{\mathbf{k}} + \hat{\mathbf{k}} \otimes \hat{\mathbf{l}},$$
$$||\varepsilon_{ij}^{y}|| = \hat{\mathbf{m}} \otimes \hat{\mathbf{k}} + \hat{\mathbf{k}} \otimes \hat{\mathbf{m}},$$
$$||\varepsilon_{ij}^{b}|| = \hat{\mathbf{l}} \otimes \hat{\mathbf{l}} + \hat{\mathbf{m}} \otimes \hat{\mathbf{m}},$$
$$||\varepsilon_{ij}^{l}|| = \hat{\mathbf{k}} \otimes \hat{\mathbf{k}}. \quad (11.38)$$

The strain measured by an interferometer detector due to the passage of a gravitational wave with the propagation direction $\hat{\mathbf{k}}$ and the polarization ψ becomes

$$h(t) = \sum_{p} h^{p}\left(t - \hat{\mathbf{k}} \cdot \mathbf{x}\right) F^{p}\left(\theta, \phi, \psi\right), \quad (11.39)$$

where $F^{p}\left(\theta, \phi, \psi\right)$ is the *antenna pattern response* for the polarization mode p and takes into account the relative orientation between the detector and the gravitational wave, as well as the detector geometry. If the arms of the interferometer detector are orthogonal each other, we have

$$F^p (\theta, \phi, \psi) = \frac{1}{2} \left(\hat{x}^i \hat{x}^j - \hat{y}^i \hat{y}^j \right) \varepsilon_{ij}^p \left(\hat{\mathbf{k}}, \psi \right) . \tag{11.40}$$

The expression of the six antenna pattern responses in terms of (θ, ϕ, ψ) is [51]

$$F^+ (\theta, \phi, \psi) = \frac{1}{2} \left(1 + \cos^2 \theta \right) \cos (2\phi) \cos (2\psi) - \cos \theta \sin (2\phi) \sin (2\psi) ,$$

$$F^\times (\theta, \phi, \psi) = -\frac{1}{2} \left(1 + \cos^2 \theta \right) \cos (2\phi) \sin (2\psi) - \cos \theta \sin (2\phi) \cos (2\psi) ,$$

$$F^x (\theta, \phi, \psi) = \sin \theta \left[\cos \theta \cos (2\phi) \cos \psi - \sin (2\phi) \sin \psi \right] ,$$

$$F^y (\theta, \phi, \psi) = - \sin \theta \left[\cos \theta \cos (2\phi) \sin \psi + \sin (2\phi) \cos \psi \right] ,$$

$$F^b (\theta, \phi) = -\frac{1}{2} \sin^2 \theta \cos (2\phi) ,$$

$$F^l (\theta, \phi) = \frac{1}{2} \sin^2 \theta \cos (2\phi) . \tag{11.41}$$

In a similar way, it is possible to derive the response of pulsar timing array experiments, see e.g. [18, 21, 41].

11.3 Matched Filtering

If we have a robust and accurate theoretical prediction of the gravitational wave signal and if the detector noise is Gaussian and stationary, the optimal detection strategy is *matched filtering*. The output of a gravitational wave detector is the sum of the (possible) signal $s(t)$ and the noise $n(t)$,

$$x(t) = s(t) + n(t) . \tag{11.42}$$

Since the noise is Gaussian and stationary, it is fully characterized by its *power spectral density* $S_n(v)$, defined by the relation

$$\langle \tilde{n}(v) \tilde{n}^*(v') \rangle = \frac{1}{2} S_n(v) \delta(v - v') , \tag{11.43}$$

where $\langle ... \rangle$ denotes an ensemble average over many noise realizations and we adopt the following conventions for the Fourier transform and its inverse

$$\tilde{y}(v) = \int_{-\infty}^{+\infty} y(t) e^{-2\pi i v t} \, dt , \qquad y(t) = \int_{-\infty}^{+\infty} \tilde{y}(v) e^{2\pi i v t} \, dv . \tag{11.44}$$

For stationary stochastic noises, the ensemble average can be replaced by a time average. From Eq. (11.43), it is easy to see that the spectral energy distribution $S_n(v)$ is related to the time average of the square of the detector noise by [49]

$$\lim_{T\to+\infty} \frac{1}{2T} \int_{-T}^{T} n(t)\,n^*(t)\,dt = \int_{0}^{+\infty} S_n(v)\,dv. \tag{11.45}$$

Since the noise $n(t)$ is real, $\tilde{n}(-v) = \tilde{n}^*(v)$, and the spectral energy distribution $S_n(v)$ is an even function, $S_n(v) = S_n(-v)$. In such a case, Fourier integrals over all frequencies can be written as integrals over positive frequencies only.

The detectability of a signal is determined by its *signal-to-noise ratio* (SNR) ρ, which is defined as

$$\rho^2 = \frac{(s|h)}{\sqrt{(h|n)\,(n|h)}}, \tag{11.46}$$

where h is a template with parameters $\{\zeta^i\}$ and $(y|w)$ denotes the overlap of the two functions $y(t)$ and $w(t)$

$$(y|w) = 4\,\mathrm{Re}\int_{0}^{+\infty} \frac{\tilde{y}\tilde{w}^*}{S_n}\,dv. \tag{11.47}$$

If the templates do not exactly match the signal, e.g. approximate families of waveforms are employed, the signal-to-noise ratio is reduced by the factor m (called the match)

$$m = \frac{(s|h)}{\sqrt{(s|s)\,(h|h)}}. \tag{11.48}$$

For any template h, the probability of measuring the signal $s(t)$ is

$$P \propto e^{-\frac{1}{2}(s-h|s-h)}. \tag{11.49}$$

The waveform h that best fits the signal is the waveform that minimizes the argument of the exponential. The 1-σ error on a given parameter ζ^i is $1/\sqrt{\Gamma_{ii}}$, where Γ_{ij} is the Fisher matrix

$$\Gamma_{ij} = \left(\frac{\partial h}{\partial \zeta^i}\,\bigg|\,\frac{\partial h}{\partial \zeta^j}\right). \tag{11.50}$$

More details on matched filtering and Fisher analysis can be found in [20, 23, 32]. It is worth noting that the Fisher method requires that the noise is stationary and Gaussian, and it is necessary that the signal-to-noise ratio is sufficiently large. If this is not the case, one may obtain incorrect results and should employ alternative parameter estimation methods. The Fisher analysis can be used to test alternative theories of gravity, see e.g. [53, 70].

Fig. 11.2 Temporal evolution of the strain, of the *black hole* separation, and of the *black hole* relative velocity in the event GW150914, which was the coalescence of two stellar-mass black holes, each of them of about 30 $M_\odot$. From [1] under the terms of the Creative Commons Attribution 3.0 License

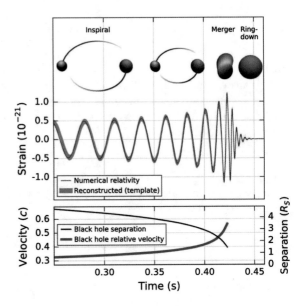

11.4   Coalescing Black Holes

Coalescing black holes are among the leading candidate sources for detection by present and future gravitational wave observatories. Ground-based laser interferometers can detect gravitational waves in the frequency range 10 Hz–10 kHz and consequently can observe the last stage of the coalescence of stellar-mass black holes. Both GW150914 and GW151226 were events generated by the coalescence of two stellar-mass black holes [1, 2]. Gravitational wave experiments sensitive at lower frequencies can detect signals from the coalescence of two supermassive black holes or from a system of a supermassive black hole and a stellar-mass compact object.

The coalescence of a system of two black holes is characterized by three stages (see Fig. 11.2):

1. *Inspiral*. The two objects rotate around each other. This causes the emission of gravitational waves. As the system loses energy and angular momentum, the separation between the two objects decreases and their relative velocity increases. The frequency and the amplitude of the gravitational waves increase (leading to the so-called "chirping" character in the waveform) until the moment of merger.
2. *Merger*. The two black holes merge into a single black hole.
3. *Ringdown*. The newly born black hole emits gravitational waves to settle down to an equilibrium configuration.

Because of the complexity of the Einstein equations, it is necessary to employ certain approximation methods to compute the gravitational wave signal. In the case of (roughly) equal mass black holes, the three stages above are treated with the following methods:

1. *Post-Newtonian (PN) methods.* They are based on an expansion in $\varepsilon \sim U \sim v^2$, where $U \sim M/R$ is the Newtonian gravitational potential and v is the black hole relative velocity. The 0PN term is the Newtonian solution of the binary system. The nPN term is the $O(\varepsilon^n)$ correction to the Newtonian solution. In general relativity, radiation backreaction shows up at 3.5PN order in g_{00}, at 3PN order in g_{0i}, and at 2.5PN order in g_{ij}.

2. *Numerical relativity.* When the PN approach breaks down because ε is not a small parameter any longer, one has to solve the field equations of the complete theory. Since the spacetime is not resolved to infinite precision, even this approach is an approximate method. The first stable simulation was presented in [57], followed by [7, 17]. For non-spinning black holes in general relativity, the stage of merger smoothly connects the stages of inspiral and of ringdown. For spinning black holes in general relativity, merger may be a more violent event, depending on the black hole spins and their alignments with respect to the orbital angular momentum.

3. *Black hole perturbation theory.* It is based on the study of small perturbations over a background metric. The method is used to describe the ringdown stage.

In the rest of this section, we briefly discuss the stage of inspiral of a binary black hole in which the two black hole masses are comparable. A recent review on the topic is [12]. Extreme-mass ratio inspirals will be briefly reviewed in Sect. 11.5. The description of merger requires numerical simulations, which is beyond the scope of the present book focused on electromagnetic approaches. The topic is discussed, for instance, in the review paper [42] or in the textbook [10]. The stage of ringdown is characterized by the emission of quasi-normal modes, which are briefly discussed in Sect. 11.6.

PN methods for the description of the inspiral stage work well until we are very close to the event of merger, even if this requires to solve the equations to high PN orders (at least 3.5PN). Within general relativity, the description of a binary system of two generic compact objects (i.e. not necessarily black holes) is relatively easy, as a consequence of the Strong Equivalence Principle [71]: if the two objects are sufficiently far each other that effects related to their finite size can be ignored, the orbital motion and the gravitational wave signal only depend on the masses and the spins of the two bodies. For example, this is not true in a typical scalar-tensor theory, where the scalar field ϕ is generated by the mass of the objects and determines the effective gravitational coupling constant. The result is that each object is characterized by several masses (e.g. gravitational mass, inertial mass, radiation mass, etc.) and the value of these masses depends on the value of the scalar field at the location of the object.

Within a PN approach, one has first to study the orbital motion of the binary. In general relativity, the equation of motion for the binary system is [12]

$$\mathbf{a} = \frac{d\mathbf{v}}{dt} = \frac{M}{R^2} \left(-\hat{\mathbf{n}} + \mathbf{A}_{1PN} + \mathbf{A}_{2PN} + \mathbf{A}_{2.5PN} + \mathbf{A}_{3PN} + \cdots \right) \qquad (11.51)$$

where $M = m_1 + m_2$, m_1 and m_2 are the masses of the two objects, $R = |\mathbf{x}_1 - \mathbf{x}_2|$ is the Euclidean distance between the two objects, $\mathbf{v} = \mathbf{v}_1 - \mathbf{v}_2$ is the relative velocity, and $\hat{\mathbf{n}} = (\mathbf{x}_1 - \mathbf{x}_2)/R$. The Newtonian term is $-(M/R^2)\hat{\mathbf{n}}$, while $(M/R^2)\mathbf{A}_{\mathrm{nPN}}$ is the correction of order ε^n to the Newtonian solution. For example, the 1PN term $\mathbf{A}_{1\mathrm{PN}}$ is

$$\mathbf{A}_{1\mathrm{PN}} = \left[(4 + 2\eta)\frac{M}{R} - (1 + 3\eta)v^2 + \frac{3}{2}\eta\dot{R}^2\right]\hat{\mathbf{n}} + (4 - 2\eta)\dot{R}\mathbf{v}, \quad (11.52)$$

where $v = |\mathbf{v}|$, $\eta = m_1 m_2/M^2$, and a dot stands for a derivative with respect to time.

In order to get the gravitational field far from the binary system, it is necessary to solve some technical issues, see e.g. [12]. Since the source $\mathcal{T}^{\mu\nu}$ contains $h^{\mu\nu}$, which is not confined to a compact region, one finds some divergent or ill-defined integrals. A few methods have been proposed to fix these problems; e.g. the Post-Minkowskian approach of [13–16]. The amplitude of the gravitational field far from the source assumes the form

$$h^{ij} = \frac{2M}{r}\left(H^{ij} + H^{ij}_{0.5\mathrm{PN}} + H^{ij}_{1\mathrm{PN}} + H^{ij}_{1.5\mathrm{PN}} + H^{ij}_{2\mathrm{PN}} + H^{ij}_{2.5\mathrm{PN}} + \cdots\right)_{t'=t-r}, \quad (11.53)$$

where r is the distance of the observer from the source and all variables must be evaluated at the retarded time $t' = t - r$. The first term on the right hand side of Eq. (11.53) gives the quadrupole formula, Eq. (11.21). In the case of a binary system

$$H^{ij} = 2\eta\left(v^i v^j - \frac{M\hat{n}^i\hat{n}^j}{R}\right). \quad (11.54)$$

The leading radiation damping terms is at 2.5PN order. Averaging over one orbit, the rate of increase of the orbital frequency due to the emission of gravitational waves is

$$\dot{f}_{\mathrm{b}} = \frac{192\pi}{5}(2\pi\mathcal{M})^{5/3}f_{\mathrm{b}}^{11/3}F(e), \quad (11.55)$$

where $\mathcal{M} = \eta^{3/5}M$ is the chirp mass and $F(e)$ is a function of the orbital eccentricity e

$$F(e) = (1 - e^2)^{-7/2}\left(1 + \frac{73}{24}e^2 + \frac{37}{96}e^4\right). \quad (11.56)$$

The gravitational waves emitted during the stage of inspiral are different from Einstein's predictions in some alternative theories of gravity and are the same as in general relativity in others. An interesting example is the case of scalar-tensor theories. For instance, [47] studied the evolution of compact binary systems of non-spinning objects in the family of the scalar-tensor theories described by the action

$$S = \frac{1}{16\pi}\int\left(\phi R - \frac{\omega}{\phi}g^{\mu\nu}(\partial_\mu\phi)(\partial_\nu\phi)\right)\sqrt{-g}d^4x + S_m, \quad (11.57)$$

where ω is a function of the scalar field ϕ, while S_m is the action of the matter sector and is independent of ϕ. In this class of theories, Eq. (11.51) becomes

$$\mathbf{a} = \frac{d\mathbf{v}}{dt} = \frac{\alpha M}{R^2} \left(-\hat{\mathbf{n}} + \mathbf{A}_{1PN} + \mathbf{A}_{1.5PN} + \mathbf{A}_{2PN} + \mathbf{A}_{2.5PN} + \mathbf{A}_{3PN} + \cdots \right) \quad (11.58)$$

where

$$\alpha = \frac{3 + 2\omega_0}{4 + 2\omega_0} + \frac{(1 - 2s_1)(1 - 2s_2)}{4 + 2\omega_0}, \quad (11.59)$$

ω_0 is the value of the function ω far from the system, while s_1 and s_2 are, respectively, the "sensitivity" of the objects 1 and 2 in the binary (see [47] for details). For neutron stars, the sensitivity depends the mass and the matter equation of state, and its numerical value should be around 0.2. For black holes, the sensitivity is exactly 1/2. The 1PN term $\mathbf{A}_{1PN}$ in Eq. (11.52) becomes

$$\mathbf{A}_{1PN} = \left[\left(4 + 2\eta + 2\bar{\gamma} + 2\bar{\beta}_+ - 2\psi\bar{\beta}_- \right) \frac{\alpha M}{R} - (1 + 3\eta + \bar{\gamma}) v^2 + \frac{3}{2}\eta \dot{R}^2 \right] \hat{\mathbf{n}}$$
$$+ (4 - 2\eta + 2\bar{\gamma}) \dot{R}\mathbf{v} \quad (11.60)$$

where $\bar{\gamma}$, $\bar{\beta}_+$, and $\bar{\beta}_-$ are other functions that exactly vanish in the case of a binary black hole, but they are different from zero in general. $\psi = (m_1 - m_2)/M$. The leading radiation reaction term is at 1.5PN order, which is associated to the emission of dipole gravitational radiation. Equation (11.54) becomes

$$h^{ij} = \frac{2(1 - \zeta)M}{r} \left(H^{ij} + H^{ij}_{0.5PN} + H^{ij}_{1PN} + H^{ij}_{1.5PN} + H^{ij}_{2PN} + \cdots \right), \quad (11.61)$$

where $\zeta = 1/(4 + 2\omega_0)$. Equation (11.54) becomes

$$H^{ij} = 2\eta \left(v^i v^j - \frac{\alpha M \hat{n}^i \hat{n}^j}{R} \right). \quad (11.62)$$

Since α only appears in the product αM, it cannot be directly measured. As long as the distortions of each black hole from the companion can be neglected, in the family of scalar-tensor theories with the action (11.57), the emission of gravitational waves from a black hole-black hole binary is the same as in general relativity, while deviations may be present if one or both the objects are not black holes.

11.5 Extreme-Mass Ratio Inspirals

An extreme-mass ratio inspiral (EMRI) is a system of a stellar-mass compact object (black hole, neutron star, or white dwarf with a mass $\mu \sim 1\text{--}10\ M_\odot$) orbiting a supermassive black hole ($M \sim 10^6\text{--}10^{10}\ M_\odot$). Since the system emits gravitational waves, the stellar-mass compact object slowly inspirals into the supermassive black hole until the final plunge. A similar system can easily form by multi-body interactions in galactic centers [62, 63]. Initially, the captured object is in a "generic" orbit, namely the orbit will have a high eccentricity and any inclination angle with respect to the black hole spin. Due to the emission of gravitational waves, the eccentricity tends to decreases (if we are far from the last stable orbit), while the inclination of the orbit remains approximately constant [24, 27].

EMRIs are an important class of detection candidates for future space-based gravitational wave interferometers. Detection will be dominated by EMRIs in which the stellar-mass compact object is a black hole rather than a neutron star or a white dwarf. There are two reasons. First, heaviest bodies more easily tend to concentrate at the center [28]. Second, the signal produced by a 10 $M_\odot$ black hole is stronger than the signal from a neutron star or a white dwarf with a mass $\mu \approx 1\ M_\odot$.

In the frequency range 1–100 mHz, it is possible to detect the last few years of inspiral into a supermassive black hole of $\sim 10^6\ M_\odot$. If the supermassive object is heavier or the inspiral is at an earlier stage, the emission of gravitational waves is at lower frequencies.

Since $\mu/M < 10^{-5}$, the evolution of the system is adiabatic; that is, the orbital parameters evolve on a timescale much longer than the orbital period of the stellar-mass compact object. A rough estimate can be obtained as follows [25]. If we are far from the last stable orbit, the Keplerian period is

$$T_K \sim 8\left(1 - e^2\right)^{-3/2} \left(\frac{M}{10^6\ M_\odot}\right) \left(\frac{p}{6M}\right)^{3/2} \text{min}, \qquad (11.63)$$

where p is the semi-latus rectum of the orbit and e is the eccentricity. The timescale of the radiation backreaction can be estimated as $T_R \sim -p/\dot{p}$ and we find

$$T_R \sim 100\left(1 - e^2\right)^{-3/2} \left(\frac{M}{\mu}\right) \left(\frac{M}{10^6\ M_\odot}\right) \left(\frac{p}{6M}\right)^4 \text{min}. \qquad (11.64)$$

As long as $M/\mu \gg 1$, we have $T_R \gg T_K$ and the evolution of the system is adiabatic. This simplifies the description, as we can neglect the radiation backreaction and assume that the small object follows the geodesics of the spacetime. EMRIs are thus relatively simple systems and offer a unique opportunity to map the metric around supermassive black holes [9, 26, 60]. Since the value of μ/M is so low, the inspiral process is slow, and it is possible to observe the signal for many ($> 10^5$) cycles. In such a case, the signal-to-noise ratio can be high and it is possible to accurately measure the parameters of the system. In particular, EMRI detections promise to provide unprecedented accurate measurements of the mass and the spin of supermassive black holes [8].

Despite that EMRIs are relatively simple systems, the calculation of inspiral waveforms is not easy, and indeed there is currently no method to compute accurate EMRI waveforms. In the past 15–20 years, there have been significant efforts to develop techniques to evaluate these waveforms. At the moment, the most reliable waveforms are the so-called Teukolsky-based waveforms, which will be briefly reviewed in the next subsection. However, this approach has been only applied to some special classes of orbits. Moreover, it is computationally time-consuming. Since the future detection of EMRIs will be based on matched-filtering and will require a huge number of waveform templates, there have been efforts to develop techniques capable of constructing approximate families of waveforms that can be generated quickly in large numbers. Section 11.5.2 briefly discusses the so-called kludge approach. More details on the calculation techniques for EMRIs can be found, e.g., in [25] and reference therein.

11.5.1 Teukolsky-Based Waveforms

Teukolsky-based waveforms are obtained from the Teukolsky black hole perturbation theory [66]; see [19] for an extended discussion of this topic and the review articles [46] or [25] for its application to EMRIs. The calculations require that the spacetime is described by the Kerr metric (i.e. the supermassive object must be a Kerr black hole) and follow three steps:

1. The orbit is parametrized by its constants of motion (specific energy E, specific axial component of the angular momentum L_z, and Carter constant $\mathcal{Q}$) or by other three orbital parameters (e.g. semi-latus rectum p, eccentricity e, and some inclination angle i). Only stable orbits are relevant.
2. The stellar-mass object is usually treated as a test-particle. In such a case, we solve the geodesic equations to find its trajectory, say $z^\mu = \{t(\tau), r(\tau), \theta(\tau), \phi(\tau)\}$, where τ is the affine parameter of the geodesic. Geodesic motion in the Kerr background has been already discussed in Sect. 3.3 and the equations of motion are Eqs. (3.52)–(3.55). If we want to take the spin of the stellar-mass object into account, the calculations are more complicated and we have deviations from geodesic motion.
3. We employ the Teukolsky formalism to study the evolution of the gravitational perturbations generated by the geodesic motion of the stellar-mass compact object. Eventually we obtain the metric perturbations h_+ and $h_\times$ far from the source, where we have the detector.

As the calculation of the emission of gravitational waves is based on the Teukoslky formalism, the approach works for the Kerr metric and can be potentially extended to other Petrov type D spacetimes. It cannot be employed to study more general spacetimes because of the absence of the counterpart of the Teukolsky equation. Without the Teukolsky equation, we should solve ten coupled metric perturbation equations, which is definitively very challenging.

In the Teukolsky formalism, one studies the perturbations of the Weyl scalars, and eventually it is possible to recover the metric perturbations far from the source. The Teukolsky formalism is based on the Newman-Penrose formalism [50]. The details can be found, e.g., in [19]. We introduce four null 4-vectors $\{l^\mu, n^\mu, m^\mu, m^{*\mu}\}$, where

l^μ and n^μ are real, while $m^{*\mu}$ is the complex conjugate of m^μ. In the Kerr metric in Boyer-Lindquist coordinates, a possible choice is

$$l^\mu = \left(\frac{r^2 + a^2}{\Delta}, 1, 0, \frac{a}{\Delta}\right), \quad n^\mu = \left(\frac{r^2 + a^2}{2\Sigma}, -\frac{\Delta}{2\Sigma}, 0, \frac{a}{2\Sigma}\right),$$

$$m^\mu = \frac{1}{\sqrt{2}\,(r + ia\cos\theta)}\left(ia\sin\theta, 0, 1, \frac{i}{\sin\theta}\right). \tag{11.65}$$

These four 4-vectors form a tetrad and any tensor can be projected onto such a tetrad. The two components of the Weyl tensor relevant for the evolution of metric perturbations are

$$\psi_0 = -C_{\mu\nu\rho\sigma}\,l^\mu m^\nu l^\rho m^\sigma, \quad \psi_4 = -C_{\mu\nu\rho\sigma}\,n^\mu m^{*\nu} n^\rho m^{*\sigma}, \tag{11.66}$$

where $C_{\mu\nu\rho\sigma}$ is the Weyl tensor. The non-vanishing spin coefficients (which will be used later) are

$$\rho = -\frac{1}{r - ia\cos\theta}, \quad \beta = -\frac{\rho^*\cot\theta}{2\sqrt{2}}, \quad \pi = \frac{ia\rho^2\sin\theta}{\sqrt{2}}, \quad \sigma = -\frac{ia\rho\rho^*\sin\theta}{\sqrt{2}},$$

$$\nu = \frac{\rho^2\rho^*\Delta}{2}, \quad \gamma = \nu + \frac{\rho\rho^*(r - M)}{2}, \quad \alpha = \pi - \beta^*. \tag{11.67}$$

The *Teukolsky equation* is [66]

$$\left[\frac{(r^2 + a^2)^2}{\Delta} - a^2\sin^2\theta\right]\frac{\partial^2\psi}{\partial t^2} + \frac{4Mar}{\Delta}\frac{\partial^2\psi}{\partial t\partial\phi} + \left(\frac{a^2}{\Delta} - \frac{1}{\sin^2\theta}\right)\frac{\partial^2\psi}{\partial\phi^2}$$

$$-\Delta^{-s}\frac{\partial}{\partial r}\left(\Delta^{s+1}\frac{\partial\psi}{\partial r}\right) - \frac{1}{\sin\theta}\frac{\partial}{\partial\theta}\left(\sin\theta\frac{\partial\psi}{\partial\theta}\right) - 2s\left[\frac{a(r - M)}{\Delta} + \frac{i\cos\theta}{\sin^2\theta}\right]\frac{\partial\psi}{\partial\phi}$$

$$-2s\left[\frac{M(r^2 - a^2)}{\Delta} - r - ia\cos\theta\right]\frac{\partial\psi}{\partial t} + \left(s^2\cot^2\theta - s\right)\psi = 4\pi\,\Sigma\mathscr{S}. \tag{11.68}$$

Here s is the *spin weight* of the field and ψ is a certain quantity that depends on the field. The Teukolsky equation can indeed be used to study the evolution of small perturbations of a scalar field ($s = 0$), of a massless fermionic field ($s = \pm 1/2$), of an electromagnetic field ($s = \pm 1$), as well as for gravitational perturbations ($s = \pm 2$). For $s = 2$, $\psi = \psi_0$, but this case is not relevant for the outgoing gravitational waves to detect far from the source, and therefore it will be ignored in what follows. For $s = -2$, ψ is given by

$$\psi = \rho^{-4}\psi_4. \tag{11.69}$$

In vacuum $\psi_4 = 0$, and in the Teukolsky equation we consider perturbations around this background. For $s = -2$, the source term $\mathscr{S}$ on the right hand side of Eq. (11.68) is $\mathscr{S} = 2\rho^{-4}T_4$, where

$$
\begin{aligned}
T_4 &= \left(\Delta + 3\gamma - \gamma^* + 4\nu + \nu^*\right)\Big[\left(\delta^* - 2\sigma^* + 2\alpha\right)\left(T_{\mu\nu}n^\mu m^{*\,\nu}\right) \\
&\qquad\qquad\qquad\qquad - \left(\Delta + 2\gamma - 2\gamma^* + \nu^*\right)\left(T_{\mu\nu}m^{*\,\mu}m^{*\,\nu}\right)\Big] \\
&\quad + \left(\delta^* - \sigma^* + \beta^* + 3\alpha + 4\pi\right)\Big[\left(\Delta + 2\gamma + 2\nu^*\right)\left(T_{\mu\nu}n^\mu m^{*\,\nu}\right) \\
&\qquad\qquad\qquad\qquad - \left(\delta^* - \sigma^* + 2\beta^* + 2\alpha\right)\left(T_{\mu\nu}n^\mu n^\nu\right)\Big].
\end{aligned}
\tag{11.70}
$$

The matter energy-momentum tensor $T^{\mu\nu}$ is that of the stellar-mass compact object. We can employ the energy-momentum tensor of a point-like particle

$$
T^{\mu\nu} = \frac{\mu}{\Sigma \sin\theta}\frac{u^\mu u^\nu}{u^0}\,\delta[r - r(\tau)]\,\delta[\theta - \theta(\tau)]\,\delta[\phi - \phi(\tau)],
\tag{11.71}
$$

where $r = r(\tau)$, $\theta = \theta(\tau)$, and $\phi = \phi(\tau)$ have been obtained at the step 2. The Teukolsky equation is separable in the frequency domain. We write

$$
\psi_4 = \rho^4 \sum_{lm}\int d\omega\, R_{lm}\, S_{lm}^{a\omega}\, e^{-i\omega t + i m\phi},
\tag{11.72}
$$

$$
4\pi\,\Sigma\,\mathcal{S} = \sum_{lm}\int d\omega\, T_{lm}\, S_{lm}^{a\omega}\, e^{-i\omega t + i m\phi},
\tag{11.73}
$$

where $R_{lm} = R_{lm}(r)$, $T_{lm} = T_{lm}(r)$, and $S_{lm}^{a\omega} = S_{lm}^{a\omega}(\theta)$. The radial Teukolsky equation for R_{lm} is

$$
\Delta^2\frac{d}{dr}\left(\frac{1}{\Delta}\frac{dR_{lm}}{dr}\right) + \left[\frac{K^2 + 4i\,(r - M)\,K}{\Delta} - 8i\omega r - \lambda\right]R_{lm} = T_{lm},
\tag{11.74}
$$

where $K = \left(r^2 + a^2\right)\omega - ma$, $\lambda = E_{lm} + a^2\omega^2 - 2am\omega$, and E_{lm} is a separation constant [66]. The angular Teukolsky equation for $S_{lm}^{a\omega}$ is

$$
\begin{aligned}
\Bigg[\frac{1}{\sin\theta}\frac{d}{d\theta}&\left(\sin\theta\frac{d}{d\theta}\right) + a^2\omega^2\cos^2\theta - \frac{m^2}{\sin^2\theta} \\
&+ 4a\omega\cos\theta + \frac{4m\cos\theta}{\sin^2\theta} - 4\cot^2\theta - 2 + E_{lm}\Bigg]S_{lm}^{a\omega} = 0.
\end{aligned}
\tag{11.75}
$$

For $r \to \infty$, the relation between the Weyl scalar ψ_4 and the two metric perturbations h_+ and $h_\times$ in the TT gauge is [66]

$$
\psi_4 = \frac{1}{2}\left(\frac{\partial^2 h_+}{\partial t^2} - i\frac{\partial^2 h_\times}{\partial t^2}\right).
\tag{11.76}
$$

Inverting this equation we can thus obtain h_+ and $h_\times$. Inspiral trajectories may be obtained by pasting together a sequence of geodesic orbits, see e.g. [30]. This is possible because the system evolves adiabatically. Plugging the resulting slowly evolving trajectory $z^\mu(\tau)$ into the Teukolsky equation, one finds the typical chirping character of the waveform.

The Teukolsky equation is traditionally solved in the frequency domain, because the equation becomes fully separable and the problem is reduced to the integration of Eqs. (11.74) and (11.75). However, when the orbit significantly deviates from an equatorial circular one, the resulting waveform is the sum of many orbital frequency harmonics, which makes it computationally time-consuming. This is an important issue, because initially the orbit is likely with high eccentricity ($e \approx 1$). In such a case, it may be computationally more convenient to solve the Teukolsky equation in the time domain. Studies on the Teukolsky equation in the time domain started in [33, 43].

11.5.2 Kludge Waveforms

While they are based on several approximations, Teukolsky-based waveforms are currently the most reliable waveforms on the market. However, they are computationally time-consuming to generate and, at the same time, matched filtering techniques require a huge number of waveform templates. This has led to the search for approximate families of waveforms that are capable of capturing the main features of the waveform and that are easy to generate quickly in large numbers. Kludge waveforms are an example of such approximate families of waveforms.

The numerical kludge waveforms of [5] are calculated in the following three steps:

1. We calculate an inspiral trajectory in terms, for instance, of (p, e, i) as a function of time [27]. Let us note that this is not a geodesic trajectory because we are taking into account the losses of energy and angular momentum.
2. We calculate the trajectory of the stellar-mass object in terms of the Boyer-Lindquist coordinates. The result is an expression like $z^\mu = \{t, r(t), \theta(t), \phi(t)\}$, where the trajectory is parametrized by the time t.
3. The gravitational waveform is calculated from the inspiral trajectory by employing some approximate formula for the generation of gravitational waves.

For the step 3, one plugs the expression of z^μ into the energy-momentum tensor of a point-like particle, i.e.

$$
\begin{aligned}
T^{\mu\nu}(t, \mathbf{x}) &= \mu \int \frac{dz^\mu}{d\tau} \frac{dz^\nu}{d\tau} \delta^4[x - z(\tau)]\, d\tau \\
&= \mu \frac{1}{u^0} \frac{dz^\mu}{d\tau} \frac{dz^\nu}{d\tau} \delta^3[\mathbf{x} - \mathbf{z}(t)].
\end{aligned}
\tag{11.77}
$$

We can then plug Eq. (11.77) into (11.14). In the slow motion limit, we can consider the quadrupole formula. Otherwise, we can consider additional terms in the expansion. For example, the quadrupole-octuple formula reads [55]

$$h^{jk} = \frac{2}{r} \left(\ddot{Q}^{jk} - 2n_i \dddot{S}^{ijk} + n_i \ddddot{M}^{ijk} \right)_{t'=t-r},$$ (11.78)

where $\mathbf{n} = \mathbf{x}/r$ and

$$S^{ijk}(t) = \int_V T^{0i}(t, \mathbf{x}) \, x^j x^k \, d^3\mathbf{x},$$

$$M^{ijk}(t) = \int_V T^{00}(t, \mathbf{x}) \, x^i x^j x^k \, d^3\mathbf{x}.$$ (11.79)

Unlike the Teukolsky approach of the previous subsection, the calculations for kludge waveforms can be more naturally extended to generic stationary and axisymmetric spacetimes [9, 26].

11.6   Quasi-normal Modes

If a stable compact object is perturbed, it returns to its equilibrium configuration through some characteristic damped oscillations, the so-called *quasi-normal modes* (QNMs). This is not a special property of black holes, and it is expected even in neutron stars or other possible strong gravity systems like boson stars, wormholes, etc. However, the frequency and the damping time of these oscillations do depend on the specific system and on its fundamental properties, while they are independent of the initial perturbations. For instance, in the case of a Schwarzschild black hole, the spectrum of the quasi-normal modes from metric perturbations are only determined by the black hole mass M. For a Kerr black hole, the spectrum depends on the mass M and the spin parameter a_*. The quasi-normal mode spectrum is thus a powerful tool to test the nature of a compact object. Moreover, quasi-normal modes are not limited to the case of metric perturbations, but they exist even for matter perturbations in the spacetime around a compact object.

The spectrum of quasi-normal modes of a stable black hole is an infinite set of complex angular frequencies. The real part is proportional to the oscillation frequency of the mode. The imaginary part is inversely proportional to the damping time of the mode. Linear stability is guaranteed when all quasi-normal modes are damped. Depending on the astrophysical scenario, some modes can receive more energy than others and dominate the ringdown stage.

The name "quasi-normal modes" was put forward in [54] to distinguish these oscillations from the normal modes present in other physical systems. Normal modes are stationary oscillations. Here, on the contrary, the system is open and loses energy by emitting some kind of radiation. For this reason, quasi-normal modes of a stable system decay with time. This is a consequence of the different boundary conditions or, equivalently, of the different behavior of the potential at the boundaries. Normal modes are confined to a finite region, while quasi-normal modes are not. There are also some other important differences. For example, it is remarkable that quasi-normal modes of black holes in asymptotically flat spacetimes do not form a complete set.

In the past decades, there has been a significant work to develop numerical and semi-analytical methods to compute the spectrum of quasi-normal modes. More details can be found in [11, 35, 36] and references therein.

Let us first consider matter perturbations in a certain black hole spacetime. Such a possibility may not have direct astrophysical implications, but it is easier to study. In the linear approximation, the matter field does not backreact on the equilibrium background metric, and the evolution of the matter perturbations turns out to be determined by the equation of motion of the matter field under consideration in the equilibrium background metric. This is indeed the way in which one obtains the equation of motion of a matter field from the Least Action Principle.

The simplest example is a scalar field Φ of mass μ. In flat spacetime, the equation of motion of Φ is the usual Klein-Gordon equation $(\Box + \mu^2)\Phi = 0$, where $\Box = \eta^{\mu\nu}\partial_\mu\partial_\nu$. In curved spacetime, the equation of motion of Φ is still given by the Klein-Gordon equation, but now $\Box = g^{\mu\nu}\nabla_\mu\nabla_\nu$, which can be rewritten as

$$\frac{1}{\sqrt{-g}}\frac{\partial}{\partial x^\mu}\left(g^{\mu\nu}\sqrt{-g}\frac{\partial\Phi}{\partial x^\mu}\right) + \mu^2\Phi = 0. \qquad (11.80)$$

In general relativity, the spectrum of the quasi-normal modes of a massless scalar field is not very different from that of the metric, so as a crude approximation one may calculate the former to get an estimate of the latter. In some alternative theories of gravity, such a statement is also true, while in other theories the two spectra may be very different.

The second natural example is the case of perturbations of the electromagnetic field A_μ. In flat spacetime, we have the Maxwell equation $\partial^\mu F_{\mu\nu} = 0$, where $F_{\mu\nu} = \partial_\mu A_\nu - \partial_\nu A_\mu$. In curved spacetime, the Maxwell equation can be rewritten as

$$\frac{\partial}{\partial x^\mu}\left(\sqrt{-g}F^{\mu\nu}\right) = 0. \qquad (11.81)$$

In the same way, one can consider other matter fields, like Dirac fields, massive vector fields, etc. In all these cases, the evolution of perturbations is governed by the equation of motion of the matter field in the curved background metric of the black hole. The evolution of the perturbations is thus only determined by the background metric and is independent of the field equation of the gravity theory. In other words, the spectrum of the quasi-normal modes of, for instance, a scalar field in the Kerr metric is the same in general relativity as in an alternative theory of gravity in which the Kerr metric is a viable black hole solution.

Metric perturbations can be generated in the spacetime around a black hole at the time of its formation, and the study of their evolution has direct astrophysical implications. The new black hole may be created by non-spherically symmetric collapse of the progenitor star or by the merger of two black holes, two neutron stars, or a black hole-neutron star system. At the time of its formation, the black hole is not in its equilibrium configuration. The system thus emits gravitational waves to settle down to its equilibrium configuration. Metric perturbations also appear when a black hole swallows a smaller object. For example, this is the case of an EMRI.

The emission of gravitational waves of a newly born, stable black hole can be divided into three stages. First, we have an outburst, which lasts for a relatively short time. The emission of gravitational waves is determined by the non-linear regime of the gravity theory and the initial perturbations. Second, we have the period dominated by the quasi-normal modes. The frequency and the damping time of each mode only depend on the fundamental properties of the system, and are independent of the initial perturbations, but the energy in each mode is determined by the initial perturbations. Last, at late time quasi-normal modes are usually suppressed by the so-called exponential late-time tails [58], which are also a characteristic of the system under consideration. Ringdown typically refers to the second stage of quasi-normal modes, but may also include the third stage of exponential late-time tails.

The gravitational waves of the outburst would be interesting, but difficult to predict theoretically and to measure with the necessary precision. The exponential late-time tails are difficult to detect, because the signal is very weak. From the observational point of view, the quasi-normal mode stage is the most interesting one.

The evolution of metric perturbations does depend on the exact field equations of the gravity theory. In 4-dimensional general relativity, we have to consider metric perturbations in the Einstein equations, namely

$$\delta R_{\mu\nu} = 8\pi \delta \left(T_{\mu\nu} - \frac{1}{2} g_{\mu\nu} T \right), \tag{11.82}$$

where $T = T_\rho^\rho$. We write the metric as $g_{\mu\nu} = g_{\mu\nu}^0 + \delta g_{\mu\nu}$, where $g_{\mu\nu}^0$ is the equilibrium background metric and $\delta g_{\mu\nu}$ is the perturbation. In the linear approximation, we plug this expression of $g_{\mu\nu}$ into Eq. (11.82) and we neglect terms of the second and higher order in $\delta g_{\mu\nu}$.

11.6.1 Calculation Methods of Quasi-normal Modes

The equations governing the evolution of matter field perturbations are relatively straightforward to obtain if we know the black hole metric. In the case of the equations for the evolution of metric perturbations, they depend on the gravity theory, but usually there are no special difficulties to face. A different story is solving these equations to find the spectrum of the perturbations. The standard strategy is to try to separate the angular variables θ and ϕ from the radial and temporal variables r and t. This is only possible when the background metric has some special symmetries and, even in this case, it may not be an easy job. The choice of suitable coordinates is also crucial.

In the case of a scalar field Φ, we can expands Φ in terms of scalar harmonics

$$\Phi\,(t,r,\theta,\phi) = \sum_{l=0}^{+\infty} \sum_{m=-l}^{l} Y_{lm}\,(\theta,\phi)\,\frac{\Phi_{lm}\,(t,r)}{r}\,. \tag{11.83}$$

In the case of vector or tensor fields, the formalism is more complicated. Any second-rank symmetric tensor, say $S_{\mu\nu}$, can be expanded in *tensor harmonics* as follows [59, 73]

$$S_{\mu\nu} = \sum_{l=0}^{+\infty} \sum_{m=-l}^{l} \sum_{i=1}^{10} s_{lm}^{i}\left(A_{lm}^{i}\right)_{\mu\nu}\,. \tag{11.84}$$

The expression of the tensor harmonics $i=1,2$, and 3 is

$$\left\|\left(A_{lm}^{1}\right)_{\mu\nu}\right\| = \begin{pmatrix} 1&0&0&0 \\ 0&0&0&0 \\ 0&0&0&0 \\ 0&0&0&0 \end{pmatrix} Y_{lm}\,, \quad \left\|\left(A_{lm}^{2}\right)_{\mu\nu}\right\| = \frac{i}{\sqrt{2}} \begin{pmatrix} 0&1&0&0 \\ 1&0&0&0 \\ 0&0&0&0 \\ 0&0&0&0 \end{pmatrix} Y_{lm}\,,$$

$$\left\|\left(A_{lm}^{3}\right)_{\mu\nu}\right\| = \begin{pmatrix} 0&0&0&0 \\ 0&1&0&0 \\ 0&0&0&0 \\ 0&0&0&0 \end{pmatrix} Y_{lm}\,. \tag{11.85}$$

These tensors have odd parity, namely they change by the factor $(-1)^{l+1}$ under space inversion $(\theta,\phi) \rightarrow (\pi-\theta,\pi+\phi)$. They are usually referred to as *axial* perturbations in the literature. The tensor harmonics $i=4,5...,$ and 10 have even parity, namely they change by the factor $(-1)^{l}$ under space inversion. They are usually referred to as *polar* perturbations. Their expression is

$$\left\|\left(A_{lm}^{4}\right)_{\mu\nu}\right\| = \frac{ir}{\sqrt{2l\,(l+1)}} \begin{pmatrix} 0&0&\partial_{\theta}&\partial_{\phi} \\ 0&0&0&0 \\ \partial_{\theta}&0&0&0 \\ \partial_{\phi}&0&0&0 \end{pmatrix} Y_{lm}\,,$$

$$\left\|\left(A_{lm}^{5}\right)_{\mu\nu}\right\| = \frac{r}{\sqrt{2l\,(l+1)}} \begin{pmatrix} 0&0&0&0 \\ 0&0&\partial_{\theta}&\partial_{\phi} \\ 0&\partial_{\theta}&0&0 \\ 0&\partial_{\phi}&0&0 \end{pmatrix} Y_{lm}\,,$$

$$\| \left(A^6_{lm} \right)_{\mu\nu} \| = \frac{r}{\sqrt{2l\,(l+1)}} \begin{pmatrix} 0 & 0 & \frac{1}{\sin\theta}\partial_\phi & -\sin\theta\,\partial_\theta \\ 0 & 0 & 0 & 0 \\ \frac{1}{\sin\theta}\partial_\phi & 0 & 0 & 0 \\ -\sin\theta\,\partial_\theta & 0 & 0 & 0 \end{pmatrix} Y_{lm},$$

$$\| \left(A^7_{lm} \right)_{\mu\nu} \| = \frac{ir}{\sqrt{2l\,(l+1)}} \begin{pmatrix} 0 & 0 & 0 & 0 \\ 0 & 0 & \frac{1}{\sin\theta}\partial_\phi & -\sin\theta\,\partial_\theta \\ 0 & \frac{1}{\sin\theta}\partial_\phi & 0 & 0 \\ 0 & -\sin\theta\,\partial_\theta & 0 & 0 \end{pmatrix} Y_{lm},$$

$$\| \left(A^8_{lm} \right)_{\mu\nu} \| = -\frac{ir^2}{\sqrt{2\,(l-1)\,l\,(l+1)\,(l+2)}} \begin{pmatrix} 0 & 0 & 0 & 0 \\ 0 & 0 & 0 & 0 \\ 0 & 0 & -\frac{1}{\sin\theta}X_{lm} & \sin\theta\,W_{lm} \\ 0 & 0 & \sin\theta\,W_{lm} & \sin\theta\,X_{lm} \end{pmatrix},$$

$$\| \left(A^9_{lm} \right)_{\mu\nu} \| = \frac{r^2}{\sqrt{2}} \begin{pmatrix} 0 & 0 & 0 & 0 \\ 0 & 0 & 0 & 0 \\ 0 & 0 & 1 & 0 \\ 0 & 0 & 0 & \sin^2\theta \end{pmatrix} Y_{lm},$$

$$\| \left(A^{10}_{lm} \right)_{\mu\nu} \| = \frac{r^2}{\sqrt{2\,(l-1)\,l\,(l+1)\,(l+2)}} \begin{pmatrix} 0 & 0 & 0 & 0 \\ 0 & 0 & 0 & 0 \\ 0 & 0 & W_{lm} & X_{lm} \\ 0 & 0 & X_{lm} & \sin^2\theta\,W_{lm} \end{pmatrix}, \quad (11.86)$$

where

$$X_{lm} = 2\frac{\partial}{\partial\phi}\left(\frac{\partial}{\partial\theta} - \cot\theta \right) Y_{lm},$$

$$W_{lm} = \left(\frac{\partial^2}{\partial\theta^2} - \cot\theta\frac{\partial}{\partial\theta} - \frac{1}{\sin^2\theta}\frac{\partial^2}{\partial\phi^2} \right) Y_{lm}. \quad (11.87)$$

Let us note that different authors often use different notations and different normalizations.

In the case of a spherically symmetric spacetime, the equation governing the evolution of the metric perturbations does not mix axial and polar perturbations, and we can thus study axial and polar perturbations separately. If the compact object is rotating – even slowly – the two types of perturbations couple. In the case of rotating black holes, it is not convenient to expand the perturbations in tensor harmonics and other techniques have been developed (see, for instance, the case of the Kerr metric in Sect. 11.6.3).

The tensor harmonics are orthonormal each other with respect to the inner product

$$(B, C) = \int_0^\pi d\theta \int_0^{2\pi} d\phi \, \sin\theta\,\eta^{\mu\nu}\eta^{\rho\sigma} B^*_{\mu\rho}C_{\nu\sigma}, \quad (11.88)$$

where $\eta_{\mu\nu}$ is the Minkowski metric and $*$ indicates the complex conjugate. The scalars $\{s_{lm}^i\}$ in Eq. (11.84) can thus be evaluated from

$$s_{lm}^i = \left(A_{lm}^i, S\right) . \tag{11.89}$$

If we can decouple the angular variables in the perturbation equation, we usually obtain a Schrödinger-like equation for the radial and temporal variables, say

$$-\frac{d^2 R}{dx^2} + V(r, \omega)R = \omega^2 R . \tag{11.90}$$

$V(r, \omega)$ is an effective potential that depends on the initial equation of motion, the background metric, and the perturbed field. The coordinate x is not the radial coordinate r. For instance, in the Schwarzschild metric x can be the *tortoise coordinate* r_*

$$r_* = r + 2M \ln \left(\frac{r}{2M} - 1\right) , \tag{11.91}$$

For $r \to 2M$, $r_* \to -\infty$, while for $r \to \infty$ we have $r_* \to \infty$.

Quasi-normal modes satisfy specific boundary conditions. In the case of an asymptotically flat black hole spacetime and in the context of astrophysical observations, it is natural to impose pure ingoing wave at the horizon and pure outgoing wave at spatial infinity; that is,

$$\begin{aligned}\Psi \sim e^{i\omega x} \quad &\text{for } x \to -\infty \quad \text{(pure ingoing wave)} ,\\ \Psi \sim e^{-i\omega x} \quad &\text{for } x \to \infty \quad \text{(pure outgoing wave)} .\end{aligned} \tag{11.92}$$

Different boundary conditions may be imposed with special potentials or in different contexts.

It is common to write the frequency of the quasi-normal mode ω as

$$\omega = \omega_R - i\omega_I . \tag{11.93}$$

$\omega_R = 2\pi\nu$ and ν is the oscillation frequency of the mode. $\omega_I = 1/\tau$ and τ is the damping time of the mode. The spacetime is linearly stable under the perturbations if all quasi-normal modes have positive imaginary part ($\omega_I > 0$). The spacetime is unstable if at least one of the modes has $\omega_I < 0$.

In general, it is not easy to solve the equation of the perturbations and find the spectrum of the quasi-normal modes. Exact solutions are rare and it is usually necessary to employ specific calculation techniques to find approximate solutions; see, for instance, [36] and references therein.

As a consequence of the incompleteness of the set of quasi-normal modes, at late time quasi-normal modes are suppressed by exponential or power-law tails; see, however, [29] for a counterexample. Late-time tails were first studied in [58]. It is not clear whether this behavior may have astrophysical implications, because the signal is very weak at late time, but it is still an interesting feature of the quasi-normal modes. For instance, massless scalar perturbations and metric perturbations in the Kerr spacetime decay as

$$|\Psi| \sim t^{-(2l+3)} . \tag{11.94}$$

In other background metrics and/or for different matter field perturbations, the decay may be different. For example, in the case of a massive scalar field of mass μ, the decay at late time $(t \gg \mu^{-2}M^{-3})$ in Schwarzschild and Reissner-Nordström spacetimes is [37–39]

$$|\Psi| \sim t^{-5/6} \sin(\mu t) . \tag{11.95}$$

11.6.2 Schwarzschild Metric

For the study of metric perturbations in the Schwarzschild spacetime, we can proceed as discussed in the previous subsection. For axial perturbation, we obtain the Schrödinger-like equation (11.90) with the *Regge–Wheeler potential* [59][2]

$$V_{\text{RW}} = \left(1 - \frac{2M}{r}\right)\left[\frac{l(l+1)}{r^2} - \frac{6M}{r^3}\right] . \tag{11.96}$$

The potential for scalar, electromagnetic, and axial perturbations in the Schwarzschild background can be written in the compact form

$$V = \left(1 - \frac{2M}{r}\right)\left[\frac{l(l+1)}{r^2} - (1-s^2)\frac{2M}{r^3}\right] , \tag{11.97}$$

where $s = 0$ for scalar field perturbations, ± 1 for electromagnetic field perturbations, and ± 2 for axial metric perturbations. For polar perturbations, we find the *Zerilli potential* [73]

$$V_Z = \left(1 - \frac{2M}{r}\right)\left[\frac{72M^3}{r^5L^2} - \frac{12M}{r^3L^2}(l-1)(l+2)\left(1 - \frac{3M}{r}\right)\right.$$
$$\left. + \frac{(l-1)l(l+1)(l+2)}{r^2L}\right] , \tag{11.98}$$

where

$$L = l(l+1) - 2 + \frac{6M}{r} . \tag{11.99}$$

---

[2]Regge and Wheeler [59] also considered polar perturbations, but there was a mistake later corrected in [73].

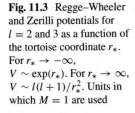

Fig. 11.3 Regge–Wheeler and Zerilli potentials for $l = 2$ and 3 as a function of the tortoise coordinate r_*. For $r_* \to -\infty$, $V \sim \exp(r_*)$. For $r_* \to \infty$, $V \sim l(l+1)/r_*^2$. Units in which $M = 1$ are used

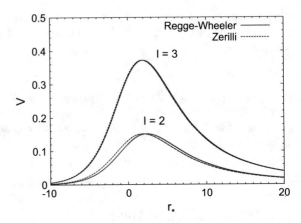

Fig. 11.4 The $l = 2$ (*diamonds*) and the $l = 3$ (*crosses*) spectra of quasi-normal modes for a Schwarzschild black hole. The 9th $l = 2$ mode and the 41st $l = 3$ mode have vanishing ω_R. From [35] following the calculations in [3] under the terms of the Creative Commons Attribution License

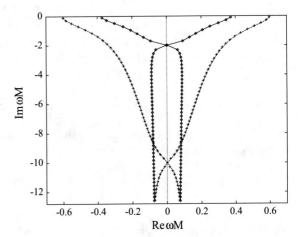

Figure 11.3 shows the Regge–Wheeler and the Zerilli potentials for $l = 2$ and 3 as a function of the tortoise coordinate r_*. Both potentials have a maximum near the photon orbit and tend to 0 when we move to the horizon or to infinity.

The quasi-normal mode spectrum in the Schwarzschild background has been studied by many authors; see, e.g., [6]. For every l ($l \geq 2$, because of the quadrupole nature of gravitational radiation), we find an infinite set of quasi-normal frequencies. For every l, the quasi-normal frequencies are labelled by the non-negative integer number n, where $n = 0$ corresponds to the fundamental mode and $n = 1, 2, 3$, etc. are used for the first, second, third, etc. modes. Figure 11.4 shows the $l = 2$ and the $l = 3$ quasi-normal mode spectra of a Schwarzschild black hole. As the number n increases, the imaginary part increases too, which means the damping time gets shorter and the mode is less and less important in a possible detection of the ringdown of a black hole. Table 11.1 reports the first four frequencies (in units in which $M = 1$) for $l = 2, 3$, and 4. The conversion factor from ω_R in Table 11.1 to ν in kHz is

Table 11.1 The first four quasi-normal frequencies $\omega = \omega_R - i\omega_I$ of a Schwarzschild black hole for $l = 2, 3$, and 4. Units in which $M = 1$ are used

n	$l = 2$	$l = 3$	$l = 4$
0	$0.37367 - 0.08896i$	$0.59944 - 0.09270i$	$0.80918 - 0.09416i$
1	$0.34671 - 0.27391i$	$0.58264 - 0.28130i$	$0.79663 - 0.28443i$
2	$0.30105 - 0.47828i$	$0.55168 - 0.47909i$	$0.77271 - 0.47991i$
3	$0.25150 - 0.70514i$	$0.51196 - 0.69034i$	$0.73984 - 0.68392i$

$$3.231 \left(\frac{10 \, M_\odot}{M} \right) \text{kHz} . \tag{11.100}$$

The fundamental mode $l = 2$ and $n = 0$ has the frequency $\nu = \omega_R / (2\pi)$ and the damping time $\tau = 1/\omega_I$

$$\nu = 1.207 \left(\frac{10 \, M_\odot}{M} \right) \text{kHz} , \quad \tau = 0.554 \left(\frac{M}{10 \, M_\odot} \right) \text{ms} . \tag{11.101}$$

Quasi-normal modes from stellar-mass black holes are in the detection band of ground-based interferometers, and they have been already observed in GW150914 and GW151226 by LIGO. Quasi-normal modes from supermassive black holes can instead be detected by space-based interferometers (e.g. eLISA).

11.6.3 Kerr Metric

In the case of the Kerr metric, the approach employed for the Schwarzschild space-time of studying the metric perturbations through an expansion in tensor harmonics does not work. As already shown in Sect. 11.5.1, if we use the Newman-Penrose formalism, we obtain the Teukolsky equation (11.68), which describes the evolution of perturbations of massless fields of various spin and is fully separable in the frequency domain. The equations to solve are (11.74) and (11.75). The quasi-normal mode spectrum has been calculated by several authors employing different techniques, see e.g. [22, 34, 40, 52, 61].

If all quasi-normal modes have $\omega_I > 0$, the background metric is stable under linear perturbations. This is the case of the Kerr spacetime [56, 67, 68].

Superradiance is the phenomenon of amplification of an incident wave and was discovered in [72]. The equation governing the scattering of a massless scalar field in the Kerr background can be written as Eq. (11.90) with the potential

$$V = \left(\omega - \frac{am}{r^2 + a^2} \right)^2 - \frac{C^2 \Delta}{\left(r^2 + a^2 \right)^2} - \frac{\Delta}{\left(r^2 + a^2 \right)^{3/2}} \frac{d}{dr} \frac{r\Delta}{\left(r^2 + a^2 \right)^{3/2}} , \tag{11.102}$$

where C is a separation constant. The asymptotic forms of the solution of the scalar field are

$$
\begin{aligned}
\Phi &= e^{-i\omega x} + \mathscr{R} e^{i\omega x} && \text{for } r \to \infty\,, \\
\Phi &= \mathscr{T} e^{-i(\omega - m\Omega_H)x} && \text{for } r \to r_+\,,
\end{aligned}
\tag{11.103}
$$

where $\mathscr{R}$ is the reflection coefficient, $\mathscr{T}$ is the transmission coefficient, and Ω_H is the angular velocity of the event horizon (see Sect. 3.5). From the Wronskian of the scalar field Φ and employing the asymptotic solutions (11.103), it is not difficult to obtain the following relation (the calculations can be found in many textbooks of quantum mechanics)

$$
|\mathscr{R}|^2 = 1 + \left(\frac{m\Omega_H}{\omega} - 1 \right) |\mathscr{T}|^2\,.
\tag{11.104}
$$

The amplitudes of the incident, reflected, and transmitted waves are, respectively, 1, $|\mathscr{R}|$, and $|\mathscr{T}|$. The incident wave is amplified (i.e. the amplitude of the reflected wave is larger than the amplitude of the incident wave, $|\mathscr{R}| > 1$) for

$$
m\Omega_H > \omega\,.
\tag{11.105}
$$

In superradiance, the incident wave "extracts" rotational energy from the black hole. There is no superradiance in the case of fermions [31, 44], while superraciance of metric perturbations is stronger than the case of a massless scalar field [64, 65].

If the reflected wave cannot escape to infinity (for instance, we put a mirror around the black hole), the system becomes unstable, as the reflected wave becomes a new incident wave, which is amplified again. The process continues and the amplitude of the wave increases. Something similar can happen, for instance, in the case of a massive scalar field. Here the effective potential has a barrier at large radii. Such a barrier can act as a mirror and the reflected wave bounces to become a new incident wave, with the result that the amplitude of the wave increases and increases. The observation of Kerr black holes with certain masses may thus be used to constrain the mass of hypothetical new scalar fields [4].

References

1. B.P. Abbott et al., LIGO Scientific and Virgo Collaborations. Phys. Rev. Lett. **116**, 061102 (2016), arXiv:1602.03837 [gr-qc]
2. B.P. Abbott et al., LIGO Scientific and Virgo Collaborations. Phys. Rev. Lett. **116**, 241103 (2016), arXiv:1606.04855 [gr-qc]
3. N. Andersson, S. Linnæus, Phys. Rev. D **46**, 10, 4179 (1992)
4. A. Arvanitaki, S. Dubovsky, Phys. Rev. D **83**, 044026 (2011), arXiv:1004.3558 [hep-th]
5. S. Babak, H. Fang, J.R. Gair, K. Glampedakis, S.A. Hughes, Phys. Rev. D **75**, 024005 (2007) (Erratum: Phys. Rev. D **77**, 04990 (2008)), arXiv:gr-qc/0607007
6. A. Bachelot, A. Motet-Bachelot, Ann. Inst. H. Poincare Phys. Theor. **59**, 3 (1993)

7. J.G. Baker, J. Centrella, D.I. Choi, M. Koppitz, J. van Meter, Phys. Rev. Lett. **96**, 111102 (2006), arXiv:gr-qc/0511103
8. L. Barack, C. Cutler, Phys. Rev. D **69**, 082005 (2004), arXiv:gr-qc/0310125
9. L. Barack, C. Cutler, Phys. Rev. D **75**, 042003 (2007), arXiv:gr-qc/0612029
10. T.W. Baumgarte, S.L. Shapiro, *Numerical Relativity: Solving Einstein's Equations on the Computer* (Cambridge University Press, Cambridge, 2010)
11. E. Berti, V. Cardoso, A.O. Starinets, Class. Quant. Grav. **26**, 163001 (2009), arXiv:0905.2975 [gr-qc]
12. L. Blanchet, Living Rev. Rel. **17**, 2 (2014), arXiv:1310.1528 [gr-qc]
13. L. Blanchet, T. Damour, Phil. Trans. Roy. Soc. Lond. A **320**, 379 (1986)
14. L. Blanchet, T. Damour, Phys. Rev. D **37**, 1410 (1988)
15. L. Blanchet, T. Damour, Ann. Inst. H. Poincare Phys. Theor. **50**, 377 (1989)
16. L. Blanchet, T. Damour, Phys. Rev. D **46**, 4304 (1992)
17. M. Campanelli, C.O. Lousto, P. Marronetti, Y. Zlochower, Phys. Rev. Lett. **96**, 111101 (2006), arXiv:gr-qc/0511048
18. S.J. Chamberlin, X. Siemens, Phys. Rev. D **85**, 082001 (2012), arXiv:1111.5661 [astro-ph.HE]
19. S. Chandrasekhar, *The Mathematical Theory of Black Holes* (Clarendon Press, Oxford, 1998)
20. C. Cutler, E.E. Flanagan, Phys. Rev. D **49**, 2658 (1994), arXiv:gr-qc/9402014
21. M.E. da Silva Alves, M. Tinto, Phys. Rev. D **83**, 123529 (2011), arXiv:1102.4824 [gr-qc]
22. S.L. Detweiler, Astrophys. J. **239**, 292 (1980)
23. L.S. Finn, D.F. Chernoff, Phys. Rev. D **47**, 2198 (1993), arXiv:gr-qc/9301003
24. J.R. Gair, K. Glampedakis, Phys. Rev. D **73**, 064037 (2006), arXiv:gr-qc/0510129
25. K. Glampedakis, Class. Quant. Grav. **22**, S605 (2005), arXiv:gr-qc/0509024
26. K. Glampedakis, S. Babak, Class. Quant. Grav. **23**, 4167 (2006), arXiv:gr-qc/0510057
27. K. Glampedakis, S.A. Hughes, D. Kennefick, Phys. Rev. D **66**, 064005 (2002), arXiv:gr-qc/0205033
28. C. Hopman, T. Alexander, Astrophys. J. **645**, L133 (2006), arXiv:astro-ph/0603324
29. G.T. Horowitz, V.E. Hubeny, Phys. Rev. D **62**, 024027 (2000), arXiv:hep-th/9909056
30. S.A. Hughes, Phys. Rev. D **64**, 064004 (2001) (Erratum: Phys. Rev. D **88**, 109902 (2013)), arXiv:gr-qc/0104041
31. B.R. Iyer, A. Kumar, Phys. Rev. D **18**, 4799 (1978)
32. P. Jaranowski, A. Krolak, Living Rev. Rel. **8**, 3 (2005) (Living Rev. Rel. **15**, 4 (2012)), arXiv:arXiv:0711.1115 [gr-qc]
33. G. Khanna, Phys. Rev. D **69**, 024016 (2004), arXiv:gr-qc/0309107
34. K.D. Kokkotas, Class. Quant. Grav. **8**, 2217 (1991)
35. K.D. Kokkotas, B.G. Schmidt, Living Rev. Rel. **2**, 2 (1999), arXiv:gr-qc/9909058
36. R.A. Konoplya, A. Zhidenko, Rev. Mod. Phys. **83**, 793 (2011), arXiv:arXiv:1102.4014 [gr-qc]
37. H. Koyama, A. Tomimatsu, Phys. Rev. D **63**, 064032 (2001), arXiv:gr-qc/0012022
38. H. Koyama, A. Tomimatsu, Phys. Rev. D **64**, 044014 (2001), arXiv:gr-qc/0103086
39. H. Koyama, A. Tomimatsu, Phys. Rev. D **65**, 084031 (2002), arXiv:gr-qc/0112075
40. E.W. Leaver, Proc. R. Soc. Lond. A **402**, 285 (1985)
41. K. Lee, F.A. Jenet, R.H. Price, N. Wex, M. Kramer, Astrophys. J. **722**, 1589 (2010), arXiv:1008.2561 [astro-ph.HE]
42. L. Lehner, Class. Quant. Grav. **18**, R25 (2001), arXiv:gr-qc/0106072
43. R. Lopez-Aleman, G. Khanna, J. Pullin, Class. Quant. Grav. **20**, 3259 (2003), arXiv:gr-qc/0303054
44. K. Maeda, Prog. Theor. Phys. **55**, 1677 (1976)
45. M. Maggiore, *Gravitational Waves: Volume 1: Theory and Experiments* (Oxford University Press, Oxford, 2007)
46. Y. Mino, M. Sasaki, M. Shibata, H. Tagoshi, T. Tanaka, Prog. Theor. Phys. Suppl. **128**, 1 (1997), arXiv:gr-qc/9712057
47. S. Mirshekari, C.M. Will, Phys. Rev. D **87**, 084070 (2013), arXiv:arXiv:1301.4680 [gr-qc]
48. C.W. Misner, K.S. Thorne, J.A. Wheeler, *Gravitation* (W.H. Freeman and Company, San Francisco, 1973)

49. C.J. Moore, R.H. Cole, C.P.L. Berry, Class. Quant. Grav. **32**, 015014 (2015), arXiv:1408.0740 [gr-qc]
50. E. Newman and R. Penrose, J. Math. Phys. **3**, 566 (1962)
51. A. Nishizawa, A. Taruya, K. Hayama, S. Kawamura, M.a. Sakagami, Phys. Rev. D **79**, 082002 (2009), arXiv:0903.0528 [astro-ph.CO]
52. H. Onozawa, Phys. Rev. D **55**, 3593 (1997), arXiv:gr-qc/9610048
53. E. Poisson, C.M. Will, Phys. Rev. D **52**, 848 (1995), arXiv:gr-qc/9502040
54. W.H. Press, Astrophys. J. **170**, L105 (1971)
55. W.H. Press, Phys. Rev. D **15**, 965 (1977)
56. W.H. Press, S.A. Teukolsky, Astrophys. J. **185**, 649 (1973)
57. F. Pretorius, Phys. Rev. Lett. **95**, 121101 (2005), arXiv:gr-qc/0507014
58. R.H. Price, Phys. Rev. D **5**, 2419 (1972)
59. T. Regge, J.A. Wheeler, Phys. Rev. **108**, 1063 (1957)
60. F.D. Ryan, Phys. Rev. D **52**, 5707 (1995)
61. E. Seidel, S. Iyer, Phys. Rev. D **41**, 374 (1990)
62. S. Sigurdsson, Class. Quant. Grav. **14**, 1425 (1997), arXiv:astro-ph/9701079
63. S. Sigurdsson, M.J. Rees, Mon. Not. R. Astron. Soc. **284**, 318 (1997), arXiv:astro-ph/9608093
64. A.A. Starobinskij, Sov. Phys. JETP **37**, 28 (1973)
65. A.A. Starobinskij, S.M Churilov, Zh. Eksp. Teor. Fiz. **65**, 3 (1973)
66. S.A. Teukolsky, Astrophys. J. **185**, 635 (1973)
67. S.A. Teukolsky, W.H. Press, Astrophys. J. **193**, 443 (1974)
68. B.F. Whiting, J. Math. Phys. **30**, 1301 (1989)
69. C.M. Will, *Theory and Experiment in Gravitational Physics* (Cambridge University Press, Cambridge, 1993)
70. C.M. Will, Phys. Rev. D **50**, 6058 (1994), arXiv:gr-qc/9406022
71. C.M. Will, Living Rev. Rel. **17**, 4 (2014), arXiv:1403.7377 [gr-qc]
72. Y.B. Zeldovich, Sov. Phys. JETP **35**, 1085 (1972)
73. F.J. Zerilli, Phys. Rev. D **2**, 2141 (1970)

Part III
Testing the Kerr Paradigm

Chapter 12
Non-Kerr Spacetimes

Techniques like the continuum-fitting and the iron line methods were originally proposed to measure black hole spins assuming that the spacetime metric around black holes is described by the Kerr solution. Under this assumption, the electromagnetic spectrum is computed in the Kerr metric and then the theoretical predictions are compared with observational data to find the best fit and estimate the values of a_* and of the other free parameters of the model.

With the same spirit, we can employ some non-Kerr metric, compute the expected spectrum in this non-Kerr spacetime, and compare the new predictions with the available data. We can then see if the non-Kerr model fits the data better than the Kerr one and constrain possible deviations from the Kerr background. Deviations from the Kerr metric may be expected, for instance, from classical extensions of general relativity [4], macroscopic quantum gravity effects [9, 10, 13], or the presence of exotic matter [16, 18].

There are two natural approaches to test the Kerr black hole hypothesis. In the *top-bottom* approach, we have a specific theoretical model in which the spacetime metric around an astrophysical black hole is not described by the Kerr solution. The key-point is that we have a theoretically motivated non-Kerr metric. Some examples will be discussed in Sect. 12.1. In the *bottom-up* approach, we consider instead a phenomenological parameterization of the metric, which ideally should be able to describe the spacetime of any possible black hole (or compact object capable of mimicking a black hole) in any possible gravity theory. A number of *deformation parameters* are used to quantify possible deviations from the Kerr metric, and we want to constrain the values of these deformation parameters. This approach will be reviewed in Sect. 12.2.

12.1 Theoretically-Motivated Spacetimes

If we have a theoretical model that predicts macroscopic deviations from the Kerr solution in the metric around an astrophysical black hole, we can compare the theoretical predictions in the Kerr and non-Kerr models with observations and see which of them can better explain some astrophysical data. The advantages of this approach is that there are some theoretical reasons to study such a scenario and the predictions are (usually) relatively precise.

There are two disadvantages. First, there are a large number of theoretical frameworks predicting deviations from the Kerr solution, and there are currently no reasons to prefer a model over the others. This would require to repeat the analysis for any model, which would be very dispersive. Actually, it may also be possible that, even assuming that the Kerr solution is not adequate to describe the metric around astrophysical black holes, we do not have the right model. Second, it is typically very difficult to find rotating black hole solutions in these models. In many gravity theories we know their non-rotating black hole solutions, in some cases we know the black hole solutions in the slow-rotation approximation, but only in a few cases we have the complete rotating black hole solutions. If we only know the non-rotating black hole solution of a certain theory, we cannot test such a theory, because the astrophysical data may not be consistent with the non-rotating solution but may be consistent when we employ the rotating one.

12.1.1 Kerr Black Holes with Scalar Hair

Kerr black holes with scalar hair were proposed in [16] and represent an example of theoretical model in which deviations from the Kerr metric are possible because of the presence of exotic matter. We have Einstein's gravity minimally coupled to a massive, complex, scalar field. The action is

$$S = \int d^4x \sqrt{-g} \left\{ \frac{R}{16\pi} - \frac{1}{2} g^{\mu\nu} \left[\left(\partial_\mu \Phi^* \right) \left(\partial_\nu \Phi \right) + \left(\partial_\nu \Phi^* \right) \left(\partial_\mu \Phi \right) \right] - \mu^2 \Phi^* \Phi \right\} ,$$

$$(12.1)$$

where μ is the mass of the scalar field. The black hole solutions can be found using the following metric and scalar field ansatz

$$ds^2 = -e^{2F_0} N dt^2 + e^{2F_1} \left(\frac{dr^2}{N} + r^2 d\theta^2 \right) + e^{2F_2} r^2 \sin^2 \theta \left(d\varphi - W dt \right)^2 , \quad (12.2)$$

$$\Phi(t, r, \theta, \phi) = e^{-iwt} e^{im\varphi} \phi(r, \theta) , \qquad (12.3)$$

where $N = 1 - r_H/r$, $r_H > 0$ is the location of the event horizon, $w \in \mathbb{R}^+$ is the scalar field frequency, and $m \in \mathbb{Z}^+$ is the azimuthal harmonic index. There is an infinite countable number of families of Kerr black holes with scalar hair, each family being obtained for different values of m and of the number of nodes, n, of the scalar field profile ϕ in the equatorial plane. In what follows, we will focus on the $n = 0$ nodeless family with $m = 1$. These are likely the most stable configurations, as increasing either of these integer numbers (roughly) increases the energy of the solutions, corresponding to more excited states.

Kerr black holes with scalar hair avoid the no-hair theorem thanks to a harmonic time-dependence in the scalar field. The spacetime metric and the energy-momentum tensor of the scalar field are stationary.

Even if the scalar field is complex, its energy-momentum tensor and the metric of the spacetime are real. The metric is described by four functions (F_0, F_1, F_2, and W) of the coordinates r and θ, whereas the scalar field introduces a fifth function of the same coordinates. These functions are found, numerically, by solving a system of five non-linear coupled partial differential equations, with appropriate boundary conditions that guarantee asymptotic flatness and regularity on and outside the horizon; see [17] for all the details. Of crucial importance is the *synchronization condition*, $w = m\Omega_H$, where Ω_H is the angular velocity of the horizon. This condition guarantees regularity of a non-trivial scalar field at the horizon, and has a very clear physical interpretation in the context of the superradiance phenomenon of Kerr black holes [16]. The vacuum Kerr metric can be written in this coordinate system, which differs from the standard Boyer–Lindquist coordinates by a radial shift. The explicit form of the coefficients of the Kerr metric in this coordinates can be found in [17, 18].

A generic Kerr black hole with scalar hair is described by three "charges". The first two can be separated in their horizon (BH) and scalar field (S) contributions. They are the ADM or total mass $M = M_{BH} + M_S$, and the total angular momentum $J = J_{BH} + J_S$.[1] The third one is a Noether charge, Q, which is conserved in a local sense (of a continuity equation), but, unlike the mass and angular momentum, has no associated Gauss law, and hence cannot be measured by an observer at infinity. This Noether charge is a consequence of the $U(1)$ global symmetry of the complex scalar field and provides a measure of the hairiness of the black hole. A given family of Kerr black holes with scalar hair (i.e. with specific values of m and n) bifurcates from a subset of vacuum Kerr black holes, see Fig. 12.1. In this limit, $M_S = 0 = J_S$ and $Q = 0$. When the horizon size vanishes, which in particular implies $r_H \to 0$, the hairy black holes reduce to rotating boson stars (with the same values of m and n, as boson stars also form an infinite countable number of families). In this limit, $M_{BH} = 0 = J_{BH}$ and $Q = m J_S$. This latter condition, which was first observed for rotating boson stars in [31, 37], also holds for Kerr black holes with scalar hair. Thus we can write, for the hairy black holes,

$$1 = \frac{J_{BH}}{J} + \frac{Q}{mJ} \quad \Leftrightarrow \quad q = 1 - \frac{J_{BH}}{J}, \qquad q \equiv \frac{Q}{mJ}, \tag{12.4}$$

---

[1] The expression total angular momentum is used to indicate the angular momentum computed as the Komar integral associated to axisymmetry at spatial infinity.

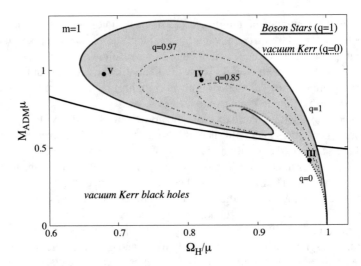

Fig. 12.1 Domain of existence of Kerr black holes with scalar hair (*blue shaded region*) in an ADM mass (here denoted M_{ADM} to avoid ambiguities) vs horizon angular velocity diagram. Vacuum Kerr black holes exist below the *black solid line* that corresponds to the extremal Kerr solution. Kerr black holes with scalar hair bifurcate from the vacuum Kerr solution at a particular existence line, corresponding to the Kerr black holes that can support stationary bound states of the massive Klein Gordon equation (*blue dotted line*) [16]. This is the $q = 0$ limit. In the $q = 1$ limit, Kerr black holes with scalar hair reduce to boson stars (*red solid line*). Constant q lines are roughly parallel to the boson star spiral. The *green dashed line* is the set of extremal (zero temperature) Kerr black holes with scalar hair. From [28], reproduced by permission of IOP Publishing. All rights reserved

where $q \in [0, 1]$ is the normalized Noether charge, which measures the fraction of angular momentum in the scalar field. This parameter provides a useful and compact measure of the hairiness, with $q = 0$ for vacuum Kerr black holes (no hair) and $q = 1$ for boson stars (only hair).

Since the hairy black hole solutions are only known numerically, for what follows we can consider three specific Kerr black holes with scalar hair. The data for these solutions are publicly available.[2] These three solutions are:

1. Configuration III ($q = 0.128$) is a Kerr-like hairy black hole. Only a relatively small fraction of the total mass and angular momentum is stored in the scalar field (5% of the mass and 13% of angular momentum). The input parameters used in obtaining this solution are: $r_H = 0.2$ and $\Omega_H = 0.975$. In units of the scalar field mass,[3] the ADM mass is $M = 0.415$ and the total angular momentum is $J = 0.172$. The black hole mass is $M_{BH} = 0.393$ and the black hole angular momentum is $J_{BH} = 0.15$. The scalar field mass is $M_S = 0.022$ and the scalar field angular momentum is $J_S = 0.022$.

---

[2]The data files are available at http://gravitation.web.ua.pt/index.php?q=node/416.
[3]That is, we have rescaled, in this discussion, $\mu r_H \to r_H$, $\Omega_H/\mu \to \Omega_H$, $\mu M \to M$ and $\mu^2 J \to J$. Let us note that any solution can have any physical mass for an appropriate choice of μ.

2. Configuration IV ($q = 0.846$) is already quite different from the vacuum Kerr solution. A large fraction of its mass and angular momentum are in the scalar field (75% of the mass and 85% of the angular momentum). The input parameters are: $r_H = 0.1$ and $\Omega_H = 0.82$. The ADM mass is $M = 0.933$ and the total angular momentum is $J = 0.739$. The black hole mass is $M_{BH} = 0.234$ and the black hole angular momentum is $J_{BH} = 0.114$. The scalar field mass is $M_S = 0.699$ and the scalar field angular momentum is $J_S = 0.625$.

3. Configuration V ($q = 0.998$) essentially describes a rotating boson star with a tiny black hole at its center. The mass and angular momentum are almost fully in the scalar field (98.2% of the mass and 97.6% of the angular momentum). The input parameters are: $r_H = 0.04$ and $\Omega_H = 0.68$. The ADM mass is $M = 0.975$ and the total angular momentum is $J = 0.85$. The black hole mass is $M_{BH} = 0.018$ and the black hole angular momentum is $J_{BH} = 0.002$. The scalar field mass is $M_S = 0.957$ and the scalar field angular momentum is $J_S = 0.848$.

Figure 12.1 shows a representation of the location of these three solutions in the domain of existence of Kerr black holes with scalar hair.

12.1.2 Black Holes in Einstein–Dilaton–Gauss–Bonnet Gravity

Black holes in Einstein–dilaton–Gauss–Bonnet gravity represent an example of non-Kerr black holes in an extension of Einstein's theory of gravity and for which we know the rotating solution numerically [22]. These solutions can evade the no-hair theorem for scalar-tensor theories [32] because of the presence of a non-trivial coupling between gravity and the dilaton field. The action of the model is the low-energy effective action for the heterotic string and reads [38]

$$S = \frac{1}{16\pi} \int d^4x \sqrt{-g} \left[R - \frac{1}{2} g^{\mu\nu} \left(\partial_\mu \Phi \right) \left(\partial_\nu \Phi \right) + \alpha e^{-\gamma \Phi} + R_{GB}^2 \right], \quad (12.5)$$

where Φ is the dilaton field, γ is the coupling constant of the dilaton field, α is another parameter of the theory, and R_{GB}^2 is the Gauss–Bonnet correction

$$R_{GB}^2 = R_{\mu\nu\rho\sigma} R^{\mu\nu\rho\sigma} - 4R_{\mu\nu} R^{\mu\nu} + R^4. \quad (12.6)$$

The equations of motion are of second order and the theory is ghost-free.

The black hole solutions can be found using the following metric ansatz [22]

$$ds^2 = -f dt^2 + \frac{m}{f} \left(dr^2 + r^2 d\theta^2 \right) + \frac{l}{f} r^2 \sin^2 \theta \left(d\phi - \frac{\omega}{r} dt \right)^2, \quad (12.7)$$

where f, m, l, and ω are functions of r and θ only. The event horizon is located at the radial coordinate $r = r_H$ defined by $f(r_H) = 0$ (independent of θ). The boundary conditions at the horizon are:

$$f|_{r=r_H} = m|_{r=r_H} = l|_{r=r_H} = 0, \quad \omega|_{r=r_H} = \Omega_H r_H, \quad \left.\frac{\partial \Phi}{\partial r}\right|_{r=r_H} = 0. \quad (12.8)$$

The boundary conditions at infinity are:

$$f|_{r\to\infty} = m|_{r\to+\infty} = l|_{r\to\infty} = 1, \quad \omega|_{r\to\infty} = \Phi|_{r\to\infty} = 0. \quad (12.9)$$

The axial symmetry of the spacetime and its regularity impose the following boundary conditions on the symmetry axis $\theta = 0$:

$$\left.\frac{\partial f}{\partial \theta}\right|_{\theta=0} = \left.\frac{\partial m}{\partial \theta}\right|_{\theta=0} = \left.\frac{\partial l}{\partial \theta}\right|_{\theta=0} = \left.\frac{\partial \omega}{\partial \theta}\right|_{\theta=0} = \left.\frac{\partial \Phi}{\partial \theta}\right|_{\theta=0} = 0. \quad (12.10)$$

The absence of conical singularities requires:

$$m|_{\theta=0} = l|_{\theta=0}. \quad (12.11)$$

The mass M, the spin angular momentum J, and the dilaton charge D are found from the asymptotic behavior of f, ω, and Φ, namely

$$f|_{r\to\infty} \to 1 - \frac{2M}{r}, \quad \omega|_{r\to\infty} \to \frac{2J}{r^2}, \quad \Phi|_{r\to\infty} \to -\frac{D}{r}. \quad (12.12)$$

Numerical solutions are obtained by solving numerically a set of five second order coupled non-linear partial differential equations for the functions f, m, l, ω, and Φ imposing the boundary conditions (12.8)–(12.11). Varying Ω_H, it is possible to obtain different rotating solutions, and the values of M, J, and D are computed a posteriori from the asymptotic behaviors of f, ω, and Φ, as shown in (12.12). Kleihaus et al. [22] only consider the case $\gamma = 1$, but otherwise γ would be another parameter of the black hole solutions.

The domain of the existence of black holes in Einstein–dilaton–Gauss–Bonnet gravity studied in [22] is shown in Fig. 12.2 by the pink shaded region. The x-axis is for the spin parameter $a_* = J/M^2$, while the y-axis is for the scaled horizon area $a_H = A_H/(16\pi M^2)$, where A_H is the area of the event horizon. The Schwarzschild solution is at $a_* = 0$ and $a_H = 1$. The Kerr solutions are all on the black solid curve. Non-rotating black holes in Einstein–dilaton–Gauss–Bonnet gravity are along the segment $a_* = 0$ and $a_H \in [0.85; 1]$. The bound $a_H = 0.85$ is determined by the existence of a critical horizon size. The latter also exists for rotating solutions and gives the lower boundary in the existence domain. It is worth noting that black holes in Einstein–dilaton–Gauss–Bonnet gravity can have a spin parameter exceeding 1. However, the extremal black holes (those with the highest spins and vanishing surface gravity) are not regular solutions, see [22] for the details.

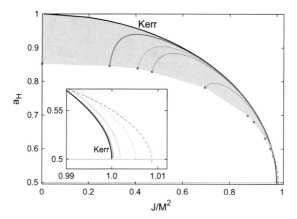

Fig. 12.2 Domain of existence of black holes in Einstein–dilaton–Gauss–Bonnet gravity studied in [22] (*pink shaded region* in the large figure) in a scaled horizon area $a_\mathrm{H} = A_\mathrm{H}/(16\pi M^2)$ (A_H is the area of the event horizon) vs spin parameter $a_* = J/M^2$. Kerr black holes are all mapped to the *black solid curve*. The other curves indicate black holes with a specific value of $\alpha^{1/2}\Omega_\mathrm{H}$: $\alpha^{1/2}\Omega_\mathrm{H} = 0.034$ (*red solid line*), 0.047 (*green dashed line*), 0.055 (*blue dotted line*), 0.082 (*magenta dotted line*), 0.110 (*cyan dashed-dotted line*), 0.126 (*black dotted line*). The *inset* shows that black holes in Einstein–dilaton–Gauss–Bonnet gravity can have a spin parameter exceeding the Kerr bound $|a_*| \leq 1$. From [22]

12.1.3 Manko–Novikov Metric

The Manko–Novikov solution is something between a theoretically-motivated metric and a phenomenological parametrization. This metric is a stationary, axisymmetric, and asymptotically flat exact solution of the vacuum Einstein equations with an infinite number of free parameters [25]. The no-hair theorem does not apply because these spacetimes have naked singularities and/or closed time-like curves exterior to their horizon. If we employ the Manko–Novikov metric to test the nature of astrophysical black holes, we are considering the possibility that the latter may not be black holes but compact bodies made of some kind of exotic matter. In other words, general relativity would hold, and the Manko–Novikov metric would describe the vacuum solution of the exterior gravitational field. The surface of the object must be at some radial coordinate larger than the null surface in the Manko–Novikov metric in order to cover the pathological region of the solution. The interior spacetime is supposed to be described by some interior solution, which is not necessary to know if we want to study the expected electromagnetic spectrum of an accretion disk in the exterior background.

The line element of the Manko–Novikov solution in quasi-cylindrical and prolate spheroidal coordinates is, respectively,

$$ds^2 = -f\,(dt - \omega d\phi)^2 + \frac{e^{2\gamma}}{f}\left(d\rho^2 + dz^2\right) + \frac{\rho^2}{f}d\phi^2 =$$

$$= -f\,(dt - \omega d\phi)^2 + \frac{k^2 e^{2\gamma}}{f}\left(x^2 - y^2\right)\left(\frac{dx^2}{x^2 - 1} + \frac{dy^2}{1 - y^2}\right)$$

$$+\frac{k^2}{f}\left(x^2 - 1\right)\left(1 - y^2\right)d\phi^2\,, \tag{12.13}$$

where

$$f = e^{2\psi}A/B\,,$$

$$\omega = 2ke^{-2\psi}CA^{-1} - 4k\alpha\left(1 - \alpha^2\right)^{-1}\,,$$

$$e^{2\gamma} = e^{2\gamma'}A\left(x^2 - 1\right)^{-1}\left(1 - \alpha^2\right)^{-2}\,, \tag{12.14}$$

and

$$\psi = \sum_{n=1}^{+\infty}\frac{\alpha_n P_n}{R^{n+1}}\,, \tag{12.15}$$

$$\gamma' = \frac{1}{2}\ln\frac{x^2 - 1}{x^2 - y^2} + \sum_{m,n=1}^{+\infty}\frac{(m+1)(n+1)\alpha_m\alpha_n}{(m+n+2)R^{m+n+2}}\left(P_{m+1}P_{n+1} - P_m P_n\right)$$

$$+\left[\sum_{n=1}^{+\infty}\alpha_n\left((-1)^{n+1} - 1 + \sum_{k=0}^{n}\frac{x - y + (-1)^{n-k}(x+y)}{R^{k+1}}P_k\right)\right]\,, \tag{12.16}$$

$$A = (x^2 - 1)(1 + ab)^2 - (1 - y^2)(b - a)^2\,, \tag{12.17}$$

$$B = [x + 1 + (x - 1)ab]^2 + [(1 + y)a + (1 - y)b]^2\,, \tag{12.18}$$

$$C = (x^2 - 1)(1 + ab)[b - a - y(a + b)]$$

$$+(1 - y^2)(b - a)[1 + ab + x(1 - ab)]\,, \tag{12.19}$$

$$a = -\alpha\exp\left[\sum_{n=1}^{+\infty}2\alpha_n\left(1 - \sum_{k=0}^{n}\frac{(x - y)}{R^{k+1}}P_k\right)\right]\,, \tag{12.20}$$

$$b = \alpha\exp\left[\sum_{n=1}^{+\infty}2\alpha_n\left((-1)^n + \sum_{k=0}^{n}\frac{(-1)^{n-k+1}(x + y)}{R^{k+1}}P_k\right)\right]\,. \tag{12.21}$$

Here $R = \sqrt{x^2 + y^2 - 1}$ and P_n are the Legendre polynomials with argument xy/R

$$P_n = P_n\left(\frac{xy}{R}\right)\,, \quad P_n(x) = \frac{1}{2^n n!}\frac{d^n}{dx^n}\left(x^2 - 1\right)^n\,. \tag{12.22}$$

It is worth noting that Eqs. (12.16), (12.20) and (12.21) correct a few typos in the original Manko–Novikov metric written in [25]: see [6, 11].

The solution has an infinite number of free parameters: k, α, and α_n ($n = 1, \ldots, \infty$). For $\alpha \neq 0$ and $\alpha_n = 0$, it reduces to the Kerr metric. For $\alpha = \alpha_n = 0$, we find the Schwarzschild solution. For $\alpha = 0$ and $\alpha_n \neq 0$, we obtain the static Weyl metric. Without loss of generality, we can put $\alpha_1 = 0$ to bring the massive object to the origin of the coordinate system. The simplest non-Kerr metric has three free parameters (k, α, α_2, and $\alpha_n = 0$ for $n \neq 2$), which are related to the mass, M, the spin parameter, $a_* = J/M^2$, and the dimensionless anomalous quadrupole moment, $q = -(Q - Q_{\text{Kerr}})/M^3$, of the object by the relations

$$\alpha = \frac{\sqrt{1 - a_*^2} - 1}{a_*}, \quad k = M\frac{1 - \alpha^2}{1 + \alpha^2}, \quad \alpha_2 = q\frac{M^3}{k^3}. \qquad (12.23)$$

q measures the deviation from the quadrupole moment of a Kerr black hole. In particular, since $Q_{\text{Kerr}} = -a_*^2 M^3$, the solution is oblate for $q > -a_*^2$ and prolate for $q < -a_*^2$. However, when $q \neq 0$, even all the higher order multipole moments of the spacetime have a different value from the Kerr ones.

It is often useful to change coordinate system. The relation between the prolate spheroidal coordinates and the quasi-cylindrical coordinates is given by

$$\rho = k\sqrt{\left(x^2 - 1\right)\left(1 - y^2\right)}, \quad z = kxy, \qquad (12.24)$$

with inverse

$$x = \frac{1}{2k}\left(\sqrt{\rho^2 + (z + k)^2} + \sqrt{\rho^2 + (z - k)^2}\right),$$

$$y = \frac{1}{2k}\left(\sqrt{\rho^2 + (z + k)^2} - \sqrt{\rho^2 + (z - k)^2}\right). \qquad (12.25)$$

The relation between the Schwarzschild coordinates and the quasi-cylindrical coordinates is given by

$$\rho = \sqrt{r^2 - 2Mr + a_*^2 M^2}\sin\theta, \quad z = (r - M)\cos\theta. \qquad (12.26)$$

The Manko–Novikov solution is written in prolate spheroidal coordinates and requires that the spin parameter does not exceed 1; that is, $|a_*| \leq 1$. There is currently no extension to describe objects with $|a_*| > 1$ available in the literature. The metrics discovered in [26, 27] are similar to the Manko–Novikov solution, but they have a small number of extra parameters. They are also written in prolate spheroidal coordinates. Such a family of solutions can be easily extended to describe objects with $|a_*| > 1$ after introducing oblate spheroidal coordinates [2].

Figure 12.3 shows the impact of the anomalous quadrupole moment q on the value of the ISCO radius r_{ISCO} for different values of the spin parameter a_*. The location of the ISCO radius is important in techniques like the continuum-fitting and the iron line methods. Every curve for a given a_* has a discontinuity at a particular value of

Fig. 12.3 ISCO radius r_{ISCO} as a function of the anomalous quadrupole moment q for different values of the spin parameter a_*. r_{ISCO} in Schwarzschild-like coordinates. Units in which $M = 1$ are used

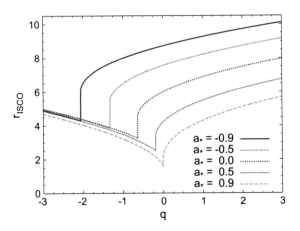

the anomalous quadrupole moment, say q_c, because the ISCO is determined by the orbital stability along the radial direction (as in the Kerr metric) for $q > q_c$ and by the orbital stability along the vertical direction for $q < q_c$.

12.2   Phenomenological Parametrizations

With the top-bottom approach discussed in the previous section, we can only compare the predictions of the standard theory with a specific alternative model. However, there are a large number of potential alternative scenarios and no one seems to be more motivated than the others. Moreover, even speculating that the metric around astrophysical black holes is not described by the Kerr solution, it is unlikely that we already have the right model. From these arguments, we may thus prefer a bottom-up approach and try to see whether it is possible to constrain possible deviations from the Kerr solution in a model-independent way.

It is remarkable that the same model-independent spirit has been successfully employed to test the Schwarzschild solution in the weak field limit with Solar System experiments. The approach is called the Parametrized Post-Newtonian (PPN) formalism [29] and can be summarized as follows. The starting point is to write the most general static, spherically symmetric, and asymptotically flat metric in terms of the expansion parameter M/r, where M is the mass of the central object and r is some radial coordinate. The PPN approach is traditionally formulated in isotropic coordinates and the line element reads

$$ds^2 = -\left(1 - \frac{2M}{r} + \beta \frac{2M^2}{r^2} + \ldots\right) dt^2$$
$$+ \left(1 + \gamma \frac{2M}{r} + \ldots\right) (dx^2 + dy^2 + dz^2) . \qquad (12.27)$$

When arranged in the more familiar Schwarzschild coordinates, the line element becomes

$$ds^2 = -\left(1 - \frac{2M}{r} + (\beta - \gamma)\frac{2M^2}{r^2} + ...\right) dt^2$$
$$+ \left(1 + \gamma\frac{2M}{r} + ...\right) dr^2 + r^2 d\theta^2 + r^2 \sin^2\theta d\phi^2 . \qquad (12.28)$$

The second term in g_{tt}, i.e. $2M/r$, is required to recover the Newtonian limit. β and γ are free parameters to be measured by experiments.

The only spherically symmetric vacuum solution of Einstein's equations is the Schwarzschild metric and it requires $\beta = \gamma = 1$. Other theories of gravity may have a different spherically symmetric vacuum solution, and in this case β and γ may not be exactly 1. Current observational data in the Solar System provide the following constraints on β and γ [5, 35]

$$|\beta - 1| < 2.3 \cdot 10^{-4}, \quad |\gamma - 1| < 2.3 \cdot 10^{-5}, \qquad (12.29)$$

confirming the validity of the Schwarzschild solution in the weak field limit within the precision of current observations.

In order to test the Kerr metric around astrophysical black holes, we would like to employ a similar approach. We would like to consider the most general stationary, axisymmetric, and asymptotically flat metric capable of describing the spacetime around a compact object. Such a metric should have a number of free parameters to be determined by observations and a posteriori we should check whether astrophysical data require that the values of these free parameters is consistent with the Kerr solution.

However, this is not so easy, and indeed there is currently no satisfactory formalism to test the Kerr metric. The crucial point is that now we want to test the strong gravity region and we cannot use an expansion in M/r. There is an infinite number of way to deform the Kerr metric and no natural way to distinguish leading order and higher order corrections. Moreover, since we are in the strong gravity regime, deformations of the Kerr metric naturally produce pathological features like naked singularities or regions with closed time-like curves. The result is that there are several proposals in the literature [7, 12, 14, 19, 20, 23, 24, 33], each of these with its advantages and disadvantages. In the next subsections, we will briefly review a few examples.

12.2.1 Johannsen–Psaltis Metric

The Johannsen–Psaltis metric [20] has been extensively employed in the literature to test the Kerr metric around astrophysical black holes. The procedure to obtain such a metric is quite ad hoc and can be summarized as follows. First, we consider the

Schwarzschild metric and we add the function $h(r)$ of the radial coordinate r only to take into account possible deviations from the Schwarzschild solution. The new line element reads

$$ds^2 = -\left(1 - \frac{2M}{r}\right)(1 + h)\,dt^2 + \frac{(1 + h)}{(1 - 2M/r)}dr^2 + r^2 d\theta^2 + r^2 \sin^2 \theta d\phi^2 . \qquad (12.30)$$

$h(r)$ is chosen to have the form

$$h = \sum_{k=0}^{+\infty} \varepsilon_{2k} \left(\frac{M}{r}\right)^k . \qquad (12.31)$$

At this point, we apply the Newman–Janis algorithm, which is essentially a trick to obtain a rotating black hole solution from the non-rotating one. After a (wrong) coordinate transformation, the line element in Boyer–Lindquist coordinates reads

$$ds^2 = -\left(1 - \frac{2Mr}{\Sigma}\right)(1 + h)\,dt^2 + \frac{\Sigma\,(1 + h)}{\Delta + ha^2 \sin^2 \theta}dr^2 + \Sigma d\theta^2$$
$$+ \left[r^2 + a^2 + \frac{2a^2 Mr \sin^2 \theta}{\Sigma} + \frac{a^2\,(\Sigma + 2Mr)\sin^2 \theta}{\Sigma}h\right]\sin^2 \theta d\phi^2$$
$$- \frac{4Mar \sin^2 \theta}{\Sigma}(1 + h)\,dt d\phi , \qquad (12.32)$$

where h is

$$h = \sum_{k=0}^{+\infty} \left(\varepsilon_{2k} + \varepsilon_{2k+1}\frac{Mr}{\Sigma}\right)\left(\frac{M^2}{\Sigma}\right)^k . \qquad (12.33)$$

The metric has an infinite number of deformation parameters $\{\varepsilon_k\}$ and it reduces to the Kerr solution when all the deformation parameters vanish. However, ε_0 must vanish in order to recover the correct Newtonian limit, while ε_1 and ε_2 are already strongly constrained by experiments in the Solar System through Eq. (12.29) [7]. The simplest non-trivial metric is thus that with ε_3 free and with all the other deformation parameters set to zero. The transformation to eliminate some off-diagonal terms and arrive at the expression in Eq. (12.32) is not correct, see the discussion in [1]. However, since the original non-rotating metric was not a solution of any equation and eventually we only want a parametrization to test the Kerr metric, this does not introduce any real mistake.

Figure 12.4 shows the contour levels of the ISCO radius $r_{\rm ISCO}$ (top panel) and of the Novikov–Thorne radiative efficiency $\eta_{\rm NT} = 1 - E_{\rm ISCO}$ (bottom panel) in the Johannsen–Psaltis metric with non-vanishing deformation parameter ε_3. The constraints that we can obtain from several measurements of the spectrum of astrophysical black holes approximately follow the contour levels of the Novikov–Thorne radiative efficiency (see next chapter).

Fig. 12.4 Contour levels of the ISCO radius r_{ISCO} (*top panel*) and of the Novikov–Thorne radiative efficiency $\eta_{\mathrm{NT}} = 1 - E_{\mathrm{ISCO}}$ (*bottom panel*) in the Johannsen–Psaltis metric with the non-vanishing deformation parameter ε_3. The *black dotted line* separates spacetimes with a regular event horizon (on the *left* of the line) from those with naked singularities (on the *right*). r_{ISCO} in Boyer–Lindquist coordinates

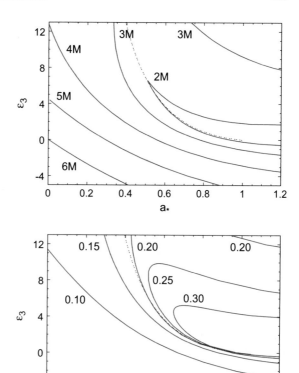

An extension of the Johannsen–Psaltis metric is the Cardoso–Pani–Rico parameterization [7]. Here the deformed Schwarzschild line element replacing Eq. (12.30) is

$$ds^2 = -\left(1 - \frac{2M}{r}\right)\left(1 + h^t\right)dt^2 + \frac{\left(1 + h^r\right)}{\left(1 - 2M/r\right)}dr^2 + r^2 d\theta^2 + r^2 \sin^2\theta d\phi^2 , \quad (12.34)$$

and we have two independent functions, $h^t(r)$ and $h^r(r)$. Then one can proceed as in the case of the Johannsen–Psaltis metric. The new line element is

$$ds^2 = -\left(1 - \frac{2Mr}{\Sigma}\right)\left(1 + h^t\right)dt^2 + \frac{\Sigma\left(1 + h^r\right)}{\Delta + h^r a^2 \sin^2\theta}dr^2 + \Sigma d\theta^2$$
$$+ \sin^2\theta \left\{\Sigma + a^2\sin^2\theta\left[2\sqrt{(1 + h^t)(1 + h^r)} - \left(1 - \frac{2Mr}{\Sigma}\right)\left(1 + h^t\right)\right]\right\}d\phi^2 ,$$
$$-2a\sin^2\theta\left[\sqrt{(1 + h^t)(1 + h^r)} - \left(1 - \frac{2Mr}{\Sigma}\right)\left(1 + h^t\right)\right]dt d\phi , \quad (12.35)$$

where

$$h^t = \sum_{k=0}^{+\infty} \left(\varepsilon_{2k}^t + \varepsilon_{2k+1}^t \frac{Mr}{\Sigma} \right) \left(\frac{M^2}{\Sigma} \right)^k , \tag{12.36}$$

$$h^r = \sum_{k=0}^{+\infty} \left(\varepsilon_{2k}^r + \varepsilon_{2k+1}^r \frac{Mr}{\Sigma} \right) \left(\frac{M^2}{\Sigma} \right)^k . \tag{12.37}$$

Now there are two infinite sets of deformation parameters, $\{\varepsilon_k^t\}$ and $\{\varepsilon_k^r\}$. The Johannsen–Psaltis metric is recovered when $\varepsilon_k^t = \varepsilon_k^r$ for all k, and the Kerr metric when $\varepsilon_k^t = \varepsilon_k^r = 0$ for all k.

12.2.2 Johannsen Metric

The Johannsen metric [19] is another phenomenological parametrization to test the Kerr metric. It is obtained by imposing that the corresponding Hamilton–Jacobi equations remain separable. While there is no theoretical reason to impose such a condition, and actually we know non-Kerr black hole solutions in alternative theories of gravity in which such a condition is not realized, it may facilitate some calculations. The most important properties of this parametrization are that: (i) the metric is regular everywhere on and outside of the event horizon, and (ii) it was explicitly shown that it is able to recover some black hole solutions in alternative theories of gravity for suitable choices of the deformation parameters.

In Boyer–Lindquist coordinates, the line element of the Johannsen parametrization reads [19]

$$ds^2 = -\frac{\tilde{\Sigma}\left(\Delta - a^2 A_2^2 \sin^2\theta\right)}{B^2} dt^2 - \frac{2a\left[\left(r^2 + a^2\right) A_1 A_2 - \Delta\right]\tilde{\Sigma}\sin^2\theta}{B^2} dt d\phi$$

$$+ \frac{\tilde{\Sigma}}{\Delta A_5} dr^2 + \tilde{\Sigma} d\theta^2 + \frac{\left[\left(r^2 + a^2\right)^2 A_1^2 - a^2 \Delta \sin^2\theta\right]\tilde{\Sigma}\sin^2\theta}{B^2} d\phi^2 \tag{12.38}$$

where

$$B = \left(r^2 + a^2\right) A_1 - a^2 A_2 \sin^2\theta , \quad \tilde{\Sigma} = \Sigma + f ,$$

$$\Sigma = r^2 + a^2 \cos^2\theta , \quad \Delta = r^2 - 2Mr + a^2 , \tag{12.39}$$

and

$$A_1 = 1 + \sum_{n=3}^{\infty} \alpha_{1n} \left(\frac{M}{r} \right)^n ,$$

$$A_2 = 1 + \sum_{n=2}^{\infty} \alpha_{2n} \left(\frac{M}{r} \right)^n ,$$

$$A_5 = 1 + \sum_{n=2}^{\infty} \alpha_{5n} \left(\frac{M}{r} \right)^n ,$$

$$f = \sum_{n=3}^{\infty} \varepsilon_n \frac{M^n}{r^{n-2}} . \tag{12.40}$$

In the expressions of A_1, A_2, A_5, and f in (12.40), the summation starts, respectively, from $n = 3, 2, 2$, and 3 after imposing that the spacetime is asymptotically flat, has the correct Newtonian limit, and the metric is consistent with Solar System experiments without constraints on its parameters; see [19] for the details. If we only consider the leading order in the expressions of A_1, A_2, A_5, and f, we have a metric with four deformation parameters, α_{13}, α_{22}, α_{52}, and ε_3. Regularity of the exterior region imposes the following constraints [19]

$$\alpha_{13}, \; \varepsilon_3 > -\frac{\left(M + \sqrt{M^2 - a^2} \right)^3}{M^3} , \tag{12.41}$$

$$\alpha_{22}, \; \alpha_{52} > -\frac{\left(M + \sqrt{M^2 - a^2} \right)^2}{M^2} . \tag{12.42}$$

12.2.3 Konoplya–Rezzolla–Zhidenko Metric

The Konoplya–Rezzolla–Zhidenko metric [23] was proposed to try to address the following issues:

1. Many parametrizations are based on an expansion in M/r. If we want to test the strong gravity region near the black hole horizon, where M/r is not a small quantity, we have an infinite number of roughly equally important parameters, which makes it impossible to isolate the dominant terms and focus the efforts on the measurement of a small number of parameters.
2. One would like to have a so general parametrization that it is possible to recover any black hole solution in any (known and unknown) alternative theories of gravity for specific values of its free parameters. While it is difficult to assert how general a parametrization can be, several proposals in the literature fail to recover the known black hole solutions in alternative theories of gravity.
3. Astrophysical black holes may have a non-negligible spin angular momentum, which plays an important rule in the features of the electromagnetic spectrum

associated to the strong gravity region, and therefore the spin cannot be ignored in any test of the Kerr metric. The Newman–Janis algorithm is a simple trick to obtain a rotating black hole solution from the non-rotating one. However, it is not guaranteed that such an algorithm works beyond general relativity. In the literature there are some examples in which the algorithm works [8, 21, 36], as well as examples in which it fails [15]. It is thus unclear whether a parametrization based on the Newman–Janis algorithm can be employed to test astrophysical black holes.

The Konoplya–Rezzolla–Zhidenko metric addresses the three issues above in the following way. First, it is not based on an expansion in M/r. There is a hierarchical structure in the deviations from the Kerr spacetime, so that higher order terms necessarily provide smaller and smaller corrections. Second, Konoplya et al. [23] explicitly show that their proposal can well approximate with a few parameters the metrics of rotating dilaton black holes and of rotating black holes in Einstein–dilaton–Gauss–Bonner gravity. For example, this is not possible with the Johannsen–Psaltis and the Cardoso–Pani–Rico metrics. Third, the Konoplya–Rezzolla–Zhidenko metric is not obtained from the Newman–Janis algorithm, because the initial ansatz is already suitable to describe rotating black holes.

Assuming reflection symmetry across the equatorial plane and neglecting coefficients of higher orders, the line element of the Konoplya–Rezzolla–Zhidenko metric reads [23]

$$ds^2 = -\frac{N^2 - W^2 \sin^2\theta}{K^2} dt^2 - 2Wr \sin^2\theta \, dt \, d\phi + K^2 r^2 \sin^2\theta \, d\phi^2$$
$$+ \frac{\Sigma B^2}{N^2} dr^2 + \Sigma r^2 d\theta^2 , \tag{12.43}$$

where

$$N^2 = \left(1 - \frac{r_0}{r}\right)\left[1 - \frac{\varepsilon_0 r_0}{r} + (k_{00} - \varepsilon_0)\frac{r_0^2}{r^2} + \frac{\delta_1 r_0^3}{r^3}\right]$$
$$+ \left[(k_{21} + a_{20})\frac{r_0^3}{r^3} + \frac{a_{21} r_0^4}{r^4}\right]\cos^2\theta ,$$

$$B = 1 + \frac{\delta_4 r_0^2}{r^2} + \frac{\delta_5 r_0^2}{r^2}\cos^2\theta ,$$

$$\Sigma = 1 + \frac{a_*^2}{r^2}\cos^2\theta ,$$

$$W = \frac{1}{\Sigma}\left[\frac{w_{00} r_0^2}{r^2} + \frac{\delta_2 r_0^3}{r^3} + \frac{\delta_3 r_0^3}{r^3}\cos^2\theta\right] ,$$

$$K^2 = 1 + \frac{a_* W}{r} + \frac{1}{\Sigma}\left\{\frac{k_{00} r_0^2}{r^2} + \frac{k_{21} r_0^3}{r^3}\left[1 - \frac{\frac{a_*^2}{r_0^2}(1 - \frac{r_0}{r})}{1 + \frac{a_*^2}{r_0^2}(1 - \frac{r_0}{r})}\right]^{-1}\cos^2\theta\right\} \tag{12.44}$$

Following [28], we can introduce six deformation parameters $\{\delta_j\}$ ($j = 1, 2, \ldots, 6$), which are related to the coefficient r_0, a_{20}, a_{21}, ε_0, k_{00}, k_{21}, and w_{00} appearing in the Konoplya–Rezzolla–Zhidenko metric by the following relations

$$r_0 = 1 + \sqrt{1 - a_*^2}\,, \qquad a_{20} = \frac{2a_*^2}{r_0^3}\,, \qquad a_{21} = -\frac{a_*^4}{r_0^4} + \delta_6\,, \qquad \varepsilon_0 = \frac{2 - r_0}{r_0}\,,$$

$$k_{00} = \frac{a_*^2}{r_0^2}\,, \qquad k_{21} = \frac{a_*^4}{r_0^4} - \frac{2a_*^2}{r_0^3} - \delta_6\,, \qquad w_{00} = \frac{2a_*}{r_0^2}\,. \tag{12.45}$$

Here the mass is $M = 1$ and a_* is the spin parameter. r_0 is the radial coordinate in the equatorial plane of the event horizon. The physical interpretation of the deformation parameters can be summarized as follows [23]:

$$
\begin{array}{rcl}
\delta_1 & \to & \text{related to deformations of } g_{tt}, \\
\delta_2, \delta_3 & \to & \text{related to rotational deformations of the metric,} \\
\delta_4, \delta_5 & \to & \text{related to deformations of } g_{rr}, \\
\delta_6 & \to & \text{related to deformations of the event horizon.}
\end{array}
$$

The mass-quadrupole moment is the same as in the Kerr metric, and deviations from the Kerr solution are only possible in the strong gravity region.

12.2.4 Ghasemi–Nodehi–Bambi Metric

The proposal in [12] is somewhat different from the other parametrized metrics. It can be used to see which parts of the Kerr solution can be tested by observations and which parts are more elusive and they have a small impact on observable quantities. The ansatz for the line element is

$$
\begin{aligned}
ds^2 = {} & -\left(1 - \frac{2b_1 Mr}{r^2 + b_2 a^2 \cos^2\theta}\right) dt^2 - \frac{4b_3 Mar\sin^2\theta}{r^2 + b_4 a^2 \cos^2\theta} dt\,d\phi \\
& + \frac{r^2 + b_5 a^2 \cos^2\theta}{r^2 - 2b_6 Mr + b_7 a^2} dr^2 + \left(r^2 + b_8 a^2 \cos^2\theta\right) d\theta^2 \\
& + \left(r^2 + b_9 a^2 + \frac{2b_{10} Ma^2 r \sin^2\theta}{r^2 + b_{11} a^2 \cos^2\theta}\right) \sin^2\theta\, d\phi^2\,.
\end{aligned}
\tag{12.46}
$$

The line element of the Kerr metric is recovered when $b_i = 1$ for all i. The asymptotic mass is $b_1 M$ and the asymptotic specific angular momentum is $b_3 a$ (if $b_1 = 1$). Solar System experiments already constrain b_6 to be very close to 1. We can thus set b_1, b_3, and b_6 equal to 1 and consider as free parameters the remaining eight: b_2, b_4, b_5, b_7, b_8, b_9, b_{10}, and b_{11}.

The line element in (12.46) is surely not a very general line element to describe black holes in alternative theories of gravity. Basically we already assume that the metric around a black hole is at least close to the Kerr solution, and we want to check whether we can constrain small deviations from it.

12.3   Important Remarks

The parametrizations discussed in the previous section can be employed to test the Kerr metric with observations of the electromagnetic spectrum of astrophysical black holes. The properties of the electromagnetic radiation are determined by the motion of the gas in the accretion disk and by the propagations of the photons from the point of emission in the disk to the point of detection in the flat faraway region. In a *metric theory of gravity* [34], test-particles follow the geodesics of the spacetime, and all the relativistic effects in the electromagnetic spectrum are determined by the metric of the spacetime.

In this context, these parametrizations can be used to test the Kerr metric as the PPN formalism can be used to test the Schwarzschild solution in the weak field limit. It is not possible to directly test the Einstein equations. For instance, we cannot distinguish a Kerr black hole of general relativity from a Kerr black hole in an alternative metric theory of gravity, because the motion of particles and photons is the same. It is indeed important to bear in mind that the Kerr metric is a black hole solution even in some alternative theories of gravity [30]. If we want to directly test the Einstein equations, we should use gravitational (rather than electromagnetic) waves, because their emission is governed by the Einstein equations [3].

The possible detection of deviations from the Kerr metric does not directly implies that general relativity is not the correct theory of gravity. As discussed in Sect. 12.1, deviations from the Kerr solution can be possible with new physics either in the gravity or in the matter sectors. In the framework of conventional physics, astrophysical black holes can only be explained as black holes and deviations from the Kerr metric should be extremely small (see Sect. 6.5). In the presence of new physics, such a conclusion does not hold. It may also happen that the metric around an astrophysical black hole is not well described by the stationary black hole solution of its gravity theory: black holes rapidly go bald in general relativity, but not necessarily in other theories of gravity.

In the case of a non-metric theory of gravity, test-particles may not follow the geodesics of the spacetime. For instance, the photon trajectory may depend on the photon energy or on the photon polarization. Photons may follow the geodesics of a certain metric, while the particle in the disk may follow the geodesics of another metric. In this context, it may still be possible to employ the phenomenological parametrizations discussed above in order to test the particle motion in the strong gravity region around a black hole. For instance, we may consider different parameterized metrics for different particle species.

References

1. M. Azreg-Ainou, Class. Quant. Grav. **28**, 148001 (2011), arXiv:1106.0970 [gr-qc]
2. C. Bambi, JCAP **1105**, 009 (2011), arXiv:1103.5135 [gr-qc]
3. E. Barausse, T.P. Sotiriou, Phys. Rev. Lett. **101**, 099001 (2008), arXiv:0803.3433 [gr-qc]
4. E. Berti et al., Class. Quant. Grav. **32**, 243001 (2015), arXiv:1501.07274 [gr-qc]
5. B. Bertotti, L. Iess, P. Tortora, Nature **425**, 374 (2003)
6. J. Brink, Phys. Rev. D **78**, 102002 (2008), arXiv:0807.1179 [gr-qc]
7. V. Cardoso, P. Pani, J. Rico, Phys. Rev. D **89**, 064007 (2014), arXiv:1401.0528 [gr-qc]
8. X. Dianyon, Class. Quant. Grav. **5**, 871 (1988)
9. G. Dvali, C. Gomez, Fortsch. Phys. **61**, 742 (2013), arXiv:1112.3359 [hep-th]
10. G. Dvali, C. Gomez, Phys. Lett. B **719**, 419 (2013), arXiv:1203.6575 [hep-th]
11. H. Fang, Ph.D. thesis, California Institute of Technology (2007)
12. M. Ghasemi-Nodehi, C. Bambi, Eur. Phys. J. C **76**, 290 (2016), arXiv:1604.07032 [gr-qc]
13. S.B. Giddings, Phys. Rev. D **90**, 124033 (2014), arXiv:1406.7001 [hep-th]
14. K. Glampedakis, S. Babak, Class. Quant. Grav. **23**, 4167 (2006), arXiv:gr-qc/0510057
15. D. Hansen, N. Yunes, Phys. Rev. D **88**, 104020 (2013), arXiv:1308.6631 [gr-qc]
16. C.A.R. Herdeiro, E. Radu, Phys. Rev. Lett. **112**, 221101 (2014),arXiv:1403.2757 [gr-qc]
17. C. Herdeiro, E. Radu, Class. Quant. Grav. **32**, 144001 (2015), arXiv:1501.04319 [gr-qc]
18. C. Herdeiro, E. Radu, H. Runarsson, Class. Quant. Grav. **33**, 154001 (2016), arXiv:1603.02687 [gr-qc]
19. T. Johannsen, Phys. Rev. D **88**, 044002 (2013), arXiv:1501.02809 [gr-qc]
20. T. Johannsen, D. Psaltis, Phys. Rev. D **83**, 124015 (2011), arXiv:1105.3191 [gr-qc]
21. H. Kim, Phys. Rev. D **59**, 064002 (1999)
22. B. Kleihaus, J. Kunz, E. Radu, Phys. Rev. Lett. **106**, 151104 (2011), arXiv:1101.2868 [gr-qc]
23. R. Konoplya, L. Rezzolla, A. Zhidenko, Phys. Rev. D **93**, 064015 (2016), arXiv:1602.02378 [gr-qc]
24. N. Lin, N. Tsukamoto, M. Ghasemi-Nodehi, C. Bambi, Eur. Phys. J. C **75**, 599 (2015), arXiv:1512.00724 [gr-qc]
25. V.S. Manko, I.D. Novikov, Class. Quant. Grav. **9**, 2477 (1992)
26. V.S. Manko, E.W. Mielke, J.D. Sanabria-Gomez, Phys. Rev. D **61**, 081501 (2000), arXiv:gr-qc/0001081
27. V.S. Manko, J.D. Sanabria-Gomez, O.V. Manko, Phys. Rev. D **62**, 044048 (2000)
28. Y. Ni, M. Zhou, A. Cardenas-Avendano, C. Bambi, C.A.R. Herdeiro, E. Radu, JCAP **1607**, 049 (2016), arXiv:1606.04654 [gr-qc]
29. K. Nordtvedt, Phys. Rev. **169**, 1017 (1968)
30. D. Psaltis, D. Perrodin, K.R. Dienes, I. Mocioiu, Phys. Rev. Lett. **100**, 091101 (2008) [erratum: Phys. Rev. Lett. **100**, 119902 (2008)], arXiv:0710.4564 [astro-ph]
31. F.E. Schunck, E.W. Mielke, Phys. Lett. A **249**, 389 (1998)
32. T.P. Sotiriou, V. Faraoni, Phys. Rev. Lett. **108**, 081103 (2012), arXiv:1109.6324 [gr-qc]
33. S. Vigeland, N. Yunes, L. Stein, Phys. Rev. D **83**, 104027 (2011), arXiv:1102.3706 [gr-qc]
34. C.M. Will, Living Rev. Rel. **17**, 4 (2014), arXiv:1403.7377 [gr-qc]
35. J.G. Williams, S.G. Turyshev, D.H. Boggs, Phys. Rev. Lett. **93**, 261101 (2004), arXiv:gr-qc/0411113
36. S. Yazadjiev, Gen. Rel. Grav. **32**, 2345 (2000)
37. S. Yoshida, Y. Eriguchi, Phys. Rev. D **56**, 762 (1997)
38. B. Zwiebach, Phys. Lett. B **156**, 315 (1985)

Chapter 13
Testing the Kerr Paradigm with X-Ray Observations

The electromagnetic radiation emitted from the inner part of the accretion disk of a black hole is significantly affected by relativistic effects (Doppler boosting, gravitational redshift, light bending) occurring in the strong gravity region and determined by the metric of the spacetime around the compact object. If we have the correct astrophysical model, the study of the properties of the electromagnetic radiation can be a tool to test the Kerr paradigm. In the previous chapters, we have already discussed the calculations of the structure of Novikov–Thorne accretion disks, of the thermal and reflection spectra of Novikov–Thorne accretion disks, and of the fundamental frequencies of a test-particle in a generic stationary, axisymmetric, and asymptotically flat spacetime. It is straightforward to consider a specific non-Kerr metric and obtain the expected properties of the electromagnetic radiation in this spacetime. We can then compare our theoretical predictions with the available data and constrain the value of some deformation parameter or check whether the data prefer the Kerr solution or the non-Kerr metric.

In this chapter, we will discuss how to test the Kerr black hole hypothesis with X-ray observations. Since this is quite a new research field and still under development, the observational constraints that can be found now in the literature will be presumably replaced by new and more accurate constraints in the near future. For this reason, here we will focus the attention on the method – what we can do with this approach – and the problems to face, rather than listing current observational constraints on certain deformation parameters.

13.1 Continuum-Fitting Method

Figure 7.3 in Chap. 7 shows the impact of the model parameters on the thermal spectrum of a thin disk around a Kerr black hole. If we compute the thermal spectrum of a thin disk in one of the phenomenological parametrizations discussed in Sect. 12.2, the spectrum will also depend on the deformation parameters of the metric. For instance,

© Springer Nature Singapore Pte Ltd. 2017
C. Bambi, *Black Holes: A Laboratory for Testing Strong Gravity*,
DOI 10.1007/978-981-10-4524-0_13

Fig. 13.1 As in Fig.7.3 for the Johannsen–Psaltis deformation parameter ε_3. The values of the other parameters are: $M = 10\ M_\odot$, $\dot{M} = 2 \cdot 10^{18}$ g s^{-1}, $D = 10$ kpc, $i = 45°$, and $a_* = 0.7$. Flux density $N_{E_{obs}}$ in photons keV^{-1} cm^{-2} s^{-1}. From [7], reproduced by permission of IOP Publishing.

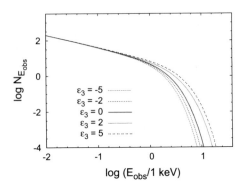

if we consider the Johannsen–Psaltis metric with the deformation parameter ε_3 (all the other deformation parameters vanish), we find the spectra shown in Fig. 13.1.

Let us now compare the panels in Figs.7.3 and 13.1. It is not necessary a quantitative analysis to understand that the impact of the spin parameter a_* and of the deformation parameter ε_3 is very similar if all the other parameters are fixed, while the impact of the other parameters (M, $\dot{M}$, i, D) seems to be quite different.

First, the thermal spectrum of a thin disk has a very simple shape: it is just a multi-color blackbody spectrum. It is not really necessary to have a sophisticated theoretical model to fit the data. If we have a theoretical model and we change its parameters, it is not difficult to find a good fit. Of course, a wrong theoretical model would lead to wrong measurements, but this is a different issue.

Second, the spacetime metric mainly affects the thermal spectrum of a thin disk because it determines the location of the inner edge of the disk (in the Novikov–Thorne model, the inner edge of the disk is at the ISCO radius, which depends on the metric). In this way, a variation of the metric can move the high energy cut-off of the spectrum to lower or higher energies. If we are not considering metrics substantially different from the Kerr one, the variation of the other relativistic effects determined by changing the metric are too weak to produce a characteristic feature in the disk's thermal spectrum.

From the above considerations, it is easy to guess that, in general, the continuum-fitting method cannot simultaneously measure both the spin and the deformation parameters. Even in the Kerr metric, the spectrum depends on five parameters (M, $\dot{M}$, i, D, and a_*). If we have independent estimates of M, i, and D, for instance from optical observations, we can fit the soft X-ray component and infer the values of a_* and $\dot{M}$. Without independent estimates of M, i, and D, it is impossible to measure the spin of the black hole because the spectrum is degenerate, namely different combinations of the values of the parameters of the model may provide good fits and it is not possible to establish a measurement. In the case of a parametrized metric with only one deformation parameter, it may not be sufficient to have independent estimates of M, i, and D to constrain a_*, $\dot{M}$, and the deformation parameter. Of course, this depends on the specific choice of the deformation parameter. Some deformation parameters may be degenerate with the spin, some deformation parameters may have no effect at all on the thermal spectrum of the disk, and some deformation parameters may produce so dramatic effects on the spectrum that they could be easily ruled out or strongly constrained.

Kong et al. [20] have studied the constraints on the Johannsen–Psaltis deformation parameter ε_3 from the available spin measurements of the continuum-fitting method. The same analysis could be repeated for the deformation parameter of another parametrized metric, as well as for a non-Kerr metric motivated by some theoretical model. In principle, we should calculate the thermal spectrum in our model, fit the data, and find the best fit and the allowed region at the confidence level of interest. However, it is often possible to proceed in a different way to get a simple estimate of the constraints on a deformation parameter from an already existing spin measurement with the continuum-fitting method of a certain black hole. The approach can be motivated a posteriori.

As an example, we can consider the Johannsen–Psaltis metric with the only possible non-vanishing deformation parameter ε_3, which is the case discussed in [20]. We choose a black hole whose spin has been already determined with the continuum-fitting method (see Table 7.1). The thermal spectrum of the disk of such a black hole depends now on six parameters, three of them are known from independent observations (M, i, D) and the other three should be measured with the analysis of the thermal spectrum of the disk $(a_*, \dot{M}, \varepsilon_3)$. Instead of working on some observational data, we adopt the theoretical spectrum of the disk around a Kerr black hole, whose spin parameter a_* and mass accretion rate $\dot{M}$ have the values of the measurement reported in the literature. Such a spectrum is then compared (for instance, by calculating the χ^2) with the theoretical spectra computed in spacetimes in which a_*, $\dot{M}$, and ε_3 vary.[1] The values of M, i, and D are always fixed to their measurements. In the end, we get the function

$$\chi^2 = \chi^2(a_*, \dot{M}, \varepsilon_3) \tag{13.1}$$

in a three-dimensional grid. After marginalizing over $\dot{M}$, we get $\chi^2(a_*, \varepsilon_3)$. The values of a_* and ε_3 that minimize $\chi^2(a_*, \varepsilon_3)$ are the measurements of the spin and the deformation parameters – by construction, the value of a_* is that of the measurement reported in the literature and $\varepsilon_3 = 0$, modulo other pairs of (a_*, ε_3) for which $\chi^2(a_*, \varepsilon_3) = 0$. If the Kerr measurement in the literature is $\tilde{a}_* \pm \Delta$, we compute $\chi^2(\tilde{a}_* \pm \Delta, 0)$, and the allowed region is now defined as the values of a_* and ε_3 for which

$$\chi^2(a_*, \varepsilon_3) \leq \chi^2(\tilde{a}_* \pm \Delta, 0). \tag{13.2}$$

A priory, this approach is not completely justified, because we are assuming that the object is a Kerr black hole and that the best fit is for the Kerr metric. However,

---

[1] The analysis reported in [20] is slightly different, because $\dot{M}$ is inferred from the measurement of the radiative efficiency. See [20] for more details.

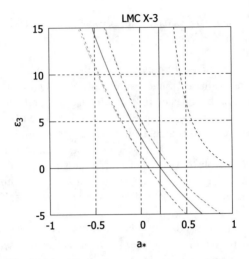

Fig. 13.2 Constraints on the spin a_* and the Johannsen–Psaltis deformation parameter ε_3 from the study of the disk thermal spectrum of the black hole in LMC X-3. The *red solid line* tracks the minimum of χ^2 for a fixed ε_3. The *dashed-dotted green lines* are the boundaries of the allowed region (1-σ error) along the *red line*. The *blue dashed lines* are the boundaries that one could infer if the best fit were exactly for $\varepsilon_3 = 0$. The *black dotted line* separates spacetimes with a regular event horizon (on the *left* of the line) from those with naked singularities (on the *right*). See the text for more details. From [20]

depending on the test-metric, it is possible to justify the method a posteriori, as explained below.

Figure 13.2 shows the constraints on the spin a_* and the Johannsen–Psaltis deformation parameter ε_3 for the black hole in LMC X-3 obtained with the above procedure in [20]. The red solid line tracks the minimum of $\chi^2(a_*, \varepsilon_3)$ for a fixed ε_3. Let us denote this quantity $\chi^2_{\min}(\varepsilon_3)$, and $a_*^{\min}(\varepsilon_3)$ the value of its spin parameter. The two blue dashed lines are the boundaries of the allowed region defined in Eq. (13.2). We remind that $\chi^2(\tilde{a}_*, 0) = 0$ by definition, as we are not introducing any kind of noise but just comparing theoretical spectra. The two green dashed-dotted curves are the boundaries of the allowed region defined as

$$\chi^2(a_*, \varepsilon_3) \leq \chi^2(\tilde{a}_* \pm \Delta, 0) + \chi^2_{\min}(\varepsilon_3) . \tag{13.3}$$

The idea behind is the following. If the spacetime of the black hole were characterized by a certain non-vanishing ε_3, assuming the correct metric the continuum-fitting method should provide the spin parameter $a_*^{\min}(\varepsilon_3)$. However, if we assume the Kerr metric, the spin measurement would be $\tilde{a}_*$. The spin uncertainty would be roughly given by Eq. (13.3).

The justification of this approach can be understood from the boundaries of the allowed regions given by Eqs. (13.2) and (13.3) in Fig. 13.2, respectively the blue dashed and the green dashed-dotted curves. The difference is negligible considering

Fig. 13.3 As in Fig. 13.2 for
the black hole in A0620-00.
From [20]

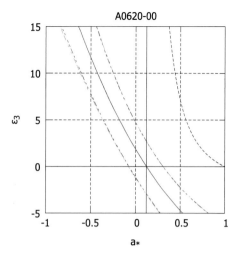

the spin uncertainty for a fixed ε_3. In other words, the spacetimes along the red
solid lines have a very similar disk's thermal spectrum and $\chi^2_{min}(\varepsilon_3) \approx 0$. It is not
important where the absolute minimum of χ^2 exactly is: the spacetimes along the red
solid line cannot be distinguished with the continuum-fitting method. If we proceeded
to constrain a_* and ε_3 from the analysis of real data, we would get constraints very
similar to those shown in Fig. 13.2, which have been instead obtained very quickly
and without much effort.

Such a simplified analysis can be repeated for other black holes with a spin measurement from the continuum-fitting method [20]. Figure 13.3 shows the constraints
on the spin a_* and the Johannsen–Psaltis deformation parameter ε_3 from the disk's
thermal spectrum of the black hole in A0620-00. There are no qualitative differences
with respect to the case of LMC X-3.

As the spin parameter of the measurement assuming the Kerr metric increases, we
find some differences. Figure 13.4 shows the constraints for the black hole in M33 X-
7. Assuming the Kerr metric, the spin is $a_* = 0.84 \pm 0.05$ [24]. The continuum-fitting
method roughly measures the radiative efficiency of Novikov–Thorne accretion disks
η_{NT}, whose contour levels were shown in Fig. 12.4 for the Johannsen–Psaltis metric
with non-vanishing ε_3. From that figure, it is clear that high values of η_{NT} cannot
be reached for large values of ε_3. As a consequence, the solid red line, the dashed
blue lines, and the dashed-dotted green lines are as in Fig. 13.4 for M33 X-7. The
allowed regions from Eqs. (13.2) and (13.3) do not coincide for large values of ε_3.
The constraint provided by the dashed blue lines should approximate the constrain
that one could obtain by reanalyzing the data. Indeed, the spacetimes along the two
red solid lines at $\varepsilon_3 \lesssim 13$ have similar disk's thermal spectra and $\chi^2_{min}(\varepsilon_3) \approx 0$, while
those along the red solid line extending from the connection point to larger ε_3 have
a higher χ^2. If the metric around the black hole in M33 X-7 were described by the
Johannsen–Psaltis metric with a larger value of the deformation parameter ε_3, the
spin measurement assuming the Kerr metric should have provided a lower value of

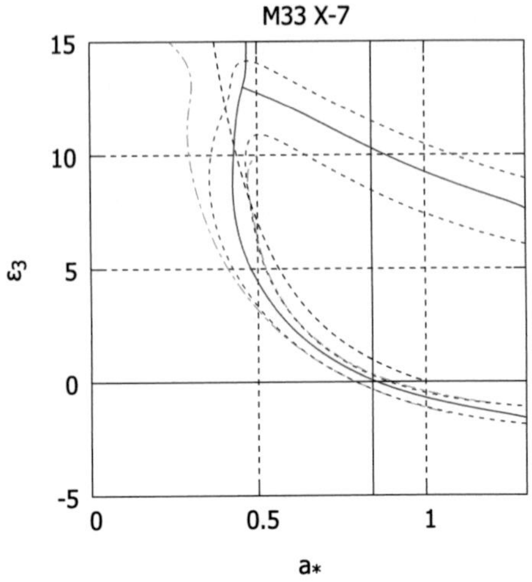

Fig. 13.4 As in Fig. 13.2 for the black hole in M33 X-7. From [20]

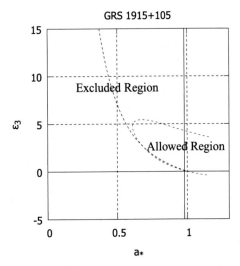

Fig. 13.5 Constraints on the spin a_* and the Johannsen–Psaltis deformation parameter ε_3 from the study of the disk thermal spectrum of the black hole in GRS 1915+105. The *black dotted line* separates spacetimes with a regular event horizon (on the *left* of the line) from those with naked singularities (on the *right*). See the text for more details. From [20]

a_*, consistently with that we would have expected from the contour levels of η_{NT} in Fig. 12.4.

Figures 13.5 and 13.6 show the constraints for GRS 1915+105 and Cygnus X-1, respectively. In these two cases, the Kerr spin measurement is $a_* > 0.98$ (see Table 7.1). In Figs. 13.5 and 13.6, the spacetimes inside the region delimited by the blue dashed curve have a disk's thermal spectrum similar to those in Kerr spacetimes with $a_* > 0.98$, while the spacetimes outside such a region have a disk's thermal spectrum similar to those in Kerr spacetimes with $a_* < 0.98$. Once again, this is consistent with the contour levels of η_{NT} in Fig. 12.4. It is worth noting that the observation of a black hole that looks like a very fast-rotating Kerr black hole can put a bound on ε_3 independently of its spin. In the case of GRS 1915+105 and

Fig. 13.6 As in Fig. 13.5 for the black hole in Cygnus X-1. From [20]

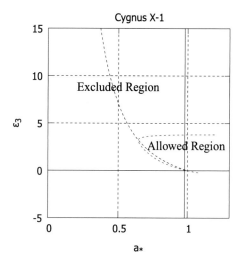

Cygnus X-1 in which the Kerr spin measurement is $a_* > 0.98$, we have $0 \lesssim \varepsilon_3 \lesssim 5$. The difference in the allowed region between GRS 1915+105 and Cygnus X-1 is mainly due to the different viewing angle, which is $i = 66°$ for GRS 1915+105 and $i = 27°$ for Cygnus X-1.

In principle, one could find a source with a thermal spectrum harder than that which is expected for a Kerr black hole with $a_* = 1$. This would essentially correspond to a spacetime in which the Novikov–Thorne radiative efficiency exceeds the Kerr black hole bound $\eta_{NT} = 0.42$ and could be an indication of deviations from the Kerr geometry. For instance, such a scenario is possible for thin disks in the spacetimes discussed in [19]. For the time being, there are no observations of this kind and therefore all the data are consistent with the Kerr metric.

The possibility of testing the Kerr black hole hypothesis from the study of the polarization of the thermal spectrum of the disk has been investigated in [22, 23]. At the moment, considering the expected sensitivity of the first generation of X-ray polarimetric detectors, this approach does not seem to be able to provide competitive constraints with respect to the other techniques.

13.2 X-Ray Reflection Spectroscopy

The reflection spectrum of a thin accretion disk has a more complicated shape than the thermal spectrum. We can thus guess that the technique is potentially more powerful to test the metric around a black hole. This is indeed what one can find with a simple quantitative analysis [15]. However, this can be achieved under two conditions: (*i*) we have the correct astrophysical model, and (*ii*) we need high quality data. The astrophysical model is more complicated than the one for the continuum-fitting

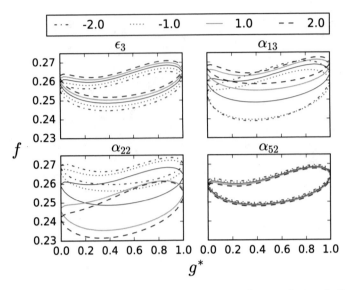

Fig. 13.7 Impact of the deformation parameters ε_3, α_{13}, α_{22}, and α_{52} on the transfer function f. The spacetime is described by the Johannsen metric with the spin parameter $a_* = 0.8$. The emission radius is $r_e = 6.855\,M$ and the viewing angle is $i = 30°$. In every plot, one of the deformation parameters assumes the values 0 (*black solid line*), ± 1, and ± 2, while the other deformation parameters vanish. From [8]

method and not fully under control at the moment. For instance, the geometry of the corona is currently unknown and therefore it is common to employ phenomenological emissivity profiles that inevitably introduce systematic errors. The flux of the disk's reflection component is weak and the intrinsic Poisson noise of the source may wash out small features determined by relativistic effects that are useful to probe the spacetime metric.

One can construct a reflection model for a certain metric with the approach discussed in Sect. 6.2 and the difference between Kerr and non-Kerr model would only be in the calculation of the transfer function. The local spectrum (with the possible exception of its emissivity profile) does not depend on the spacetime. A similar model has been presented in [8] for the Johannsen metric, but it would be straightforward to consider a different background metric. Figure 13.7 shows the impact of the deformation parameters ε_3, α_{13}, α_{22}, and α_{52} of the Johannsen metric on the transfer function f. Integrating the transfer function over the emission radius and the relative redshift factor for a single iron line, we obtain the profiles shown in Fig. 13.8.

The next subsections discuss some issues related to the use of this technique to test the nature of astrophysical black holes. It is worth stressing – once again – that precise tests of the Kerr metric with this technique will only be possible in the presence of the correct astrophysical model. In this context, one of the most crucial ingredients is the possibility of having the correct theoretical prediction of the emissivity profile. For the time being, the iron line method seems to be the most promising techniques, or

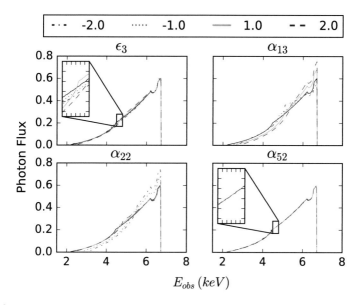

Fig. 13.8 Impact of the deformation parameters ε_3, α_{13}, α_{22}, and α_{52} on the iron line shape. The spacetime is described by the Johannsen metric with the spin parameter $a_* = 0.8$. The viewing angle is $i = 30°$. The profile of the emissivity is modeled with a simple power law with emissivity index $q = 3$, namely $I_e \propto 1/r_e^3$. The inner edge of the disk is at the ISCO radius $r_{in} = r_{ISCO}$, and the outer edge is at $r_{out} = 400\ M$. From [8]

at least one of the most promising techniques, to test the Kerr black hole hypothesis with electromagnetic radiation. However, we cannot yet assert that precise tests of the Kerr metric, as shown by some simulations for the next generation of X-ray missions (see e.g. Sect. 13.2.2), will be possible.

13.2.1 Fitting a Non-Kerr Model with Kerr Models

The shape of the iron Kα line determined by the relativistic effects occurring in the strong gravity region around a black hole is discussed in Sect. 8.2 and, in particular, illustrated in Fig. 8.7. While the iron line profiles in many non-Kerr spacetimes are similar to the Kerr predictions and can be understood with Fig. 8.7, there are also scenarios in which the expected profile is quite different. Considering that current X-ray data are normally fitted with Kerr models and there is no tension between observations and theoretical models, in the sense that the fits are acceptable, spacetimes with very different predictions can be ruled out. Such a statement also depends on the quality of the available data.

Let us now illustrate such a strategy to rule out scenarios with exotic iron lines with an example. We consider the Kerr black holes with scalar hair discussed in

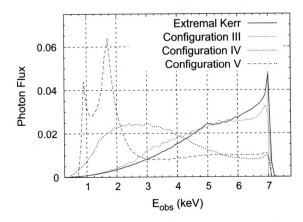

Fig. 13.9 Iron line profiles of an extremal Kerr black hole (*red solid curve*) and of the three Kerr black holes with scalar hair named configurations III, IV, and V in Fig. 12.1. The "wiggles" appearing in the shape of these lines, in particular for the configurations III and IV, are due to resolution effects of the numerical metric. From [28], reproduced by permission of IOP Publishing. All rights reserved

Sect. 12.1.1 and, in particular, the three configurations called, respectively, III, IV, and V. Here we follow the study presented in [28]. Assuming that the inclination angle of the disk with respect to the line of sight of the observer is $i = 45°$ and employing the emissivity profile $1/r^3$, the iron Kα lines in the reflection spectrum of the accretion disk in the spacetimes III, IV, and V are shown in Fig. 13.9. The same figure also shows the iron line profile of an extremal Kerr black hole (red solid curve).

The iron line shape of Configuration III is not too different from that expected from a fast-rotating Kerr black hole. The dimensionless parameter q of Configuration III is indeed much smaller than those of the other two configurations; that is, the scalar field contribution is small and the spacetime is mainly determined by the central Kerr black hole. On the contrary, the iron line of Configuration V is definitively different from those studied so far and its shape seems also to be difficult to understand in terms of the contributions from different annuli of Fig. 8.7. In the case of Configuration IV, its iron line is surely different from that studied in the Kerr metric, but it is not as weird as that of Configuration V.

As done in Sect. 8.2, the best way to understand the shape of the iron line is to see the contributions from different annuli. This is shown in Figs. 13.10, 13.11, and 13.12, respectively for the spacetimes III, IV, and V. Figure 13.10 is not qualitatively different from Fig. 8.7 and therefore the interpretation is the same.[2]

Figure 13.11 already presents some important differences. The contribution from the inner annulus is very large and it is responsible for the "bump" in the total iron

---

[2]Since the metric of the spacetimes III, IV, and V is given numerically, the precision of the calculations is lower and a smaller accretion disk is employed. The "wiggles" in the shape of these lines are due to resolution effects of the numerical metric.

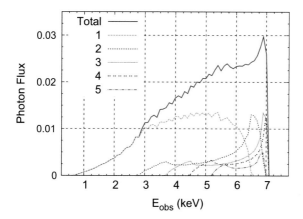

Fig. 13.10 Total iron Kα line (*red solid line*) and contributions to the total iron line from different annuli (lines 1, 2, 3, 4, and 5) for Configuration III. The annulus 1 has the inner edge at the ISCO radius and the outer edge at the radius $r = r_{\rm ISCO} + 1$. The annulus 2 has the inner edge at the same radius as the outer edge of the annulus 1 and the outer edge at the radius $r = r_{\rm ISCO} + 2$. The annulus 3 has the inner edge at the same radius as the outer edge of the annulus 2 and the outer edge at the radius $r = r_{\rm ISCO} + 4$. The annulus 4 has the inner edge at the same radius as the outer edge of the annulus 3 and the outer edge at the radius $r = r_{\rm ISCO} + 10$. The annulus 5 has the inner edge at the same radius as the outer edge of the annulus 4 and the outer edge at the radius $r = r_{\rm ISCO} + 25$, which corresponds to the outer edge of the disk. Units in which $1/\mu = 1$ are used. See the text, Sect. 12.1.1, and the original paper for more details. From [28], reproduced by permission of IOP Publishing.

line profile. Since we are assuming an intensity profile $\propto 1/r^3$, the emission at very small radii is always high, especially if the inner edge of the disk is at a smaller radius. However, in the case of a fast-rotating Kerr black hole a large fraction of photons is captured by the black hole. Here gravity is "weaker" and it is easier for photons emitted at small radii to escape to infinity.

In the case of Configuration V, such an effect is amplified, as shown in Fig. 13.12. Now most of the mass of the object comes from the scalar field cloud and the central black hole only provides a small contribution. The two peaks at 1–2 keV are produced near the inner edge of the disk, where the gravitational redshift is strong but the light bending is much weaker than the one around a black hole (and indeed the radius of the event horizon is very small). This permits to a large fraction of photons to escape to infinity. The two horns in the iron line profile of the inner annulus are produced by Doppler redshift and blueshift, like the two horns expected from annuli at large radii in the Kerr metric.

How can we decide if a certain iron line is so different from the Kerr predictions that can be excluded by current observations? A simple and fast selection criterion can be the following. We simulate an observation with a past or current X-ray mission using the iron line calculated in the exotic spacetime. We then treat the simulation as real data and we fit the spectrum with a Kerr model. If the Kerr model is clearly unable to provide at least an acceptable fit, we can conclude that similar iron lines

Fig. 13.11 As in Fig. 13.10 for Configuration IV. From [28], reproduced by permission of IOP Publishing. All rights reserved

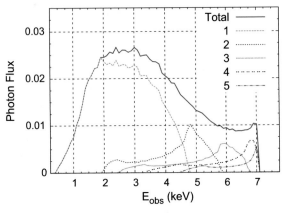

Fig. 13.12 As in Fig. 13.10 for Configuration V. From [28], reproduced by permission of IOP Publishing. All rights reserved

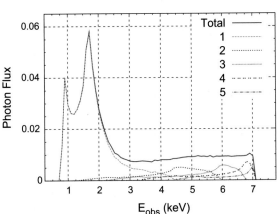

have never been seen so far, as there is currently no tensions between observational data and theoretical predictions in the Kerr spacetime. It is important to check that this is true for any viewing angle and even changing the emissivity profiles. This approach was employed in [28] for the case of Ker black holes with scalar hair. See also [12, 33] for similar studies to test other exotic objects.

Figure 13.13 shows the results of a simulation of the spectrum of the solution V Kerr black hole with scalar hair. The theoretical model is a simple power-law with a single iron line. Such a spectrum is definitively simple, but it is enough to rule out exotic scenarios. For this kind of tests, it is important to have bright sources, so the simulation is done assuming the typical parameters for a bright black hole binary with a strong iron line: the total energy flux in the 0.7–10 keV range is about $4 \cdot 10^{-9}$ erg/s/cm^2 and the iron line has an equivalent width of about 200 eV. The simulation is done using XSPEC[3] with the background, the ancillary, and the response

---

[3]XSPEC is an X-ray spectral-fitting software commonly used in X-ray astronomy. See [2] and http://heasarc.gsfc.nasa.gov/docs/xanadu/xspec/index.html for more details.

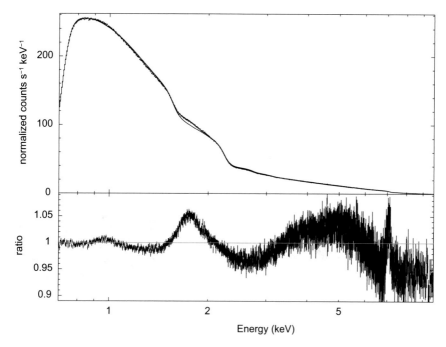

Fig. 13.13 Simulation of the spectrum of Configuration V Kerr black hole with scalar hair in a binary with XIS/Suzaku. *Top panel* folded spectra of the simulated data and of the best fit. *Bottom panel* ratio between the simulated data and the best fit. The minimum of the reduced χ^2 is $\chi^2_{min,red} = 6.08$. See the text for more details. Figure courtesy of Yueying Ni

matrix files of XIS/Suzaku assuming an exposure time of 100 ks (see Sect. 5.1 for the explanation of these concepts). The photon count turns out to be about $3 \cdot 10^7$. The data are then treated as real data and fitted with XSPEC with a power-law and an iron line for a Kerr model. The top panel in Fig. 13.13 shows the folded spectra of the simulated data and of the best fit. The bottom panel is the key-plot for our selection criterion: it is the ratio between the simulated data and the best fit. It is clear that we are missing important features, as the spectrum cannot be fitted with the Kerr model. The minimum of the reduced χ^2 is another indication of a bad fit ($\chi^2_{min,red} = 6.08$ in the simulation in Fig. 13.13). Similar iron lines are not observed in real data and therefore we can argue that the metric around astrophysical black holes is not described by the solution V.

Figure 13.14 is the counterpart of Fig. 13.13 for Configuration III. While the metric is not equivalent to the Kerr one, the fit obtained from the Kerr model is good. In this case, we can conclude that we cannot distinguish Configuration III from a Kerr black hole with a similar observation. As the photon count increases (for a specific instrument, this requires a longer exposure time), the error bars of the data become smaller and it is possible to resolve smaller differences. However, such an approach employing the analysis of simulations rather than real data is a simple criterion to

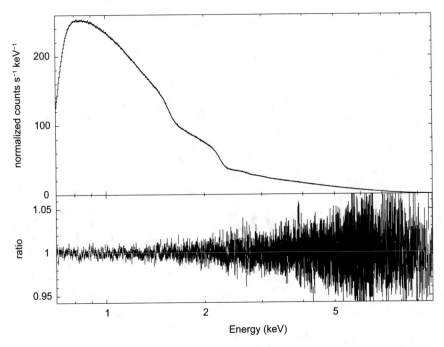

Fig. 13.14 As in Fig. 13.13 for Configuration III. The minimum of the reduced χ^2 is $\chi^2_{min,red} = 1.01$. See the text for more details. Figure courtesy of Yueying Ni

distinguish spacetimes with predictions significantly different from Kerr from those with similar predictions, but it cannot be used to perform precise tests. For this reason, we can adopt typical parameters of an observation and of a source, without looking at the possible best observations. If we use this approach and we find a simulation that does not provide a good fit, but it is neither so horrible, we cannot arrive at any conclusion: we should perform a more detailed analysis with real observations of specific sources.

13.2.2 Constraining Deviations from the Kerr Metric

The natural way to test the Kerr black hole hypothesis using X-ray reflection spectroscopy is not with the approach of the previous subsection, but fitting simulations or real data with a model in which the metric of the spacetime is either a certain solution of a well specified theoretical framework or a parametrized metric. Eventually we will obtain a measurement of the parameters of the model, including those related to the metric of the spacetime.

As a simple example of a similar measurement, we briefly review the study in [27] based on some simulated observations with NuSTAR and LAD/eXTP. The results obtained with NuSTAR can be seen as the typical constraints that can be obtained today with current X-ray missions, while those with LAD/eXTP can be seen as the constraints that will be possible to obtain with the next generation of satellites. See Sect. 5.1 and references therein for more details about these two missions.

Figures 13.15 and 13.16 show the constraints on the deformation parameters of the version of the Konoplya–Rezzolla–Zhidenko metric discussed in Sect. 12.2.3. These simulations do not consider a specific source, but simply employ the typical parameters for a bright black hole binary, which is likely the most suitable type of source for this kind of measurements. The model of the simulated observation is simple: a power-law component with a single iron Kα line. The power-law continuum, representing the spectrum from the hot corona, is generated with the photon index $\Gamma = 2$. The iron line used in the simulations is generated assuming the Kerr metric ($\delta_j = 0$ for all j) with the spin parameter $a_* = 0.7$ and the inclination angle $i = 45°$. The intensity profile is $\propto 1/r_e^3$. The energy flux of the source is $6 \cdot 10^{-9}$ ergs/s/cm^2 in the range 1–9 keV, and the equivalent width of the iron line is 230 eV.

At 6 keV, the effective area of NuSTAR is about 800 cm^2, and the energy resolution is about 400 eV. Among the expected four instruments on board of eXTP, LAD is the most suitable for the study of bright sources. It has an unprecedented effective area of more than 30,000 cm^2 at 6 keV, which is a significant advantage with respect to current X-ray missions for testing the Kerr metric. The energy resolution at 6 keV is expected to be better than 200 eV. For both NuSTAR and LAD/eXTP, the exposure time of the simulations is $\tau = 100$ ks.

The simulated observations are treated as real data and fitted with XSPEC with the model

$$\texttt{powerlaw + KRZ.} \tag{13.4}$$

`powerlaw` is a model already in XSPEC to describe a power-law component. `KRZ` is a table model with the iron line in the Konoplya–Rezzolla–Zhidenko metric with three parameters (a_*, i, and the non-vanishing deformation parameter under investigation). In the fit, there are six free parameters: the photon index of the power-law continuum Γ, the normalization of the continuum, the spin a_*, the inclination angle i, one of the six deformation parameters δ_j, and the normalization of the iron line. The results are the constraints in Figs. 13.15 and 13.16, which show the contour levels of χ^2 in the plane spin parameter versus deformation parameter.

The most remarkable result is the significant difference in the constraining power between NuSTAR and LAD/eXTP. In the case of the NuSTAR observation, it is impossible to constrain the deformation parameter. The much higher photon count number with LAD/eXTP permits to measure both the spin and the deformation parameters, and the constraints look impressive. A particular case is the deformation parameter δ_6. It does not appreciably affect the shape of the iron line, and even in the case of LAD/eXTP it is impossible to get a meaningful bound on its value.

The difference between NuSTAR and eXTP is mainly due to the very large effective area of the LAD instrument. Actually, the constraints from LAD/eXTP are so stringent that systematics effects may be dominant. For this purpose, it will be very

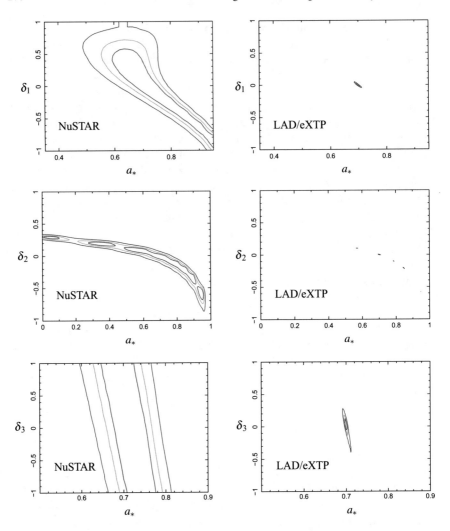

Fig. 13.15 $\Delta\chi^2$ contours from simulations with NuSTAR (*left panels*) and LAD/eXTP (*right panels*) of a bright black hole binary, assuming an exposure time of 100 ks. The reference model is a Kerr black hole in which the spin parameter is $a_* = 0.7$ and the inclination angle is $i = 45°$. It is compared with the predictions for spacetimes with non-vanishing Konoplya–Rezzolla–Zhidenko deformation parameters δ_1 (*top panels*), δ_2 (*central panels*), and δ_3 (*bottom panels*). The *red*, *green*, and *blue curves* indicate, respectively, the 1-, 2-, and 3-σ confidence level limits. See the text and the original paper for more details. From [27], reproduced by permission of IOP Publishing. All rights reserved

important to fit the future X-ray data with sophisticated theoretical models. Current uncertainties, in particular concerning the behavior of the emissivity profile, may prevent or limit the possibility of performing accurate tests of the Kerr metric.

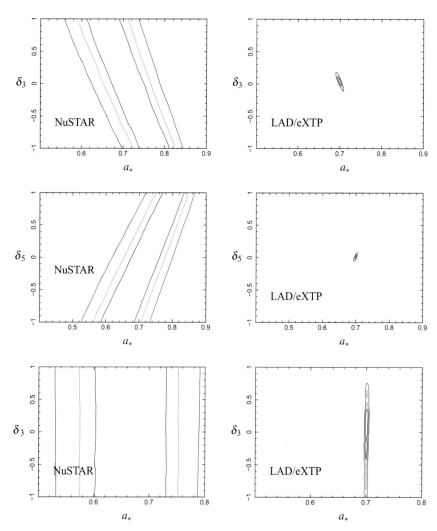

Fig. 13.16 As in Fig. 13.15 for the Konoplya–Rezzolla–Zhidenko deformation parameters δ_4 (*top panels*), δ_5 (*central panels*), and δ_6 (*bottom panels*). The *red*, *green*, and *blue curves* indicate, respectively, the 1-, 2-, and 3-σ confidence level limits. See the text and the original paper for more details. From [27], reproduced by permission of IOP Publishing. All rights reserved

13.2.3 Iron Line Reverberation Mapping

As discussed in Sect. 8.5, the time information in the iron line signal can be a valuable tool to better probe the spacetime geometry around a black hole [14, 16]. In particular, this technique can constrain certain deformation parameters that cannot be constrained, or whose constraints are weak, by the standard time-integrated

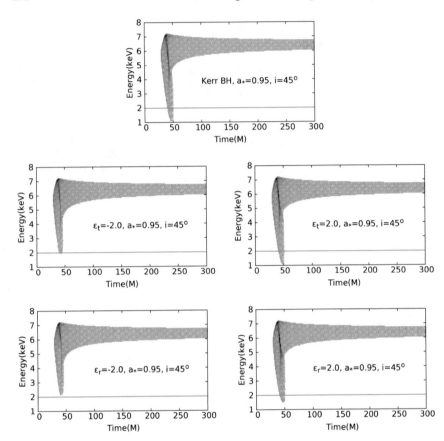

Fig. 13.17 Impact of the Cardoso–Pani–Rico deformation parameters ε_3^t and ε_3^r on the 2D transfer function: Kerr spacetime with $\varepsilon_3^t = \varepsilon_3^r = 0$ (*top panel*) and non-Kerr spacetimes with $\varepsilon_3^t = -2$ and $\varepsilon_3^r = 0$ (*central left panel*), $\varepsilon_3^t = 2$ and $\varepsilon_3^r = 0$ (*central right panel*), $\varepsilon_3^t = 0$ and $\varepsilon_3^r = -2$ (*bottom left panel*), $\varepsilon_3^t = 0$ and $\varepsilon_3^r = 2$ (*bottom right panel*). In all the plots, the spin parameter is $a_* = 0.95$, the viewing angle is $i = 45°$, the height of the source is $h = 10\ M$, and the intensity profile is $\propto 1/r^3$. The color of the 2D transfer function indicates the photon number density and ranges from *light gray to black* as the density increases (in arbitrary units). From [16]

measurement. As a simple example to understand this point, we can consider a deformation parameter that only alters the photon propagation time. If Doppler boosting, gravitational redshift, and light bending do not change, a standard time-integrated measurement cannot constrain the value of such a deformation parameter. A reverberation measurement, which is sensitive to the time of arrival of photons, may do it.

Jiang et al. [16] have studied the constraining power of iron line reverberation measurements within the Cardoso–Pani–Rico metric with the deformation parameters ε_3^t and ε_3^r. Figure 13.17 shows the impact of ε_3^t and ε_3^r on the 2D transfer function. From these plots, we see that deviations from the Kerr background affect the 2D

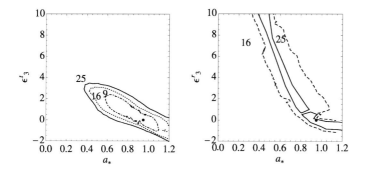

Fig. 13.18 $\Delta\chi^2$ contours obtained from a simulation with $N_{\text{line}} = 10^4$ photons in the iron line with the analysis of the time-integrated spectrum. The simulated observation of the spectrum of a Kerr black hole with the spin parameter $a_* = 0.95$ and the viewing angle $i = 70°$ is fitted with the spectra calculated in the Cardoso–Pani–Rico metric in which the only non-vanishing deformation parameter is either ε_3^t (*left panel*) or ε_3^r (*right panel*). The free parameters in the fit are: the spin parameter, the deformation parameter (ε_3^t in the *left panel* and ε_3^r in the *right panel*), the viewing angle, the ratio between the continuum and the iron line photon flux, and the photon index of the continuum. From [16]; see the original paper for more details

transfer function. However, it is necessary a quantitative analysis to figure out the difference between the constraining power of a time-integrated and of a reverberation measurement.

Figure 13.18 shows the constraints from a simulation of a time-integrated measurement in which the spacetime is described by the Kerr metric. The spin parameter is $a_* = 0.95$ and the viewing angle is $i = 70°$. The simulated spectrum is fitted with the spectra calculated in the Cardoso–Pani–Rico metric in which the only non-vanishing deformation parameter is either ε_3^t (left panel) or ε_3^r (right panel). Here the key-point is that the deformation parameter ε_3^r cannot be constrained by the time-integrated measurement, even assuming a favorable source with a high spin and a large inclination angle, two ingredients that should maximize the relativistic effects affecting the photons. In this simulation, the photon count in the iron line is $N_{\text{line}} = 10^4$. If we increase this number, we reduce the intrinsic noise of the source and the measurement (typically) becomes more precise, but we only make the width of the allowed region thinner. As we can see in the right panel in Fig. 13.18, the allowed region extends to very large values of ε_3^r.

Figure 13.19 illustrates the constraints from the same simulation when the time information is taken into account. The allowed region is now very small and ε_3^r cannot assume large values. The possibility of getting very strong constraints as those shown in Fig. 13.19 clearly relies on the possibility of having the correct astrophysical model. If this were not the case, the constraints might be stringent, but wrong. As discussed in [16], a crucial role is also played by the signal to noise. In the case of a low photon count in the iron line, the time sampling of the reverberation measurement effectively dilutes the signal to noise by apportioning the signal into additional time bins. The measurement is thus significantly affected by Poisson noise of the source.

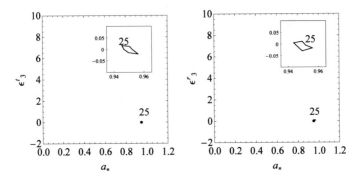

Fig. 13.19 As in Fig. 13.18, but from the analysis of the 2D transfer function. The free parameters in the fit are: the spin parameter, the deformation parameter (ε_3^t in the *left panel* and ε_3^r in the *right panel*), the viewing angle, the height of the source h, the ratio between the continuum and the iron line photon flux, and the photon index of the continuum. From [16]; see the original paper for more details

Reverberation measurements can provide stronger and stronger constraints than time-integrated observations when N_{line} increases. With the X-ray data available today, the signal is too diluted, and reverberation mapping has no advantages with respect to a measurement of the time-integrated spectrum.

13.3   Quasi-periodic Oscillations

As discussed in Chap. 9, QPOs posses a few interesting properties that make them a promising technique to get very precise measurements of the metric around astro-physical black holes. In particular, their frequencies can be determined with high accuracy, and therefore they can potentially provide more precise measurements than other techniques like the continuum-fitting and the iron line methods. However, there is currently no consensus on which mechanism is responsible for their produc-tion, or even if it is a single mechanism or there are multiple mechanisms. Different models provide a different measurement of the parameters of the background metric, which means that QPO data cannot be used to test fundamental physics at this time.

Despite the current uncertainty on the exact mechanism responsible for QPOs, several authors have already explored the possibility of using QPOs to test the Kerr metric [1, 5, 6, 9, 17, 25, 32]. Even if QPO data can potentially provide very accurate measurements, there is a fundamental degeneracy among the estimate of the spin parameter and possible deviations from the Kerr solution. In other words, the fact that the QPO frequencies can be measured with high precision permits one to obtain a narrow allowed region on the spin parameter versus deformation parameter plane, but (depending on the specific deviation from the Kerr solution) it may be impossible to break the degeneracy and constrain deviations from Kerr. This appears to be a fundamental characteristic of this approach and is easy to understand.

The QPO frequencies are just a few numbers and even if they are measured with infinite precision, the same few numbers can be obtained in many different scenarios by properly adjusting the metric.

Let us consider an explicit example [9]. We employ a variant of the Konoplya–Rezzolla–Zhidenko metric of Sect. 12.2.3. The line element is still given by Eq. (12.43), but now [21]

$$N^2(r, \theta) = \frac{r^2 - 2Mr + a^2}{r^2} - \frac{\eta}{r^3},$$

$$B^2(r, \theta) = 1,$$

$$\Sigma(r, \theta) = \frac{r^2 + a^2 \cos^2 \theta}{r^2},$$

$$K^2(r, \theta) = \frac{\left(r^2 + a^2\right)^2 - a^2 \sin^2 \theta \left(r^2 - 2Mr + a^2\right)}{r^2 \left(r^2 + a^2 \cos^2 \theta\right)} + \frac{\eta a^2 \sin^2 \theta}{r^3 \left(r^2 + a^2 \cos^2 \theta\right)},$$

$$W(r, \theta) = \frac{2Ma}{r^2 + a^2 \cos^2 \theta} + \frac{\eta a}{r^2 \left(r^2 + a^2 \cos^2 \theta\right)}, \tag{13.5}$$

where $a = J/M$. This metric can be obtained from the Kerr solution by adding a static deformation η such that

$$M \rightarrow M + \frac{\eta}{2r^2}. \tag{13.6}$$

It is convenient to rewrite η as

$$\eta = r_0 \left(r_0^2 - 2Mr_0 + a^2\right), \tag{13.7}$$

where r_0 is the radial coordinate of the black hole event horizon. If we write

$$r_0 = r_{\text{Kerr}} + \delta r = M + \sqrt{M^2 - a^2} + \delta r, \tag{13.8}$$

we can use δr as the deformation parameter to quantify possible deviations from the Kerr spacetime. If $\delta r = 0$, r_0 reduces to the radial position of the event horizon of a Kerr black hole. In the general case, δr measures the difference of the radial coordinate of the event horizon with respect to that of a Kerr black hole with the same mass and spin.

Let us now constrain the quantity $\delta r / r_{\text{Kerr}}$ from the current QPO data of GRO J1655-40 assuming the relativistic precession model discussed in Sect. 9.3. Presently, we have an observation in which we can measure all the three frequencies (ν_U, ν_L, and ν_C) and one in which we observe two frequencies (ν_U and ν_C) [26]. Moreover, we have an independent dynamical measurement of the mass of the black hole [11]. In summary, we have the following six measurements:

$$(\nu_U, \nu_L, \nu_C) = (441 \pm 2, \; 298 \pm 4, \; 17.3 \pm 0.1) \text{ Hz},$$
$$(\nu_U, \nu_L, \nu_C) = (451 \pm 5, \; -, \; 18.3 \pm 0.1) \text{ Hz},$$
$$M_{dyn} = 5.4 \pm 0.3 \, M_\odot. \tag{13.9}$$

The free parameters are five: the mass M, the spin parameter a_*, the radius of the observation with three frequencies, the radius of the observation with two frequencies, and the deformation parameter. Figure 13.20 shows the constraints on a_* and $\delta r/r_{Kerr}$ [9].

Now we want to investigate if the constraints of Fig. 13.20 can be improved if, in the future, we can get better measurements of the frequencies than those available today and reported in (13.9). Assuming that GRO J1655-40 is indeed a Kerr black hole with $a_* = 0.29$ and $M = 5.31 \, M_\odot$, we employ the following set of seven measurements

$$(\nu_U, \nu_L, \nu_C) = (441.4 \pm 1.0, \; 298.0 \pm 1.0, \; 17.59 \pm 0.05) \text{ Hz},$$
$$(\nu_U, \nu_L, \nu_C) = (451.0 \pm 1.0, \; 313.1 \pm 1.0, \; 18.36 \pm 0.05) \text{ Hz},$$
$$M_{dyn} = 5.4 \pm 0.3 \, M_\odot. \tag{13.10}$$

The QPO frequencies are now six and their uncertainty is smaller than in the measurements in (13.9). The measurement of the mass is the same, because it is not obvious that we can have better dynamical measurements in the near future. We also note that (assuming the Kerr metric) the radial coordinate of the two sets of QPOs is, respectively, $r_1 = 5.67 \, M$ and $r_2 = 5.59 \, M$. The new constraints are shown in Fig. 13.21. It is remarkable that the allowed region is now very thin. However, if we consider the allowed range of $\delta r/r_{Kerr}$, it is almost the same as in Fig. 13.20. The frequency measurements in (13.10), despite being much better than in (13.9), do not help to break the degeneracy between the spin and possible deviations from the Kerr metric.

As discussed in [9], in order to break the parameter degeneracy between a_* and $\delta r/r_{Kerr}$ we should have two or more three-frequency observations at very different radii. This would be equivalent to measure the spacetime metric at several radii and, in the case of many three-frequency observations, to be able to get the radial profile of the metric. It is currently unclear whether this can be a realistic possibility. For sure, it is natural to expect that better observations/measurements are possible when the QPO is at smaller radii.

13.4 Violation of the Kerr Bound $|a_*| \leq 1$

In the Kerr spacetime, the condition for the existence of the event horizon is $|a_*| \leq 1$. If $|a_*| > 1$, there is no horizon, and the Kerr metric describes the spacetime of a naked singularity. According to the cosmic censorship conjecture, naked singularities cannot be created by gravitational collapse [30], even if we know some counterexamples

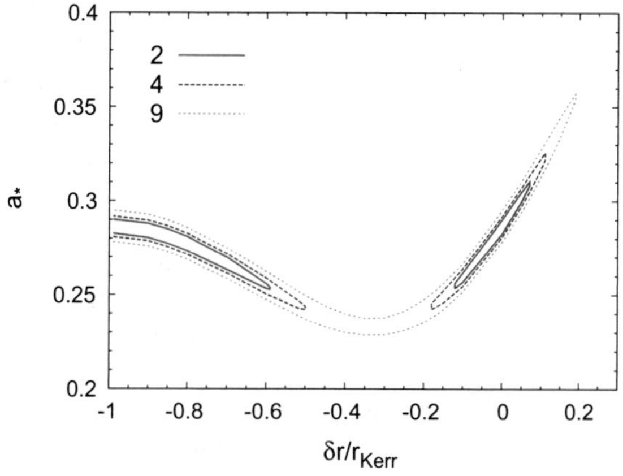

Fig. 13.20 Constraints on the spin parameter a_* and the deformation $\delta r/r_{\text{Kerr}}$ for the black hole in GRO J1655-40 using current QPO observations within the relativistic precession model. The *red-solid line*, *blue-dashed line*, and *green-dotted line* represent, respectively, the contour levels $\Delta\chi^2 = 2, 4$, and 9

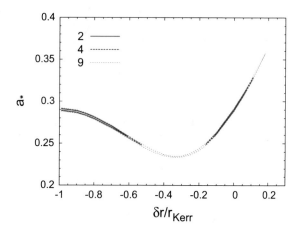

Fig. 13.21 As in Fig. 13.20, but assuming the set of measurements in Eq. (13.10), where the QPO frequencies are six and their uncertainties are smaller than those of current data

in which, starting from regular initial data, a naked singularity can be created for an infinitesimal time [18]. Can an astrophysical black hole have a spin parameter $|a_*| > 1$? Can we measure with the techniques discussed in the previous sections a spin parameter $|a_*| > 1$? And in the case we had such a measurement, how should it be interpreted? These questions can be addressed as follows:

1. First, we should bear in mind that an object with $|a_*| \sim 1$ is a fast-rotating body if it is very compact. For instance, the spin parameter of Earth is about 10^3 and there is no violation of any principle because the vacuum solution holds up to the Earth surface. The Earth radius is $r_{\text{Earth}} \approx 6,400$ km, which is much larger than its gravitational radius $r_g = M \approx 4.4$ mm.
2. The Kerr metric with $|a_*| > 1$ is not a viable astrophysical scenario. For instance, Giacomazzo et al. [13] found that in general relativity stellar models with $|a_*| > 1$ do not collapse without losing angular momentum. If we consider an existing Kerr black hole and we try to overspin it up to $|a_*| > 1$, we fail [10]. The same

negative result is found by considering the collision of two black holes at the speed of light in full non-linear general relativity [31]. None of these studies proves that it is impossible to create a Kerr spacetime with $|a_*| > 1$, but at least they suggest that naked Kerr singularities may not be physical merely because they may not be created in an astrophysical process. Assuming it is somehow possible to create a similar object, Pani et al. [29] have shown that the spacetime would be unstable independently of the boundary conditions imposed at the excision radius r, which means that possible unknown quantum gravity corrections at the singularity cannot change these conclusions. Very compact objects described by the Kerr solution with $|a_*| > 1$ can unlikely exist in Nature.

3. If the measurement of the spin parameter of a black hole gave a value larger than 1 assuming the Kerr metric, the measurement would be presumably wrong, but it would be a clear indication of new physics. As discussed at the point 2, the Kerr metric with $|a_*| > 1$ is not a viable astrophysical scenario. This means that: (i) the metric around the objects is not described by the Kerr solution, and (ii) we do not know the value of the spin parameter. Since spin measurements strongly depend on the exact background metric, they can provide the correct value only if we employ the right metric. A non-Kerr object with $|a_*| < 1$ may be measured to have $|a_*| > 1$ when it is assumed the Kerr metric. The contrary may also be possible, namely a non-Kerr object with $|a_*| > 1$ could be interpreted with $|a_*| < 1$ if we incorrectly adopt the Kerr metric.

4. It is worth noting that $|a_*| = 1$ is a critical value only in the Kerr metric. In the case of other black hole solutions, the critical bound is typically different, and it may be either larger or smaller than 1, depending on the spacetime geometry. For instance, Fig. 12.2 shows that black holes in Einstein–dilaton–Gauss–Bonnet gravity may have a spin parameter slightly larger than 1. If the metric is not described by the Kerr solutions, the results in [10, 13, 31] mentioned at the point 1 do not hold and there are indeed examples showing that it is possible to spin a non-Kerr compact object up to $|a_*| > 1$ [3, 4].

References

1. A.N. Aliev, G.D. Esmer, P. Talazan, Class. Quant. Grav. **30**, 045010 (2013), arXiv:1205.2838 [gr-qc]
2. K.A. Arnaud, *Astronomical Data Analysis Software and Systems V*, vol. 101 (1996), p. 17
3. C. Bambi, JCAP **1105**, 009 (2011), arXiv:1103.5135 [gr-qc]
4. C. Bambi, Phys. Lett. B **705**, 5 (2011), arXiv:1110.0687 [gr-qc]
5. C. Bambi, JCAP **1209**, 014 (2012), arXiv:1205.6348 [gr-qc]
6. C. Bambi, Eur. Phys. J. C **75**, 162 (2015), arXiv:1312.2228 [gr-qc]
7. C. Bambi, J. Jiang, J.F. Steiner, Class. Quant. Grav. **33**, 064001 (2016), arXiv:1511.07587 [gr-qc]
8. C. Bambi, A. Cardenas-Avendano, T. Dauser, J.A. Garcia, S. Nampalliwar, arXiv:1607.00596 [gr-qc]
9. C. Bambi, S. Nampalliwar, Europhys. Lett. **116**, 30006 (2016), arXiv:1604.02643 [gr-qc]
10. E. Barausse, V. Cardoso, G. Khanna, Phys. Rev. Lett. **105**, 261102 (2010), arXiv:1008.5159 [gr-qc]

11. M.E. Beer, P. Podsiadlowski, Mon. Not. Roy. Astron. Soc. **331**, 351 (2002), arXiv:astro-ph/0109136
12. Z. Cao, A. Cardenas-Avendano, M. Zhou, C. Bambi, C.A.R. Herdeiro, E. Radu, JCAP **1610**, 003 (2016), arXiv:1609.00901 [gr-qc]
13. B. Giacomazzo, L. Rezzolla, N. Stergioulas, Phys. Rev. D **84**, 024022 (2011), arXiv:1105.0122 [gr-qc]
14. J. Jiang, C. Bambi, J.F. Steiner, JCAP **1505**, 025 (2015), arXiv:1406.5677 [gr-qc]
15. J. Jiang, C. Bambi, J.F. Steiner, Astrophys. J. **811**, 130 (2015), arXiv:1504.01970 [gr-qc]
16. J. Jiang, C. Bambi, J.F. Steiner, Phys. Rev. D **93**, 123008 (2016), arXiv:1601.00838 [gr-qc]
17. T. Johannsen, D. Psaltis, Astrophys. J. **726**, 11 (2011), arXiv:1010.1000 [astro-ph.HE]
18. P.S. Joshi, D. Malafarina, Int. J. Mod. Phys. D **20**, 2641 (2011), arXiv:1201.3660 [gr-qc]
19. P.S. Joshi, D. Malafarina, R. Narayan, Class. Quant. Grav. **31**, 015002 (2014), arXiv:1304.7331 [gr-qc]
20. L. Kong, Z. Li, C. Bambi, Astrophys. J. **797**, 78 (2014), arXiv:1405.1508 [gr-qc]
21. R. Konoplya, A. Zhidenko, Phys. Lett. B **756**, 350 (2016), arXiv:1602.04738 [gr-qc]
22. H. Krawczynski, Astrophys. J. **754**, 133 (2012), arXiv:1205.7063 [gr-qc]
23. D. Liu, Z. Li, Y. Cheng, C. Bambi, Eur. Phys. J. C **75**, 383 (2015), arXiv:1504.06788 [gr-qc]
24. J. Liu, J. McClintock, R. Narayan, S. Davis, J. Orosz, Astrophys. J. **679**, L37 (2008) [Erratum: Astrophys. J. **719**, L109 (2010)], arXiv:0803.1834 [astro-ph]
25. A. Maselli, L. Gualtieri, P. Pani, L. Stella, V. Ferrari, Astrophys. J. **801**, 115 (2015), arXiv:1412.3473 [astro-ph.HE]
26. S.E. Motta, T.M. Belloni, L. Stella, T. Muoz-Darias and R. Fender. Mon. Not. Roy. Astron. Soc. **437**, 2554 (2014), arXiv:1309.3652 [astro-ph.HE]
27. Y. Ni, J. Jiang, C. Bambi, JCAP **1609**, 014 (2016), arXiv:1607.04893 [gr-qc]
28. Y. Ni, M. Zhou, A. Cardenas-Avendano, C. Bambi, C.A.R. Herdeiro, E. Radu, JCAP **1607**, 049 (2016), arXiv:1606.04654 [gr-qc]
29. P. Pani, E. Barausse, E. Berti, V. Cardoso, Phys. Rev. D **82**, 044009 (2010), arXiv:1006.1863 [gr-qc]
30. R. Penrose, Riv. Nuovo Cim. **1**, 252 (1969) [Gen. Rel. Grav. **34**, 1141 (2002)]
31. U. Sperhake, V. Cardoso, F. Pretorius, E. Berti, T. Hinderer, N. Yunes, Phys. Rev. Lett. **103**, 131102 (2009), arXiv:0907.1252 [gr-qc]
32. Z. Stuchlik, A. Kotrlova, Gen. Rel. Grav. **41**, 1305 (2009), arXiv:0812.5066 [astro-ph]
33. M. Zhou, A. Cardenas-Avendano, C. Bambi, B. Kleihaus, J. Kunz, Phys. Rev. D **94**, 024036 (2016), arXiv:1603.07448 [gr-qc]

Chapter 14
Tests with Other Approaches

This chapter provides, for completeness, a quick overview on a number of approaches to test the Kerr metric that are alternative to those discussed in Chap. 13 based on X-ray observations. Some topics have been only very briefly discussed in Part II, while other techniques are based on physics not mentioned before in this book. Some of these measurements are sensitive to different relativistic effects with respect to the X-ray observations in Chap. 13, and they can thus be seen as complementary tests to investigate the strong gravity region around astrophysical black holes.

14.1 The Special Case of SgrA*

SgrA* is the name of the radio source at the center of the Galaxy and is identified with the supermassive black hole in the nucleus of the Milky Way. Its mass is about $4 \cdot 10^6 \, M_\odot$ and its distance from us is about 8 kpc. It is a peculiar object among all the black holes in the Universe, because it is the only nearby supermassive black hole. The next subsections list a number of possible techniques potentially capable of probing the spacetime metric around SgrA* and test the Kerr black hole hypothesis with this object. Some of these techniques may not be used in the case of other black holes, or at least they do not seem to be so promising as in the case of SgrA*.

14.1.1 Accretion Structure Imaging

As already discussed in Chap. 10, it is widely believed that sub-mm VLBI facilities will be soon able to image SgrA* and detect its shadow. In the ideal case, the boundary of the shadow of a black hole surrounded by an optically thin emitting medium corresponds to the apparent image of the photon capture sphere, which is only

© Springer Nature Singapore Pte Ltd. 2017
C. Bambi, *Black Holes: A Laboratory for Testing Strong Gravity*,
DOI 10.1007/978-981-10-4524-0_14

determined by the spacetime metric and is independent of the particular properties of the accretion flow. In the reality, the situation is more complicated, and at present it is not clear whether it will be possible to measure the apparent photon capture sphere of SgrA* with a sufficient precision to perform tests of the Kerr metric. Despite that, several authors have calculated the boundary of the shadow of a number of black holes in alternative theories of gravity and in parametrized metrics [4–8, 12–14, 16–18, 29, 30, 41, 46, 50, 63, 69, 73, 74].

Figure 14.1 shows the impact of the Cardoso–Pani–Rico deformation parameters ε_3^t (left panels) and ε_3^r (right panels) on the apparent boundary of the photon capture sphere. The small inset in every panel is for the function $R(\phi)/R(0)$ discussed in Sect. 10.3. The deformation parameter ε_3^t mainly alters the size of the shadow for any value of the spin parameter a_*. This is understandable, because ε_3^t describes a deformation in g_{tt} (and only in g_{tt} if $a_* = 0$), which regulates the intensity of the gravitational field. In other words, varying ε_3^t we change the value of the photon capture radius. On the contrary, the deformation parameter ε_3^r can affect the shape of the shadow and the effect is more important as the spin increases. This is because ε_3^r can alter the boundary of the photon capture sphere mainly through its presence in $g_{t\phi}$, which vanishes when $a_* = 0$. As shown in Sect. 10.1.1, for a static and spherically symmetric spacetime the boundary of the photon capture sphere is only determined by g_{tt} and is completely independent of g_{rr}.

Considering the current uncertainties on the possibility of being able to get a very precise detection of the photon capture sphere, tests of the Kerr metric may require the analysis of the complete image of the accretion structure around SgrA*. However, the image of the accretion structure does not depend on the metric only, but also on the astrophysical model. SgrA* is thought to have a radiatively inefficient accretion flow, or RIAF, but there are several models in the literature [81]. The uncertainty in the choice of the correct model to describe the accretion structure around SgrA* is clearly a weak point of this approach.

Johannsen et al. [47] constrain the metric around SgrA* with the available mm-VLBI data of the Event Horizon Telescope (EHT) obtained with the current three-station array, and simulate future constraints obtainable with the forthcoming eight-station array. The Event Horizon Telescope[1] is a project involving mm and sub-mm observatories equipped with VLBI instrumentation to get high resolution images of the accretion flow around supermassive black holes at 230 and 345 GHz. One of the main goals of this project is the observation of the shadow of SgrA*. The current three-station array consists of the James Clerk Maxwell Telescope and the Sub-Millimeter Array in Hawaii (Hawaii), the Submillimeter Telescope Observatory (SMT) in Arizona, and some dishes of the Combined Array for Research in Millimeter-wave Astronomy (CARMA) in California. Observations of SgrA* with the three-station array Hawaii-SMT-CARMA were conducted from 2008 to 2013 [33, 36, 37]. With the forthcoming eight-station array, the sensitivity and the resolution will be significantly increased, especially thanks to the better sensitivity of the Atacama Large Millimeter/submillimeter Array (ALMA) in Chile and the long baselines

---

[1]http://www.eventhorizontelescope.org/.

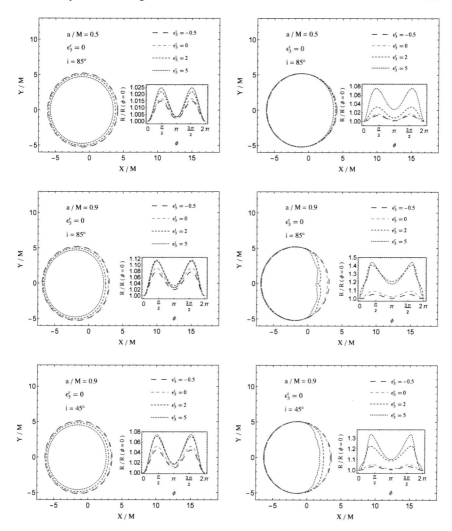

Fig. 14.1 Impact of the Cardoso–Pani–Rico deformation parameters ε_3^t (*left panels*) and ε_3^r (*right panels*) on the boundary of the photon capture sphere as seen by a distant observer. The spin parameter is $a_* = 0.5$ (*top panels*) and 0.9 (*central* and *bottom panels*). The viewing angle is $i = 85°$ (*top* and *central panels*) and 45° (*bottom panels*). The small inset in every panel shows the function $R(\phi)/R(0)$ discussed in Sect. 10.3. From [41], under the terms of the Creative Commons Attribution License

from the stations in the northern hemisphere and the South Pole Telescope (SPT) in Antarctica.

Johannsen et al. [47] employ the quasi-Kerr metric of [42], which reads

$$g_{\mu\nu}^{QK} = g_{\mu\nu}^{K} + \varepsilon h_{\mu\nu}, \qquad (14.1)$$

where $g_{\mu\nu}^{K}$ is the Kerr metric in Boyer–Lindquist coordinates, ε is the deformation parameter, and $h_{\mu\nu}$ has the following four non-vanishing components

$$
h_{tt} = \frac{5\,[1 + 3\cos{(2\theta)}]}{32M^2 r^2}\left[2M\left(3r^3 - 9Mr^2 + 4M^2 r + 2M^3\right)\right.
$$
$$
\left. - 3r^2\,(r - 2M)^2\ln\left(\frac{r}{r - 2M}\right)\right],
$$

$$
h_{rr} = -\frac{5\left(1 - 3\cos^2\theta\right)}{16M^2\,(r - 2M)^2}\left[2M\,(r - M)\left(3r^2 - 6Mr - 2M^2\right)\right.
$$
$$
\left. - 3r^2\,(r - 2M)^2\ln\left(\frac{r}{r - 2M}\right)\right],
$$

$$
h_{\theta\theta} = -\frac{5r\,[1 + 3\cos{(2\theta)}]}{32M^2}\left[-2M\left(3r^2 + 3Mr - 2M^2\right)\right.
$$
$$
\left. + 3r\left(r^2 - 2M^2\right)\ln\left(\frac{r}{r - 2M}\right)\right],
$$

$$
h_{\phi\phi} = -\frac{5r\,[1 + 3\cos{(2\theta)}]}{32M^2}\left[-2M\left(3r^2 + 3Mr - 2M^2\right)\right.
$$
$$
\left. + 3r\left(r^2 - 2M^2\right)\ln\left(\frac{r}{r - 2M}\right)\right]\sin^2\theta. \quad (14.2)
$$

For sufficiently small values of the spin and of the deformation parameter ε, the physical interpretation of the quasi-Kerr metric is that the mass-quadrupole moment of the central object is

$$
Q = -\left(a^2 + \varepsilon M^2\right) M, \quad (14.3)
$$

which reduces to the Kerr one for $\varepsilon = 0$. Within a specific radiatively inefficient accretion flow model, Johannsen et al. [47] obtain the constraints on the spin parameter a_*, the inclination angle θ, and the deformation parameter of the quasi-Kerr metric ε in Fig. 14.2 from the available data with the three-station array Hawaii-SMT-CARMA. The red square indicates the maximum of the 2D probability density. The cyan, blue, and violet regions denote, respectively, the 1-, 2-, and 3-σ confidence level areas. From current observations, the metric around SgrA* is consistent with the Kerr one at 2-σ. As pointed out in [47], these constraints are actually dominated by systematic model uncertainties, which neglect several effects.

Figure 14.3 shows the constraints obtained from a simulated observation with the forthcoming eight-station array, assuming a single 24 h observing run at 230 GHz [47]. The difference between the constraining power of the three-station array Hawaii-SMT-CARMA and the eight-station array is impressive. It is important to remark that precise tests of the Kerr metric as shown in Fig. 14.3 will be only possible by

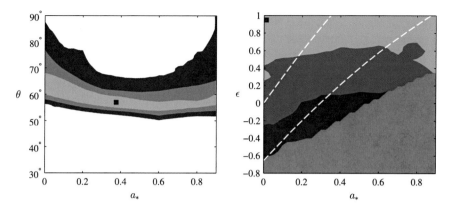

Fig. 14.2 Constraints on the spin parameter a_*, the deformation parameter of the quasi-Kerr metric ε, and the inclination angle θ for SgrA* from the observations of the three-station array Hawaii-SMT-CARMA. The *red square* indicates the maximum of the 2D probability density. The *cyan*, *blue*, and *violet regions* denote, respectively, the 1-, 2-, and 3-σ confidence level areas. The *gray region* is ignored in the analysis, because the spacetimes there have an ISCO radius close to regions with pathological properties of the metric (naked singularities and closed time-like curves). The *dashed white lines* mark the spacetimes with ISCO radius $r = 6\,M$ (*upper line*) and $5\,M$ (*lower line*). See the text for more details. From [47]

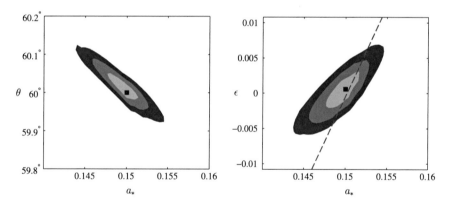

Fig. 14.3 As in Fig. 14.2, but from a simulated observation with the forthcoming eight-station array, assuming a 24 h observing run at 230 GHz. See the text for more details. From [47]

employing a sophisticated astrophysical model capable of describing the accretion flow of SgrA*, while a similar model is not available at the moment.

14.1.2 Accretion Structure Spectrum

If we do not have a sufficiently high angular resolution to image the accretion flow, we can still measure the spectrum of the accretion structure around SgrA*. Of course, this causes a loss of information, just like iron line reverberation mapping in Sect. 13.2.3

is more informative than the standard time-integrated measurement. The possibility of testing the Kerr metric around SgrA* from the analysis of the spectrum of its accretion structure was explored in [53] within the ion torus model of [65, 72]. In practice, however, it seems to be difficult to test the Kerr metric with this approach. At present, there are several models in the literature that have been proposed to describe the accretion flow around SgrA* and we do not know which one, if any, is the right one [81]. Unlike the Novikov–Thorne model for thin accretion disks, these models are not very constrained, namely they have several free parameters that should be inferred from observations. It seems unlikely that we can, at the same time, both determine with a sufficient precision the astrophysical parameters and constrain possible deviations from the Kerr solution.

14.1.3 Orbiting Hot Spots

SgrA* exhibits powerful flares in the X-ray, near-infrared, and sub-millimeter bands [32, 39, 67]. During a flare, the flux increases up to a factor 10. There are a few flares per day. Every flare lasts 1-3 hours and has a quasi-periodic substructure with a timescale of about 20 min. These flares may be associated with blobs of plasma orbiting near the ISCO of the supermassive black hole [44], even if current observations cannot exclude other explanations [57, 66, 85]. Temporary clumps of matter should indeed be common in the region near the ISCO radius [31] and, if so, they may be studied by the GRAVITY instrument for the ESO Very Large Telescope Interferometer (VLTI) [35].

The radiation emitted by a blob of plasma orbiting the strong gravity region of a black hole is affected by the metric of the spacetime and can potentially be used to test the Kerr metric [52]. Figure 14.4 shows some images of a monochromatic

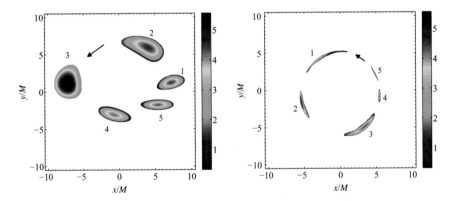

Fig. 14.4 Specific intensity of the primary image (*left panel*) and of the secondary image (*right panel*) of a hot spot of radius $R_{\rm spot} = 0.5\,M$ orbiting a Schwarzschild black hole at the ISCO radius. The viewing angle of the observer is $i = 60°$ and the time interval between two adjacent spot images is $T/5$, where T is the orbital period of the hot spot. From [52]

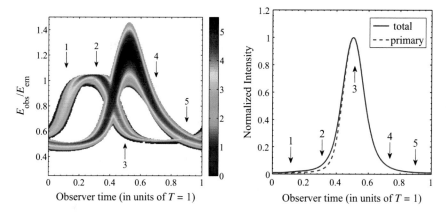

Fig. 14.5 Spectrogram (*left panel*) and light curve (*right panel*) of the hot spot in Fig. 14.4. In the panel of the spectrogram, we can clearly distinguish the spectra of the primary and of the secondary images. The panel of the light curve shows the total light curve and that generated by the primary image. In both panels, the *five arrows* refer to the five images shown in Fig. 14.4. From [52]

blob of plasma orbiting the ISCO of a Schwarzschild black hole and observed at a viewing angle $i = 60°$. The left panel is for the primary images and the right panel is for the secondary images (caused by the strong light bending in the vicinity of the black hole). The color of the images indicate the intensity of the radiation. The flux of higher order images is weaker and weaker. The model employed for describing the blob of plasma is discussed in [52].

Figure 14.5 shows the spectrogram, namely the spectrum as a function of time, of the blob of plasma in Fig. 14.4. An accurate measurement of the spectrogram would be surely an ideal tool to test the metric around SgrA*. However, in the reality the situation is much more complicated. The astrophysical model (shape and size of the blob of plasma, spectrum of the blob of plasma in its rest-frame, etc.) is likely more important than the small features associated with the relativistic effects characterizing the background metric [52].

At the moment, it is not clear if tests of the Kerr metric are possible with this approach. In some spacetimes, the photon capture radius can be significantly different from that of Kerr black holes, and in this case it is possible to identify specific signatures of these metrics [51, 54]. In general, it seems more likely that relativistic effects cannot really be identified and eventually the radiation from a blob of plasma can (at best) only provide an estimate of the orbital frequency. Since the observed period of the quasi-periodic substructure of the flares of SgrA* ranges from 13 to about 30 min, the orbital radius of these hot spots should change and be at a radius larger than the ISCO. For a Kerr black hole of $4 \cdot 10^6 \, M_\odot$, the ISCO period ranges from about 30 min ($a_* = 0$) to 4 min ($a_* = 1$ and corotating orbit). The shortest period ever measured is 13 ± 2 min, and it may be an upper bound for the ISCO period. In the Kerr metric, such a measurement translates into the spin estimate $a_* \geq 0.70 \pm 0.11$ [68]. In the case of a metric with a non-vanishing deformation parameter, there is clearly a degeneracy between the estimate of the spin and possible deviations from the Kerr solution. Such a degeneracy may be broken with another measurement, for instance a precise estimate of the spin parameter of SgrA* by the observation of a radio pulsar.

14.1.4 Accurate Measurements in the Weak Field

The nature of SgrA* can be potentially investigated even with accurate measurements of the spacetime metric at relatively large radii. In this case, the gravitational field is weak, $M/r \ll 1$, and we can adopt an approach similar to the PPN formalism of Solar System experiments. If a spacetime is stationary, axisymmetric, asymptotically flat, Ricci-flat outside the source, and analytic about the point at infinity, its metric in the region outside the source can be expanded in terms of the mass moments $\{M_\ell\}$ and the current moments $\{S_\ell\}$ [40, 45].[2] In the case of reflection symmetry, the odd M-moments and the even S-moments are identically zero, so the leading order terms are the mass $M_0 = M$, the spin angular momentum $S_1 = J$, and the mass quadrupole moment $M_2 = Q$. In the case of a Kerr black hole, the metric is completely determined by M and J, and all the moments $\{M_\ell\}$ and $\{S_\ell\}$ are locked to the mass and the spin by the following relation [40, 45]

$$
M_\ell + i S_\ell = M \left(i \frac{J}{M} \right)^\ell ,
\tag{14.4}
$$

where in this expression i is the unit imaginary number. In particular, the mass quadrupole term is $Q = -J^2/M$.

This approach has the advantage that it is quite general and relies on a small number of assumptions. Even the requirement of the Ricci-flat spacetime may approximatively hold in many cases. Here the spin measurement is really a spin measurement, and it should not be correlated to possible deviations from the Kerr solution in the near horizon region because the latter corresponds to higher order corrections. Such a measurement could be combined with a measurement in the strong gravity regime, which is usually a constraint on the spin and possible deviations from the Kerr geometry [15]. In some favorable cases, it could be possible to determine also the mass

---

[2]The expansion in multipole moments is also possible when the spacetime is not axisymmetric, but in this case the mass and the current moments of order ℓ are tensors. If the spacetime is axisymmetric, there are some simplifications, and the mass and the current moments of any order ℓ are completely determined by two scalars, namely M_ℓ and S_ℓ. In the case of tests of the Kerr metric in the weak field, it is common to assume that the spacetime is axisymmetric, because it sounds a physically plausible hypothesis and simplifies the problem. Let us also note that some assumptions may not hold in some relevant cases. For instance, black hole solutions in alternative theories of gravity may not be Ricci-flat, and an example is the case of black holes in Einstein–dilaton–Gauss–Bonnet gravity of Sect. 12.1.2. The analyticity assumption may also not hold, as in the case of the presence of a massive scalar field with a Yukawa type solution.

quadrupole moment Q. This could permit us to test the Kerr metric at the quadrupole term, because, in the case of a Kerr black hole, we must recover that $Q = -J^2/M$. Higher order terms can unlikely be tested, but clean measurements of J and Q would instantly be very helpful if combined with observations in the strong gravity field.

14.1.4.1 Radio Pulsars

It is thought that a population of radio pulsars is orbiting SgrA* with a short orbital period and there is already an intense work to detect these objects [56]. For instance, Chennamangalam and Lorimer[25] argue that there may be ~200 pulsars in the inner parsec region (orbital period $\lesssim 10^4$ yrs). Accurate timing observations of a radio pulsar orbiting SgrA* in a very close orbit ($\lesssim 1$ yr) would allow a precise measurement of the mass, the spin, and – in exceptional cases – even of the mass quadrupole moment of the supermassive black hole at the center of our Galaxy if the system is sufficiently free of external perturbations [55].

Because of the high electron density in the ionized gas at the center of the Galaxy, this kind of observations must be made at much higher frequencies than those normally used for pulsar timing, which further challenges these measurements. A more serious problem is the possible presence of a population of stars or black holes orbiting very close to SgrA*. The presence of these bodies may strongly affect, or even prevent, a clean measurement of the spin and of the quadrupole moment of SgrA* with the radio pulsar. At the moment it is impossible to make predictions about the potentialities of future observations because we do not know the actual situation near SgrA*.

Assuming a population of 10^3 objects with a mass $M = M_\odot$ isotropically distributed within 1 mpc around SgrA*, Liu et al. [55] have estimated the necessary orbital period of the pulsar to get a measurement of the mass, the spin, and the quadrupole moment of SgrA*. Figure 14.6 shows the timescales of secular orbital precession for

Fig. 14.6 Precession timescale from the mass (M), the spin (S), the quadrupole moment (Q), and stellar perturbation (P) for a pulsar orbiting SgrA* as a function of orbital period P_b, assuming an orbital eccentricity of 0.5 and 10^3 objects, all with mass $M = M_\odot$, within 1 mpc around SgrA*. From [55]. © AAS. Reproduced with permission

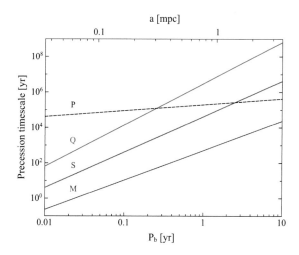

a pulsar orbiting SgrA* as a function of its orbital period: the contributions are from the mass monopole M (pericenter advance), the spin S (frame dragging), the quadrupole moment Q, and stellar perturbation P. The orbital eccentricity is assumed to be 0.5. The precession timescale of the pericenter advance is already lower than that of stellar perturbation for a 10 year orbital period, which means that the observation of a radio pulsar with an orbital period less than 10 years can already be used to estimate the black hole mass M. The measurement of the spin would require an orbital period less than 0.5 years to have the contribution from frame dragging significantly above that from stellar perturbation. The measurement of the quadrupole moment Q requires an orbital period less than 0.1 years. These estimates have to be taken as a general guide and the actual situation may be different. For instance, a population of 10 $M_\odot$ black holes in this region may completely spoil the measurement of the parameters of the metric around SgrA*, as well as a significant anisotropy in the distribution of these bodies may challenge it.

Assuming that we can observe a radio pulsar in an orbit of several months, timing observations could measure the mass and the spin of SgrA*. Since the pulsar is in the weak field of SgrA*, this would be a clean measurement of the spin parameter a_* (with the restrictions listed at the beginning of the subsection), namely independent of the higher order multipole moments of the spacetime. First of all, such a measurement should satisfy the Kerr bound $|a_*| \leq 1$, because otherwise SgrA* could not be a Kerr black hole. Second, the spin measurement could be combined with other observations of the strong field (shadow, hot spot, etc.) in which there is typically a strong correlation between the estimate of the spin and possible deviations from the Kerr solution. The combination of different measurements may break the parameter degeneracy and permit to test the Kerr metric [15].

14.1.4.2   Normal Stars

Even normal stars in compact orbits can be used to probe the weak gravitational field of SgrA* and measure its spin parameter and, possibly, its mass quadrupole moment. The idea was proposed in [77] and further explored in [9, 58].

If SgrA* is rotating fast, the observation of at least two stars with an orbital period of 0.1 years or less and in orbits with a high eccentricity, say ∼0.9, may provide a measurement of the mass, the spin, and the mass quadrupole moment of SgrA* and thus test the Kerr nature of this object at the level of the quadrupole term. Today we know stars with an orbital period as short as 10 years. These objects are still too far from SgrA* and currently no relativistic effects in their orbit are observed (but they should be observed in the near future). However, observational facilities like GRAVITY [35], with the capability of accurate astrometric measurements at the level of 10 μas (micro arc second) close to SgrA*, may observe stars with a sufficiently short orbital period to test the Kerr metric.

Will [77] has proposed to test the Kerr metric by measuring the precession of the orbital plane of these stars. The advance of the pericenter of these stars is dominated by the mass term of the supermassive object, while the contribution of the spin and the quadrupole moment would be subdominant and difficult to estimate. On the contrary, in the weak field limit, the precession of the orbital plane is only determined by the spin (through the frame-dragging) and the mass quadrupole moment; see [77] for the details. Here the dominant contribution comes from the spin, but in the case of sufficiently compact orbits it is also possible to infer the mass quadrupole moment Q. As in the pulsar case, a measurement of the spin could already be useful in combination with other measurements probing the strong gravity field. In the case a quadrupole moment measurement is also available, one can check a posteriori whether it satisfies the Kerr relation $Q = -J^2/M$.

Recently, Zhang et al. [86] have shown that measurements of the spin of SgrA* might be possible even with stars with orbital configurations similar to those already known, in the case of long-term high precision observations.

14.2 Testing the Kerr Paradigm in the Weak Gravity Region

SgrA* is not the only black hole for which we may get accurate measurements at relatively large radii. However, in the case of other black holes we may not have independent measurements in the strong gravity field. In order to test the Kerr metric with these objects, we need to have measurements good enough to determine M, J, and Q. One can then check a posteriori whether $Q = -J^2/M$, as it is expected in the case of a Kerr black hole.

The ideal candidate for this kind of tests is a pulsar binary in which the companion is a stellar-mass black hole [75]. A similar system is not known at the moment. The chances of a discovery are not clear. Stellar evolution studies suggest that in our Galaxy there are about 10^8 neutron stars. Only a small fraction of them, when they have a beam of electromagnetic radiation pointing toward Earth, can be see as pulsars by us. The expected number of pulsars in the Galaxy is about 4,000, and we currently know about 2,500 pulsars. If the estimate of the total number of pulsars is correct, the probability of the existence of a pulsar binary with a stellar-mass black hole companion among the objects not yet discovered is not high.

Even if the uncertainty is large in comparison with what could be possible with a pulsar binary, the measurement of the mass quadrupole moment of a black hole has been done in [70] and it could be relatively improved in the future [71]. The object is the supermassive black hole in the quasar OJ287. Optical observations show a quasi-periodic light curve characterized by two timescales, one of ~ 12 years and another of ~ 60 years. The interpretation is that the system is a binary black hole, with the secondary black hole orbiting the more massive primary one with an orbital period of ~ 12 years and a periastron precession of ~ 60 years [49]. The observed major outbursts occurring every ~ 12 years are thought to be due to the secondary black hole that crosses the accretion disk of the primary. Within this interpretation, Valtonen et al. [70] have employed a 2.5PN (Post-Newtonian) accurate orbital dynamics to fit current observational data. Since the mass quadrupole moment interaction term enters at the 2PN order, it is possible to constrain the mass quadrupole moment of the primary black hole. Writing the quadrupole moment as $Q = -q(J^2/M)$, where $q = 1$ for a Kerr black hole, the current observational data provide the measurement [70]

$$q = 1 \pm 0.3, \tag{14.5}$$

namely the mass quadrupole moment is tested at the level of 30%. In the next few years, this test could be improved at the level of 10% [71].

14.3   Gravitational Waves

As it has been already stressed throughout the book, there is at least an important difference on how electromagnetic radiation and gravitational waves can test astrophysical black holes. The study of the electromagnetic spectrum can essentially test the motion of particles around the black hole. If we assume that particles follow the geodesics of the spacetime, we can measure the metric. Otherwise, it is also possible to test some specific framework with deviations from geodesic motion, but still we test the motion of particles. In the case of gravitational waves, the signal is determined by the evolution of the spacetime metric, so it depends on both the metric and the field equations of the gravity theory. Each approach has its own advantages and disadvantages. In general, it is difficult to say which one is more promising, because it depends on the specific scenario we want to test. In the case of model-independent tests, it is also difficult to compare the constraints from the two approaches because there is no direct link outside a well defined theoretical model. We can generically say that the two approaches can be complementary.

There are a large number of studies on the possibility of testing astrophysical black holes and general relativity in the strong gravity regime with gravitational waves. The two main review articles on the capabilities of gravitational wave detectors to constrain new physics are probably [83] (for ground-based detectors and pulsar timing-arrays) and [38] (for space-based detectors). A more recent review paper listing the expected gravitational wave signals in some specific theories of gravity is [78].

In the case of gravitational waves, the most natural approach is to consider a specific alternative theory of gravity, compute its gravitational wave signal from the kind of events we want to study, and then check whether the observational data (if available) prefer general relativity or the alternative theory of gravity. However, it is also possible to employ an approach in which the gravitational wave signal is parametrized in order to take into account generic deviations from general relativity [10, 11, 28, 82].

As discussed in Chap. 11, the typical event is the coalescence of two black holes or of a black hole with another compact object. Particularly suitable sources for testing the spacetime metric around black holes with space-based detectors are EMRIs, in which a stellar-mass compact object (a black hole, but potentially also a neutron star or a white dwarf) slowly inspirals into a supermassive black hole. Assuming the emission of gravitational waves as in general relativity, we can map the metric of the spacetime around the supermassive black hole [20, 42, 62]. Otherwise, we can perform the calculations within a theoretical framework [23, 60, 64]. While EMRIs are often thought to be the best class of events to test astrophysical black holes, accurate tests may be prevented by mismodelling due to unknown self-force effects already in general relativity [19, 61]. Tests within specific theoretical models can be performed by studying the inspiral phase of binary black holes in which the two objects have roughly the same mass [76, 79, 80]. We can also study the ringdown phase and see whether the final black hole oscillates as it is predicted in general relativity or there are deviations from it [21, 22, 26, 34, 43, 59].

14.3.1 Constraints from GW150914

In the events GW150914 [1] and GW151226 [3], the signal-to-noise ratio was, respectively, 24 and 13. The statistics is good enough to be confident that we really observed gravitational waves from the coalesce of a binary system. GW150914 was probably quite a fortuitous event, in the sense it was relatively close to Earth and the masses of the two black holes were high, thus producing a strong signal. The signal was also at the most sensitive frequency band of LIGO.

The simplest consistency check is to measure the mass and the spin parameter of the final black hole from different parts of the signal of GW150914 [2]. Figure 14.7 shows the constraints that we can obtain from the analysis of the inspiral stage (dashed dark-violet curve), the post-inspiral stage (dashed-dotted violet curve), and the whole signal (black solid curve) assuming the coalescence of two black holes in general relativity. The inspiral and post-inspiral analysis provide consistent results, while a priori this would not be guaranteed either in the presence of different compact objects or in alternative theories of gravity.

Chirenti and Rezzolla [27] have studied the ringdown signal of GW150914 to check whether the final object could be a rotating gravastar rather than a Kerr black hole. The quasi-normal modes of a rotating gravastar are different and not consistent with the signal of GW150914.

Konoplya and Zhidenko [48] study the quasi-normal frequencies of a scalar field in the Konoplya–Rezzolla–Zhidenko metric in Eq. (13.5) in order to constrain the deformation parameter $\delta r / r_{\text{Kerr}}$. In general relativity and in other theories of gravity, the frequencies of a scalar field are not very different from those of the gravitational waves derived from the field equations. However, this is not always true and there are also examples of gravity theories in which the scalar field quasi-normal frequencies can be quite different from those of the gravitational waves. The analysis in [48] shows that it is difficult to constrain deviations from the Kerr solution from GW150914 because the measurement of the spin and the deformation parameter are correlated. This would remain true even in the presence of the detection of additional modes.

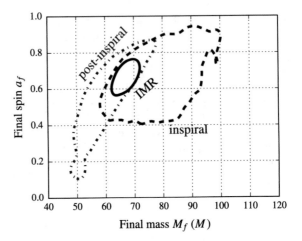

Fig. 14.7 Consistency test with GW150914. The *dashed dark-violet curve* represents the 90% confidence level constraints on the mass and the spin parameter of the final black hole in GW150914 as they are inferred from the analysis of the inspiral signal assuming general relativity. The *dashed-dotted violet curve* shows the 90% confidence level constraints as inferred from the post-inspiral signal. The *black solid curve* provides the constraints obtained from the analysis of the inspiral, merger, and ringdown signals of GW150914. From [2]

Yunes et al. [84] point out several weak points in the study presented in [48]. One is the lack of a sound theoretical framework to discuss the emission of gravitational waves. Unlike the properties of the electromagnetic radiation emitted by the accretion flow, we need to know the dynamics of the system. The beginning of the ringdown signal is not suitable to distinguish objects that posses similar photon capture spheres. The frequency and the damping time of the quasi-normal modes may be related, respectively, to the orbital frequency and the instability timescale of circular null geodesics, and therefore to the photon capture sphere. The beginning of the ringdown signal cannot be seen as a conclusive proof for the formation of an event horizon after the merger, because we can only probe the photon capture sphere. See [24] for more details.

The most detailed analysis on the constraints that can be obtained from GW150914 on the nature of the remnant and on general relativity in the strong field regime is presented in [84]. The authors show that the data of GW150914 can constrain:

1. The emission mechanism of gravitational waves. For instance, the activation of scalar fields, gravitational leakage into large extra dimensions, the variability of Newton's constant, etc.
2. The propagation of gravitational waves. For instance, the speed of gravity, modified dispersion relations, gravitational Lorentz violation, etc.

3. The nature of the compact object, namely if it is a Kerr black hole or something else.

Yunes et al. [84] point out that the limitation to constrain new physics with GW150914 is mainly related to our poor knowledge of the coalescence regime in alternative theories of gravity.

The observed drop in the signal of GW150914 after reaching peak amplitude is consistent with the quick hair-loss expected for Kerr black holes in general relativity and is not what one would expect in the case of compact objects made of exotic matter. This fact can constrain the effective viscosity of the putative exotic matter the remnant would be made of [84].

References

1. B.P. Abbott et al., LIGO Scientific and Virgo Collaborations. Phys. Rev. Lett. **116**, 061102 (2016), arXiv:1602.03837 [gr-qc]
2. B.P. Abbott et al., LIGO Scientific and Virgo Collaborations. Phys. Rev. Lett. **116**, 221101 (2016), arXiv:1602.03841 [gr-qc]
3. B.P. Abbott et al., LIGO Scientific and Virgo Collaborations. Phys. Rev. Lett. **116**, 241103 (2016), arXiv:1606.04855 [gr-qc]
4. A. Abdujabbarov, F. Atamurotov, Y. Kucukakca, B. Ahmedov, U. Camci, Astrophys. Space Sci. **344**, 429 (2013), arXiv:1212.4949 [physics.gen-ph]
5. A. Abdujabbarov, M. Amir, B. Ahmedov, S.G. Ghosh, Phys. Rev. D **93**, 104004 (2016), arXiv:1604.03809 [gr-qc]
6. L. Amarilla, E.F. Eiroa, Phys. Rev. D **85**, 064019 (2012), arXiv:1112.6349 [gr-qc]
7. L. Amarilla, E.F. Eiroa, Phys. Rev. D **87**, 044057 (2013), arXiv:1301.0532 [gr-qc]
8. L. Amarilla, E.F. Eiroa, G. Giribet, Phys. Rev. D **81**, 124045 (2010), arXiv:1005.0607 [gr-qc]
9. R. Angelil, P. Saha, D. Merritt, Astrophys. J. **720**, 1303 (2010), arXiv:1007.0007 [astro-ph.GA]
10. K.G. Arun, B.R. Iyer, M.S.S. Qusailah, B.S. Sathyaprakash, Class. Quantum Gravity **23**, L37 (2006), arXiv:gr-qc/0604018
11. K.G. Arun, B.R. Iyer, M.S.S. Qusailah, B.S. Sathyaprakash, Phys. Rev. D **74**, 024006 (2006), arXiv:gr-qc/0604067
12. F. Atamurotov, A. Abdujabbarov, B. Ahmedov, Astrophys. Space Sci. **348**, 179 (2013)
13. F. Atamurotov, A. Abdujabbarov, B. Ahmedov, Phys. Rev. D **88**, 064004 (2013)
14. C. Bambi, Phys. Rev. D **87**, 107501 (2013), arXiv:1304.5691 [gr-qc]
15. C. Bambi, Class. Quantum Gravity **32**, 065005 (2015), arXiv:1409.0310 [gr-qc]
16. C. Bambi, K. Freese, Phys. Rev. D **79**, 043002 (2009), arXiv:0812.1328 [astro-ph]
17. C. Bambi, N. Yoshida, Class. Quantum Gravity **27**, 205006 (2010), arXiv:1004.3149 [gr-qc]
18. C. Bambi, F. Caravelli, L. Modesto, Phys. Lett. B **711**, 10 (2012), arXiv:1110.2768 [gr-qc]
19. L. Barack. Quantum Gravity **26**, 213001 (2009), arXiv:0908.1664 [gr-qc]
20. L. Barack, C. Cutler, Phys. Rev. D **75**, 042003 (2007), arXiv:gr-qc/0612029
21. E. Berti, V. Cardoso, Int. J. Mod. Phys. D **15**, 2209 (2006), arXiv:gr-qc/0605101
22. E. Berti, V. Cardoso, C.M. Will, Phys. Rev. D **73**, 064030 (2006), arXiv:gr-qc/0512160
23. P. Canizares, J.R. Gair, C.F. Sopuerta, Phys. Rev. D **86**, 044010 (2012), arXiv:1205.1253 [gr-qc]
24. V. Cardoso, E. Franzin, P. Pani, Phys. Rev. Lett. **116**, 171101 (2016) (Erratum: Phys. Rev. Lett. **117**, 089902 (2016)), arXiv:1602.07309 [gr-qc]
25. J. Chennamangalam, D.R. Lorimer, Mon. Not. R. Astron. Soc. **440**, 86 (2014), arXiv:1311.4846 [astro-ph.HE]
26. C.B.M.H. Chirenti, L. Rezzolla, Class. Quantum Gravity **24**, 4191 (2007), arXiv:0706.1513 [gr-qc]

27. C. Chirenti, L. Rezzolla, arXiv:1602.08759 [gr-qc]
28. N. Cornish, L. Sampson, N. Yunes, F. Pretorius, Phys. Rev. D **84**, 062003 (2011), arXiv:1105.2088 [gr-qc]
29. P.V.P. Cunha, C.A.R. Herdeiro, E. Radu, H.F. Runarsson, Phys. Rev. Lett. **115**, 211102 (2015), arXiv:1509.00021 [gr-qc]
30. P.V.P. Cunha, C.A.R. Herdeiro, E. Radu, H.F. Runarsson, Int. J. Mod. Phys. D **25**, 1641021 (2016), arXiv:1605.08293 [gr-qc]
31. J.P. De Villiers, J.F. Hawley, J.H. Krolik, Astrophys. J. **599**, 1238 (2003), arXiv:astro-ph/0307260
32. K. Dodds-Eden, P. Sharma, E. Quataert, R. Genzel, S. Gillessen, F. Eisenhauer, D. Porquet, Astrophys. J. **725**, 450 (2010), arXiv:1005.0389 [astro-ph.GA]
33. S. Doeleman et al., Nature **455**, 78 (2008), arXiv:0809.2442 [astro-ph]
34. O. Dreyer, B.J. Kelly, B. Krishnan, L.S. Finn, D. Garrison, R. Lopez-Aleman, Class. Quantum Gravity **21**, 787 (2004), arXiv:gr-qc/0309007
35. F. Eisenhauer et al., Proc. SPIE Int. Soc. Opt. Eng. **7013**, 2A (2008), arXiv:0808.0063 [astro-ph]
36. V.L. Fish et al., Astrophys. J. **727**, L36 (2011), arXiv:1011.2472 [astro-ph.GA]
37. V.L. Fish et al., Astrophys. J. **820**, 90 (2016), arXiv:1602.05527 [astro-ph.GA]
38. J.R. Gair, M. Vallisneri, S.L. Larson, J.G. Baker, Living Rev. Relativ. **16**, 7 (2013), arXiv:1212.5575 [gr-qc]
39. R. Genzel, R. Schödel, T. Ott, A. Eckart, T. Alexander, F. Lacombe, D. Rouan, B. Aschenbach, Nature **425**, 934 (2003), arXiv:astro-ph/0310821
40. R.P. Geroch, J. Math. Phys. **11**, 2580 (1970)
41. M. Ghasemi-Nodehi, Z. Li, C. Bambi, Eur. Phys. J. C **75**, 315 (2015), arXiv:1506.02627 [gr-qc]
42. K. Glampedakis, S. Babak, Class. Quantum Gravity **23**, 4167 (2006), arXiv:gr-qc/0510057
43. S. Gossan, J. Veitch, B.S. Sathyaprakash, Phys. Rev. D **85**, 124056 (2012), arXiv:1111.5819 [gr-qc]
44. N. Hamaus, T. Paumard, T. Muller, S. Gillessen, F. Eisenhauer, S. Trippe, R. Genzel, Astrophys. J. **692**, 902 (2009), arXiv:0810.4947 [astro-ph]
45. R.O. Hansen, J. Math. Phys. **15**, 46 (1974)
46. T. Johannsen, D. Psaltis, Astrophys. J. **718**, 446 (2010), arXiv:1005.1931 [astro-ph.HE]
47. T. Johannsen, C. Wang, A.E. Broderick, S.S. Doeleman, V.L. Fish, A. Loeb, D. Psaltis, Phys. Rev. Lett. **117**, 091101 (2016), arXiv:1608.03593 [astro-ph.HE]
48. R. Konoplya, A. Zhidenko, Phys. Lett. B **756**, 350 (2016), arXiv:1602.04738 [gr-qc]
49. H.J. Lehto, M.J. Valtonen, Astrophys. J. **460**, 207 (1996)
50. Z. Li, C. Bambi, JCAP **1401**, 041 (2014), arXiv:1309.1606 [gr-qc]
51. Z. Li, C. Bambi, Phys. Rev. D **90**, 024071 (2014), arXiv:1405.1883 [gr-qc]
52. Z. Li, L. Kong, C. Bambi, Astrophys. J. **787**, 152 (2014), arXiv:1401.1282 [gr-qc]
53. N. Lin, Z. Li, J. Arthur, R. Asquith, C. Bambi, JCAP **1509**, 038 (2015), arXiv:1505.05329 [gr-qc]
54. D. Liu, Z. Li, C. Bambi, JCAP **1501**, 020 (2015), arXiv:1411.2329 [gr-qc]
55. K. Liu, N. Wex, M. Kramer, J.M. Cordes, T.J.W. Lazio, Astrophys. J. **747**, 1 (2012), arXiv:1112.2151 [astro-ph.HE]
56. D.R. Lorimer, M. Kramer, *Handbook of Pulsar Astronomy* (Cambridge University Press, Cambridge, 2005)
57. S. Markoff, H. Falcke, F. Yuan, P.L. Biermann, Astron. Astrophys. **379**, L13 (2001), arXiv:astro-ph/0109081
58. D. Merritt, T. Alexander, S. Mikkola, C.M. Will, Phys. Rev. D **81**, 062002 (2010), arXiv:0911.4718 [astro-ph.GA]
59. P. Pani, E. Berti, V. Cardoso, Y. Chen, R. Norte, Phys. Rev. D **80**, 124047 (2009), arXiv:0909.0287 [gr-qc]
60. P. Pani, V. Cardoso, L. Gualtieri, Phys. Rev. D **83**, 104048 (2011), arXiv:1104.1183 [gr-qc]
61. E. Poisson, A. Pound, I. Vega, Living Rev. Relativ. **14**, 7 (2011), arXiv:1102.0529 [gr-qc]
62. F.D. Ryan, Phys. Rev. D **52**, 5707 (1995)
63. J. Schee, Z. Stuchlik, Int. J. Mod. Phys. D **18**, 983 (2009), arXiv:0810.4445 [astro-ph]

64. C.F. Sopuerta, N. Yunes, Phys. Rev. D **80**, 064006 (2009), arXiv:0904.4501 [gr-qc]
65. O. Straub, F.H. Vincent, M.A. Abramowicz, E. Gourgoulhon, T. Paumard, Astron. Astrophys. **543**, A83 (2012), arXiv:1203.2618 [astro-ph.GA]
66. M. Tagger, F. Melia, Astrophys. J. **636**, L33 (2006), arXiv:astro-ph/0511520
67. G. Trap et al., Astron. Astrophys. **528**, A140 (2011), arXiv:1102.0192 [astro-ph.HE]
68. S. Trippe, T. Paumard, T. Ott, S. Gillessen, F. Eisenhauer, F. Martins, R. Genzel, Mon. Not. R. Astron. Soc. **375**, 764 (2007), arXiv:astro-ph/0611737
69. N. Tsukamoto, Z. Li, C. Bambi, JCAP **1406**, 043 (2014), arXiv:1403.0371 [gr-qc]
70. M.J. Valtonen et al., Astrophys. J. **709**, 725 (2010), arXiv:0912.1209 [astro-ph.HE]
71. M.J. Valtonen, S. Mikkola, H.J. Lehto, A. Gopakumar, R. Hudec, J. Polednikova, Astrophys. J. **742**, 22 (2011), arXiv:1108.5861 [astro-ph.CO]
72. F.H. Vincent, W. Yan, M.A. Abramowicz, A.A. Zdziarski, O. Straub, Astron. Astrophys. **574**, A48 (2015), arXiv:1406.0353 [astro-ph.GA]
73. S.W. Wei, Y.X. Liu, JCAP **1311**, 063 (2013), arXiv:1311.4251 [gr-qc]
74. S.W. Wei, P. Cheng, Y. Zhong, X.N. Zhou, JCAP **1508**, 004 (2015), arXiv:1501.06298 [gr-qc]
75. N. Wex, S. Kopeikin, Astrophys. J. **514**, 388 (1999), arXiv:astro-ph/9811052
76. C.M. Will, Phys. Rev. D **50**, 6058 (1994), arXiv:gr-qc/9406022
77. C.M. Will, Astrophys. J. **674**, L25 (2008), arXiv:0711.1677 [astro-ph]
78. K. Yagi, L.C. Stein, Class. Quantum Gravity **33**, 054001 (2016), arXiv:1602.02413 [gr-qc]
79. K. Yagi, L.C. Stein, N. Yunes, T. Tanaka, Phys. Rev. D **85**, 064022 (2012) (Erratum: Phys. Rev. D **93**, 029902 (2016)), arXiv:1110.5950 [gr-qc]
80. K. Yagi, N. Yunes, T. Tanaka, Phys. Rev. Lett. **109**, 251105 (2012) (Erratum: Phys. Rev. Lett. **116**, 169902 (2016)), arXiv:1208.5102 [gr-qc]
81. F. Yuan, R. Narayan, Annu. Rev. Astron. Astrophys. **52**, 529 (2014), arXiv:1401.0586 [astro-ph.HE]
82. N. Yunes, F. Pretorius, Phys. Rev. D **80**, 122003 (2009), arXiv:0909.3328 [gr-qc]
83. N. Yunes, X. Siemens, Living Rev. Relativ. **16**, 9 (2013), arXiv:1304.3473 [gr-qc]
84. N. Yunes, K. Yagi, F. Pretorius, arXiv:1603.08955 [gr-qc]
85. F. Yusef-Zadeh, D. Roberts, M. Wardle, C.O. Heinke, G.C. Bower, Astrophys. J. **650**, 189 (2006), arXiv:astro-ph/0603685
86. F. Zhang, Y. Lu, Q. Yu, Astrophys. J. **809**, 127 (2015), arXiv:1508.06293 [astro-ph.HE]

Appendix A
Stationary and Axisymmetric Spacetimes

The aim of this appendix is to collect together some useful expressions for generic stationary and axisymmetric spacetimes and for the special case of the Kerr metric. More details can be found in Chap. 3 and in [2].

A.1 General Case

The so-called *canonical form* of the line element of a generic stationary, axisymmetric, and asymptotically flat spacetime is[1]

$$ds^2 = -e^{2\nu}dt^2 + e^{2\psi}(d\phi - \omega dt)^2 + e^{2\lambda}dr^2 + e^{2\mu}d\theta^2 , \tag{A.1}$$

where ν, ψ, ω, λ, and μ are functions of the coordinates r and θ and are independent of t and ϕ.

In this class of spacetimes, it is always possible to introduce the so-called *locally non-rotating observers* [1]. These observers carry an orthonormal tetrad of 4-vectors which represent their locally Minkowskian coordinate basis vectors [3]. The orthonormal tetrad at any point of the spacetime is given by

$$E^\mu_{(t)} = \left(e^{-\nu}, 0, 0, \omega e^{-\nu}\right), \qquad E^\mu_{(r)} = \left(0, e^{-\lambda}, 0, 0\right),$$
$$E^\mu_{(\theta)} = \left(0, 0, e^{-\mu}, 0\right), \qquad E^\mu_{(\phi)} = \left(0, 0, 0, e^{-\psi}\right), \tag{A.2}$$

and the corresponding basis of one-forms is

---

[1]As pointed out in Sect. 3.1, this is not "the most general" stationary and axisymmetric spacetime, because even the g_{tr} metric coefficient may be non-vanishing in the most general case.

© Springer Nature Singapore Pte Ltd. 2017
C. Bambi, *Black Holes: A Laboratory for Testing Strong Gravity*,
DOI 10.1007/978-981-10-4524-0

$$E_\mu^{(t)} = \left(e^\nu, 0, 0, 0\right), \qquad E_\mu^{(r)} = \left(0, e^\lambda, 0, 0\right),$$

$$E_\mu^{(\theta)} = \left(0, 0, e^\mu, 0\right), \qquad E_\mu^{(\phi)} = \left(-\omega e^\psi, 0, 0, e^\psi\right). \qquad (A.3)$$

The metric of any locally non-rotating observer is $\eta_{(a)(b)}$ and is related to the metric $g_{\mu\nu}$ by

$$\eta_{(a)(b)} = E_{(a)}^\mu g_{\mu\nu} E_{(b)}^\nu, \qquad \eta^{(a)(b)} = E_\mu^{(a)} g^{\mu\nu} E_\nu^{(b)}. \qquad (A.4)$$

The relation between the components of vectors and dual vectors in the two reference systems is

$$V^{(a)} = E_\mu^{(a)} V^\mu, \qquad V_{(a)} = E_{(a)}^\mu V_\mu. \qquad (A.5)$$

For instance, the 3-velocity of a massive particle with respect to a locally non-rotating observer is

$$v^{(i)} = \frac{E_\mu^{(i)} u^\mu}{E_\nu^{(t)} u^\nu}, \qquad (A.6)$$

where u^μ is the particle 4-velocity. In the case of time-like equatorial circular orbits (e.g. the motion of particles in a Novikov–Thorne accretion disk), one finds

$$v^{(r)} = v^{(\theta)} = 0, \qquad v^{(\phi)} = e^{\psi - \nu} (\Omega - \omega), \qquad (A.7)$$

where $\Omega = u^\phi / u^t$ is the angular velocity of the particle.

A.2   Kerr Metric

In Boyer–Lindquist coordinates, the line element of the Kerr metric reads

$$ds^2 = -\left(1 - \frac{2Mr}{\Sigma}\right) dt^2 - \frac{4Mar\sin^2\theta}{\Sigma} dt d\phi$$
$$+ \frac{\Sigma}{\Delta} dr^2 + \Sigma d\theta^2 + \sin^2\theta \left(r^2 + a^2 + \frac{2Ma^2 r\sin^2\theta}{\Sigma}\right) d\phi^2, \quad (A.8)$$

with contravariant form

$$\left(\frac{\partial}{\partial s}\right)^2 = -\frac{A}{\Sigma\Delta}\left(\frac{\partial}{\partial t}\right)^2 - \frac{4Mar}{\Sigma\Delta}\left(\frac{\partial}{\partial t}\right)\left(\frac{\partial}{\partial\phi}\right)$$
$$+ \frac{\Delta}{\Sigma}\left(\frac{\partial}{\partial r}\right)^2 + \frac{1}{\Sigma}\left(\frac{\partial}{\partial\theta}\right)^2 + \frac{\Delta - a^2\sin^2\theta}{\Sigma\Delta\sin^2\theta}\left(\frac{\partial}{\partial\phi}\right)^2, \quad (A.9)$$

where

$$\Sigma = r^2 + a^2 \cos^2 \theta \,,$$
$$\Delta = r^2 - 2Mr + a^2 \,,$$
$$A = \left(r^2 + a^2\right)^2 - a^2 \Delta \sin^2 \theta \,. \tag{A.10}$$

With the notation of Eq. (A.1), we have

$$e^{2\nu} = \frac{\Sigma \Delta}{A} \,, \qquad e^{2\psi} = \frac{A \sin^2 \theta}{\Sigma} \,,$$
$$e^{2\lambda} = \frac{\Sigma}{\Delta} \,, \qquad e^{2\mu} = \Sigma \,, \qquad \omega = \frac{2Mar}{A} \,. \tag{A.11}$$

The orthonormal tetrad carried by locally non-rotating observers is given by

$$E^{\mu}_{(t)} = \left(\sqrt{\frac{A}{\Sigma \Delta}}, 0, 0, \frac{2Mar}{\sqrt{A\Sigma\Delta}}\right), \qquad E^{\mu}_{(r)} = \left(0, \sqrt{\frac{\Delta}{\Sigma}}, 0, 0\right),$$

$$E^{\mu}_{(\theta)} = \left(0, 0, \frac{1}{\sqrt{\Sigma}}, 0\right), \qquad E^{\mu}_{(\phi)} = \left(0, 0, 0, \sqrt{\frac{\Sigma}{A}}\frac{1}{\sin\theta}\right). \tag{A.12}$$

The associated basis of one-forms is

$$E^{(t)}_{\mu} = \left(\sqrt{\frac{\Sigma \Delta}{A}}, 0, 0, 0\right), \qquad E^{(r)}_{\mu} = \left(0, \sqrt{\frac{\Sigma}{\Delta}}, 0, 0\right),$$

$$E^{(\theta)}_{\mu} = \left(0, 0, \sqrt{\Sigma}, 0\right), \qquad E^{(\phi)}_{\mu} = \left(-\frac{2Mar\sin\theta}{\sqrt{A\Sigma}}, 0, 0, \sqrt{\frac{A}{\Sigma}}\right). \tag{A.13}$$

A.3 Kerr Metric in Kerr–Schild Coordinates

In the Kerr–Schild coordinates (t, r, θ, ϕ), the metric is regular at the event horizon. In these coordinates, the line element of the Kerr solution reads

$$ds^2 = -\left(1 - \frac{2Mr}{\Sigma}\right)dt^2 + \frac{4Mr}{\Sigma}dtdr - \frac{4Mar\sin^2\theta}{\Sigma}dtd\phi$$
$$+ \left(1 + \frac{2Mr}{\Sigma}\right)dr^2 - 2a\left(1 + \frac{2Mr}{\Sigma}\right)\sin^2\theta drd\phi + \Sigma d\theta^2$$
$$+ \sin^2\theta\left(r^2 + a^2 + \frac{2Ma^2r\sin^2\theta}{\Sigma}\right)d\phi^2 \,, \tag{A.14}$$

where $\Sigma = r^2 + a^2 \cos^2 \theta$ as before.

If we indicate with $(t^{KS}, r^{KS}, \theta^{KS}, \phi^{KS})$ the Kerr–Schild coordinates and with $(t^{BL}, r^{BL}, \theta^{BL}, \phi^{BL})$ the Boyer–Lindquist ones, the relation between the two coordinate systems is

$$dt^{KS} = dt^{BL} + \frac{2Mr}{\Delta} dr^{BL},$$
$$dr^{KS} = dr^{BL},$$
$$d\theta^{KS} = d\theta^{BL},$$
$$d\phi^{KS} = \frac{Ma}{\Delta} dr^{BL} + d\phi^{BL}. \tag{A.15}$$

The radial coordinate of the even horizon is still

$$r_H = M + \sqrt{M^2 - a^2}, \tag{A.16}$$

but now neither g_{rr} nor other metric coefficients diverge there.

References

1. J.M. Bardeen, Astrophys. J. **162**, 71 (1970).
2. J.M. Bardeen, W.H. Press, S.A. Teukolsky, Astrophys. J. **178**, 347 (1972).
3. M. Nakahara, *Geometry, Topology, and Physics* (Institute of Physics Publishing, Bristol, 1990).

Appendix B
(r, θ)-Motion in the Kerr Spacetime

We want to solve Eq. (6.34)

$$s_r \int_{r_e}^{D} \frac{dr'}{\sqrt{\tilde{R}}} = s_\theta \int_{\pi/2}^{i} \frac{d\theta'}{\sqrt{\tilde{\Theta}}}, \tag{B.1}$$

where

$$\tilde{R}(r) = r^4 + \left(a^2 - \lambda^2 - q^2\right) r^2 + 2M \left[q^2 + (\lambda - a)^2\right] r - a^2 q^2, \tag{B.2}$$

$$\tilde{\Theta}(\theta) = q^2 + a^2 \cos^2 \theta - \lambda^2 \cot^2 \theta. \tag{B.3}$$

$s_r = \pm 1$ and $s_\theta = \pm 1$ according to the direction of propagation. The point of emission in the disk is at $(r_e, \theta_e = \pi/2)$. The observer is at (D, i), where $D \to \infty$. The goal is to obtain r_e in terms of the constants of motion λ and q:

$$r_e = r_e(\lambda, q). \tag{B.4}$$

Equation (B.4) is necessary to calculate the Jacobian (6.30) in the transfer function. More details on the integration of photon orbits in the Kerr spacetime can be found, for instance, in [3–5], and references therein.

B.1 θ-Motion

With the change of variable $\mu = \cos \theta$, we have to solve the integral

$$I_\mu = \int_0^{\mu_{obs}} \frac{d\mu}{\sqrt{\tilde{U}}}, \tag{B.5}$$

where the disk is at $\mu = 0$, the observer at $\mu = \mu_{obs}$, and

© Springer Nature Singapore Pte Ltd. 2017
C. Bambi, *Black Holes: A Laboratory for Testing Strong Gravity*,
DOI 10.1007/978-981-10-4524-0

$$\tilde{U} = q^2 - \left(\lambda^2 + q^2 - a^2\right)\mu^2 - a^2\mu^4 . \tag{B.6}$$

Here we are interested in photons emitted from the disk in the equatorial plane ($\mu_e = 0$) and detected by the distant observer at $\mu_{obs} \neq 0$. For $\mu = 0$, $\tilde{U} = q^2$, so $q^2 \geq 0$ or the square root in the integral is not real. We can thus restrict our attention to the orbits with $q^2 > 0$. Orbits with $q^2 \leq 0$ do not connect the disk with the plane of the distant observer.

$\tilde{U}$ can be written as

$$\tilde{U} = a^2 \left(\mu_-^2 + \mu^2\right)\left(\mu_+^2 - \mu^2\right) , \tag{B.7}$$

where

$$\mu_\pm^2 = \frac{\mp\left(\lambda^2 + q^2 - a^2\right) + \sqrt{4q^2a^2 + \left(\lambda^2 + q^2 - a^2\right)^2}}{2a^2} . \tag{B.8}$$

For $\mu = 0$, $\tilde{U} = q^2 = a^2\mu_+^2\mu_-^2$, and therefore both μ_+^2 and μ_-^2 are non-negative (positive in our case with $q^2 > 0$).

μ^2 cannot exceed μ_+^2 or from Eq. (B.7) $\tilde{U}$ would become negative. The integral of $1/\sqrt{\tilde{U}}$ from some μ to μ_+ can be written as (see Eq. 213.00 in [2])

$$\int_\mu^{\mu_+} \frac{d\mu'}{\sqrt{\tilde{U}}} = \frac{1}{\sqrt{a^2\left(\mu_+^2 + \mu_-^2\right)}} \text{cn}^{-1}(\cos\psi_\mu, \kappa_\mu) , \tag{B.9}$$

where cn^{-1} is the inverse of the Jacobian elliptic function cn, and ψ_μ and κ_μ are, respectively, the argument and the modulus of the Jacobian elliptic function

$$\cos\psi_\mu = \frac{\mu}{\mu_+} , \qquad \kappa_\mu^2 = \frac{\mu_+^2}{\mu_+^2 + \mu_-^2} . \tag{B.10}$$

The Jacobian elliptic functions can be defined as $\text{cn}(x, k) = \cos\varphi$ and $\text{sn}(x, k) = \sin\varphi$, where

$$x = \int_0^\varphi \frac{d\theta}{\sqrt{1 - k^2\sin^2\theta}} . \tag{B.11}$$

Let us note that $\text{cn}^{-1}(\cos\psi_\mu, \kappa_\mu) = F(\psi_\mu, \kappa_\mu)$, where $F(\psi_\mu, \kappa_\mu)$ is the incomplete elliptic integral of the first kind. The latter can be defined as

$$F(\varphi, k) = \int_0^\varphi \frac{d\theta}{\sqrt{1 - k^2\sin^2\theta}} , \tag{B.12}$$

or, after a change of variables, as

$$F(x, k) = \int_0^x \frac{dt}{\sqrt{(1 - t^2)(1 - k^2 t^2)}}. \tag{B.13}$$

It follows that $F(x, k) = \mathrm{sn}^{-1}(x, k)$.

It is easy to see that photons hitting the image plane of the distant observer with $Y > 0$ must have a turning point, which means that μ starts from 0, and its value first increases to reach the maximum value μ_+ and then decreases to μ_{obs}. If $Y < 0$, there is no turning point, so μ starts from 0 and monotonically increases to μ_{obs}. In the presence of a turning point, the integral is

$$I_\mu = \int_0^{\mu_+} \frac{d\mu}{\sqrt{\tilde{U}}} + \int_{\mu_{\mathrm{obs}}}^{\mu_+} \frac{d\mu}{\sqrt{\tilde{U}}} \quad (Y > 0). \tag{B.14}$$

In the absence of a turning point, the integral is

$$I_\mu = \int_0^{\mu_+} \frac{d\mu}{\sqrt{\tilde{U}}} - \int_{\mu_{\mathrm{obs}}}^{\mu_+} \frac{d\mu}{\sqrt{\tilde{U}}} \quad (Y < 0). \tag{B.15}$$

Equations (B.14) and (B.15) can be rewritten in a compact form as

$$I_\mu = \frac{1}{\sqrt{a^2 (\mu_+^2 + \mu_-^2)}} \left[K(\kappa_\mu) + s_Y \, \mathrm{cn}^{-1}(\cos \psi_\mu, \kappa_\mu) \right], \tag{B.16}$$

where $s_Y = 1$ (respectively $0, -1$) for $Y > 0$ (respectively $= 0, < 0$), ψ_μ is evaluated at $\mu = \mu_{\mathrm{obs}}$, and $K(\kappa_\mu)$ is the complete elliptic integral of the first kind, namely

$$K(\kappa_\mu) = \mathrm{cn}^{-1}(\cos \psi_\mu = 0, \kappa_\mu). \tag{B.17}$$

B.2   r-Motion

Let r_1, r_2, r_3, and r_4 be the four roots of the equation $\tilde{R}(r) = 0$. We write

$$\tilde{R}(r) = (r - r_1)(r - r_2)(r - r_3)(r - r_4). \tag{B.18}$$

The procedure for finding the roots of a quartic equation are discussed, for instance, in [1]. Since $\tilde{R}(r \to \pm\infty) = \infty$ and $\tilde{R}(r = 0) = -a^2 q^2 \le 0$, at least two roots are real. We can distinguish two cases: (i) there are two complex roots and two real roots, or (ii) there are four real roots.

B.2.1 $\tilde{R}(r) = 0$ Has Two Complex Roots and Two Real Roots

Let r_1 and r_2 be the two complex roots ($r_1 = r_2^*$, where r_2^* is the complex conjugate of r_2) and r_3 and r_4 the two real roots (assuming $r_3 > r_4$). Since

$$\tilde{R}(0) = r_1 r_2 r_3 r_4 = |r_1|^2 r_3 r_4 = -a^2 q^2 \le 0, \tag{B.19}$$

r_3 must be non-negative and r_4 must be non-positive. Since $\tilde{R} \ge 0$, the motion of the photon is in the region $r \ge r_3$.

Let us write the four roots as $r_1 = u + iw$, $r_2 = u - iw$, $r_3 = -u + v$, and $r_4 = -u - v$, where u, v, and w are real numbers and v is positive. The integration from r_3 to $r > r_3$ can be written as the inverse Jacobian elliptic function cn^{-1}, see Eq. 260.00 in [2]

$$\int_{r_3}^{r} \frac{dr'}{\sqrt{\tilde{R}}} = \frac{1}{\sqrt{AB}} \, \text{cn}^{-1} (\cos \psi_2, \kappa_2), \tag{B.20}$$

where

$$\cos \psi_2 = \frac{(A - B)r - r_4 A + r_3 B}{(A + B)r - r_4 A - r_3 B}, \tag{B.21}$$

$$\kappa_2^2 = \frac{(A + B)^2 - (r_3 - r_4)^2}{4AB}, \tag{B.22}$$

and

$$A = \sqrt{(v - 2u)^2 + w^2}, \qquad B = \sqrt{(v + 2u)^2 + w^2}. \tag{B.23}$$

B.2.2 $\tilde{R}(r) = 0$ Has Four Real Roots

Let us assume that $r_1 \ge r_2 \ge r_3 \ge r_4$. From

$$\tilde{R}(r = 0) = r_1 r_2 r_3 r_4 = -a^2 q^2 \le 0, \tag{B.24}$$

it follows that $r_4 \le 0$. Since $\tilde{R}(r) \ge 0$, motion is allowed in the region $r \ge r_1$, $r_2 \ge r \ge r_3$, and $r \le r_4$. In our case, we are interested in the region $r \ge r_1$, because the observer is far from the black hole. For $r_1 \ne r_2$, the integral of $1/\sqrt{\tilde{R}}$ from r_1 to some r can be written in terms of the inverse Jacobian elliptic function sn^{-1}, see Eq. 258.00 in [2]

$$\int_{r_1}^{r} \frac{dr'}{\sqrt{R}} = \frac{2}{\sqrt{(r_1 - r_3)(r_2 - r_4)}} \, \mathrm{sn}^{-1}(\sin \psi_4, \kappa_4), \tag{B.25}$$

where

$$\sin \psi_4 = \sqrt{\frac{(r_2 - r_4)(r - r_1)}{(r_1 - r_4)(r - r_4)}}, \qquad \kappa_4^2 = \frac{(r_2 - r_3)(r_1 - r_4)}{(r_1 - r_3)(r_2 - r_4)}. \tag{B.26}$$

Again, $\mathrm{sn}^{-1}(\sin \psi_4, \kappa_4) = F(\psi_4, \kappa_4)$ where $F(\psi_4, \kappa_4)$ is the incomplete elliptic integral of the first kind. For the special case $r_1 = r_2$, we have

$$\int_{r_1}^{r} \frac{dr'}{\sqrt{R}} = \frac{1}{\sqrt{(r_1 - r_3)(r_1 - r_4)}}$$
$$\cdot \ln \left[\frac{\sqrt{(r - r_3)(r - r_4)}}{r - r_1} + \frac{r_1^2 + r_3 r_4 + 2 r r_1}{(r - r_1)\sqrt{(r_1 - r_3)(r_1 - r_4)}} \right]. \tag{B.27}$$

B.2.3   Integration over r

If $\tilde{R}(r) = 0$ has two complex roots and two real roots, the allowed region of interest for us is $r \geq r_3$. In the case of four real roots, the allowed region from the emission point in the disk to the detection point is $r \geq r_1$. The possible turning point will thus be at $r_t = r_3$ in the first case, and at $r_t = r_1$ in the second case. Let us define

$$I_r^\infty = \int_{r_t}^{\infty} \frac{dr'}{\sqrt{\tilde{R}}}, \qquad I_r^e = \int_{r_1}^{r_e} \frac{dr'}{\sqrt{\tilde{R}}}. \tag{B.28}$$

Necessary and sufficient condition for the existence of a turning point is

$$I_r^\infty < I_\mu, \tag{B.29}$$

where I_μ is the integral defined in Eq. (B.16). In the presence of a turning point, r starts from r_e, first decreases to r_t, and then increases to ∞. The integration over r along the photon path is

$$I_r = -\int_{r_e}^{r_t} \frac{dr'}{\sqrt{\tilde{R}}} + \int_{r_t}^{\infty} \frac{dr'}{\sqrt{\tilde{R}}} = I_r^\infty + I_r^e. \tag{B.30}$$

In the absence of a turning point, r monotonically increases from r_e to ∞ and we have

$$I_r = \int_{r_e}^{\infty} \frac{dr'}{\sqrt{\tilde{R}}} = I_r^\infty - I_r^e. \tag{B.31}$$

B.3   Determination of the Emission Point r_e

We have to distinguish three cases: (i) $\tilde{R}(r) = 0$ has two complex roots and two real roots, (ii) $\tilde{R}(r) = 0$ has four real roots and $r_1 \neq r_2$, and (iii) $\tilde{R}(r) = 0$ has four real roots and $r_1 = r_2$.

In the case (i), we have

$$I_r^\infty = \frac{1}{\sqrt{AB}} \, \mathrm{cn}^{-1}(\cos \psi_\infty, \kappa_2) \,, \qquad I_r^e = \frac{1}{\sqrt{AB}} \, \mathrm{cn}^{-1}(\cos \psi_e, \kappa_2) \,, \quad \text{(B.32)}$$

where

$$\cos \psi_\infty = \frac{A - B}{A + B} \,, \qquad \cos \psi_e = \frac{(A - B)r_e - r_4 A + r_3 B}{(A + B)_e - r_4 A - r_3 B} \,. \qquad \text{(B.33)}$$

If we plug these two expressions into Eqs. (B.30) or (B.31) and we impose $I_\mu = I_r$, we obtain r_e

$$r_e = \frac{r_4 A - r_3 B - (r_4 A + r_3 B)\,\mathrm{cn}(\cos \psi_2, \kappa_2)}{(A - B) - (A + B)\,\mathrm{cn}(\cos \psi_2, \kappa_2)} \,, \qquad \text{(B.34)}$$

where

$$\cos \psi_2 = \sqrt{AB} \left(I_\mu - I_r^\infty \right) \,. \qquad \text{(B.35)}$$

Since in $\mathrm{cn}(\cos \psi_2, \kappa_2)$ the sign of ψ_2 does not matter, it is not necessary to distinguish the cases with $(I_\mu > I_r^\infty)$ and without $(I_\mu < I_r^\infty)$ turning point.

In the case (ii), I_r^∞ and I_r^e are, respectively,

$$I_r^\infty = \frac{2}{\sqrt{(r_1 - r_3)(r_2 - r_4)}} \, \mathrm{sn}^{-1}(\sin \psi_\infty, \kappa_4) \,,$$

$$I_r^e = \frac{2}{\sqrt{(r_1 - r_3)(r_2 - r_4)}} \, \mathrm{sn}^{-1}(\sin \psi_e, \kappa_4) \,, \qquad \text{(B.36)}$$

with

$$\sin \psi_\infty = \sqrt{\frac{r_2 - r_4}{r_1 - r_4}} \,, \qquad \sin \psi_e = \sqrt{\frac{(r_2 - r_4)(r_e - r_1)}{(r_1 - r_4)(r_e - r_2)}} \qquad \text{(B.37)}$$

Imposing $I_\mu = I_r$, we obtain r_e

$$r_e = \frac{r_1(r_2 - r_4) - r_2(r_1 - r_4)\,\mathrm{sn}^2(\sin \psi_4, \kappa_4)}{(r_2 - r_4) - (r_1 - r_4)\,\mathrm{sn}^2(\sin \psi_4, \kappa_4)} \,, \qquad \text{(B.38)}$$

where

$$\sin \psi_4 = \frac{1}{2} \left(I_\mu - I_r^\infty \right) \sqrt{(r_1 - r_3)(r_2 - r_4)}. \tag{B.39}$$

Since the Jacobian elliptic function sn compares at the square, it does not matter if $\sin \psi_4$ is positive or negative, so if there is or there is no turning point.

Lastly, there is the case (iii), in which $\tilde{R}(r) = 0$ has four real roots and $r_1 = r_2$. Here, we do not have turning point, so

$$I_r = \int_{r_e}^{\infty} \frac{dr}{\sqrt{\tilde{R}}} = \frac{1}{\sqrt{(r_1 - r_3)(r_1 - r_4)}} \left\{ -\ln \left[1 + \frac{2r_1}{\sqrt{(r_1 - r_3)(r_1 - r_4)}} \right] \right.$$
$$\left. + \ln \left[\frac{\sqrt{(r_e - r_3)(r_e - r_4)}}{r_e - r_1} + \frac{r_1^2 + r_3 r_4 + 2 r_1 r_e}{(r_e - r_1)\sqrt{(r_1 - r_3)(r_1 - r_4)}} \right] \right\}. \tag{B.40}$$

From $I_\mu = I_r$, the solution for the emission radius r_e is

$$r_e = \frac{1}{r_1} \left\{ r_3 r_4 - \left[\gamma r_1 + \frac{r_1^2 + r_3 r_4}{\sqrt{(r_1 - r_3)(r_1 - r_4)}} \right]^2 \right\}$$
$$\cdot \left\{ 1 - \left[\gamma - \frac{2 r_1}{\sqrt{(r_1 - r_3)(r_1 - r_4)}} \right]^2 \right\}^{-1}, \tag{B.41}$$

where γ is

$$\gamma = \left[1 + \frac{2 r_1}{\sqrt{(r_1 - r_3)(r_1 - r_4)}} \right] \exp \left[I_\mu \sqrt{(r_1 - r_3)(r_1 - r_4)} \right]. \tag{B.42}$$

References

1. G. Birkhoff, S.S. Mac Lane, *A Survey of Modern Algebra* (Macmillan, New York, 1965).
2. P.F. Byrd, M.D. Friedman, *Handbook of Elliptic Integrals for Engineers and Physicists* (Springer, Berlin, 1954).
3. S. Chandrasekhar, *The Mathematical Theory of Black Holes* (Clarendon Press, Oxford, 1998).
4. J. Dexter, E. Agol, Astrophys. J. **696**, 1616 (2009), arXiv:0903.0620 [astro-ph.HE].
5. L.X. Li, E.R. Zimmerman, R. Narayan, J.E. McClintock, Astrophys. J. Suppl. **157**, 335 (2005), arXiv:astro-ph/0411583.

Appendix C
AGN Classification

Active galactic nuclei (AGN) are very bright galactic nuclei, powered by the mass accretion onto their central supermassive black hole. Figure C.1 shows the AGN classification, groups and subgroups, and the corresponding fraction of members. While it is thought that any normal galaxy has a central supermassive black hole at its center, only a small fraction of them host an AGN. In most galaxies, the central supermassive object is "dormient", like SgrA*, which has a luminosity of the order of 10^{-7} in Eddington units.

About 93% of the galaxies are non-active. Among the 7% of the active galaxies, most of them are star-forming galaxies or low-ionization nuclear emission-line regions (LINERs). The latter are sometimes considered AGN. Proper AGN are relatively rare: they are in 0.5% of the active galaxies, which means only in 0.035% of all galaxies.

AGN are mainly classified according to their luminosity and spectral features. It is thus useful to briefly review their possible spectral components:

1. *Radio emission* from jets with the typical spectrum from synchrotron radiation.
2. *IR emission* from the thermal spectrum of the accretion disk, which is reprocessed by gas and dust around the nucleus. This occurs when the accretion disk is obscured by gas and dust.
3. *Optical continuum* mainly from the thermal spectrum of the accretion disk, and in part from possible jets.
4. *Narrow optical lines* from cold material orbiting relatively far from the supermassive black hole. The orbital velocity of this material is 500–1,000 km/s.
5. *Broad optical lines* from cold material orbiting close to the supermassive black hole. The orbital velocity of this material is 1,000–5,000 km/s. The lines are broad due to Doppler boosting.
6. *X-ray continuum* from a hot corona and possible jets.
7. *X-ray lines* from fluorescence emission of the gas in the accretion disk illuminated by the X-ray continuum. The iron Kα line at 6.4 keV is usually one of the most prominent lines.

© Springer Nature Singapore Pte Ltd. 2017
C. Bambi, *Black Holes: A Laboratory for Testing Strong Gravity*,
DOI 10.1007/978-981-10-4524-0

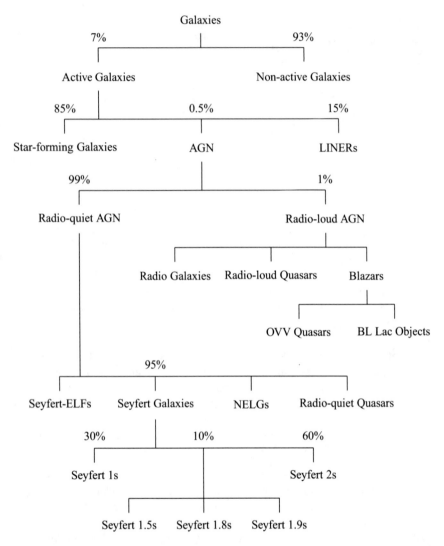

Fig. C.1 Sketch of the AGN family and of its subgroups. This classification has to be taken with caution, because different authors may use slightly different classifications. The diagram shows also the fraction of members in each subgroup. AGN represent only 0.035% of the galactic nuclei. Most of the AGN are radio-quiet and belong to the class of Seyfert galaxies

The AGN classification is sometimes confusing, some objects may not be easily associated to a specific group, and different authors may use different classifications. With reference to Fig. C.1, we see that AGN can be grouped into two classes, radio-quiet and radio-loud AGN. In radio-quiet AGN, the jet component is absent or negligible, so the radio luminosity is low. Radio-loud AGN have powerful jets. These jets may be powered by the black hole spin.

Radio-loud AGN (Powerful Jet)

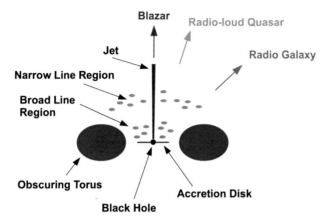

Fig. C.2 Sketch of a radio-loud AGN according to the unified AGN model [1]. The *black hole* is surrounded by an accretion disk, which may be obscured by a dusty torus. The *broad line region* is close to the *black hole* and there are clouds orbiting with high velocity. The *narrow line region* is relatively far from the *black hole* and there are clouds moving at lower velocity. The *arrows* represent the lines of sight of the observer. Depending on the angle between the jet and the line of sight of the observer, the AGN can appear as a blazar, as a radio-loud quasar, or as a radio galaxy

Radio-quiet AGN may be grouped into four classes: Seyfert extremely luminous far IR galaxies (Seyfert-ELFs), Seyfert galaxies, narrow emission line galaxies (NELGs), and radio-quiet quasars. The classification is based on a number of properties. For instance, Seyfert galaxies have an optical continuum and emission lines. Seyfert 1s have both narrow and broad emission lines, while Seyfert 2s have only narrow emission lines. Seyfert galaxies of type 1.5, 1.8, and 1.9 are grouped according to their spectral appearance.

Radio-loud AGN can be grouped into three classes: radio galaxies, radio-loud quasars, and blazars. Blazars are characterized by rapid variability and by polarized optical, radio and X-ray emission. They are divided into BL Lacertae objects (BL Lac objects) and optically violent variable quasars (OVV quasars). OVV quasars have stronger broad emission lines than BL Lac objects.

According to the unified AGN model [1], all AGN would essentially belong to the same class of objects. They look different because they are observed from different viewing angles. Figures C.2 and C.3 illustrate the idea of the unified AGN model. In the case of blazars, the jet would be along our line of sight. As the angle between the jet and our line of sight increases, we would have radio-loud quasars and then radio galaxies. In the case of radio-quiet AGN (Fig. C.3), we have a similar situation, even if there is no jet: depending on the viewing angle of the observer, the AGN can appear as a radio-quiet quasar, as a Seyfert 1 galaxy, or as a Seyfert 2 galaxy.

Radio-quiet AGN (No or Weak Jet)

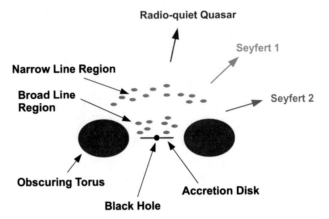

Fig. C.3 As in Fig. C.2, in the case of a radio-quiet AGN. Depending on the viewing angle of the observer, the AGN can appear as a radio-quiet quasar, as a Seyfert 1 galaxy, or as a Seyfert 2 galaxy

Reference

1. C.M. Urry, P. Padovani, Publ. Astron. Soc. Pac. **107**, 803 (1995), arXiv:astro-ph/9506063.

Appendix D
Jets

Jets are a common feature of several astrophysical objects, including protostars, stars, neutron stars, and black holes. Jets are observed both from stellar-mass black holes in X-ray binaries and from supermassive black holes in galactic nuclei, see e.g. [5, 11, 21]. The exact mechanism responsible for the formation of black hole jets is currently unknown.

D.1 Formation Mechanisms

The two most popular mechanisms for the formations of black hole jets are the Blandford–Znajek model [4] and the Blandford–Payne model [3], both with a number of variants and extensions. There are also proposals of hybrid models, in which the two mechanisms can coexist [10].

In the *Blandford–Znajek mechanism*, magnetic fields thread the black hole horizon and can extract the rotational energy of the compact object [4]. This mechanism exploits the existence of the ergoregion via the Penrose process [15]. However, strictly speaking, the extraction of the rotational energy of a compact object may be possible even in the case of neutron stars in the presence of magnetic fields anchored on the surface of the body.

In the original paper by Blandford and Znajek, the jet power P_{BZ} was derived assuming $a_* \ll 1$. In this case, one finds $P_{BZ} \propto a_*^2$. A more detailed analysis provides the following formula [18]

$$P_{BZ} = \frac{\kappa}{16\pi} \Phi_B^2 \Omega_H^2 f(\Omega_H).$$ (D.1)

κ is a numerical constant which depends on the magnetic field configuration; for instance, in the Kerr metric $\kappa = 0.053$ for a split monopole geometry and 0.044 for a parabolic geometry. Ω_H is the angular frequency at the black hole horizon

© Springer Nature Singapore Pte Ltd. 2017
C. Bambi, *Black Holes: A Laboratory for Testing Strong Gravity*,
DOI 10.1007/978-981-10-4524-0

$$\Omega_H = \left(-\frac{g_{t\phi}}{g_{\phi\phi}}\right)_{r=r_H} \quad \text{(General case)} \tag{D.2}$$

$$= \frac{a_*}{2r_H} \quad \text{(Kerr metric)}. \tag{D.3}$$

Equation (D.2) assumes that the spacetime is stationary and axisymmetric and holds when the metric is written in the canonical form (see Sect. 3.1). Φ_B is the magnetic flux threading the black hole horizon

$$\Phi_B = \int_0^\pi \int_0^{2\pi} |B^r| \sqrt{-g} \, d\theta d\phi. \tag{D.4}$$

$f(\Omega_H)$ is a dimensionless function that takes into account higher order terms in Ω_H, so

$$f(\Omega_H) \approx 1 + c_1 M^2 \Omega_H^2 + c_2 M^4 \Omega_H^4 + \cdots \tag{D.5}$$

where $\{c_i\}$ are numerical coefficients. For a Kerr black hole with a thin accretion disk, $c_1 = 1.38$ and $c_2 = -9.2$ [18]. The jet power eventually depends on the background metric, the angular frequency of the horizon Ω_H, and the magnetic field.

In the *Blandford–Payne mechanism*, magnetic fields thread the accretion disk, corrotating with it [3]. Now the energy is provided by the gravitational potential energy of the accretion flow. The power of the jet can be written as

$$P_{BZ} \sim \varepsilon \, L_{acc} \, \ln\left(\frac{r_{out}}{r_{in}}\right), \tag{D.6}$$

where ε is the efficiency of the transformation of the binding energy of the accreting matter into jet power at the inner radius of the disk r_{in}, and r_{out} is the outer radius of the disk.

D.2   Observations

D.2.1   Stellar-Mass Black Holes

In the case of black hole binaries, we observe two kinds of jets [5]. *Steady jets* manifest when a source is in the low-hard state. The jet is steady, typically not very relativistic, and may extend up to a few tens of AU. *Transient jets* are instead observed when a source switches from the hard to the soft state and crosses the "jet line" (see Sect. 4.5). These pc-scale jets appears as blobs of plasma emitting mainly in the radio band and are relativistic. They have features similar to the kpc-scale jets observed in AGN and for this reason the black hole binaries producing transient jets are also called microquasars [11].

If the mechanism responsible for the formation of jets were the Blandford–Znajek model, one may expect a correlation between black hole spin measurements and estimates of the jet power. Fender et al. [6] have studied spin measurements of black hole binaries reported in the literature and inferred from the continuum-fitting and the iron line methods. Their plots do not show any correlation between black hole spin and jet power. Narayan and McClintock [12] have proposed that the Blandford–Znajek mechanism is responsible for the formation of transient jets. They selected some spin measurements obtained via the continuum-fitting method and used a different proxy for the jet power. They find a correlation between black hole spin measurements and jet power. For the moment, this is a controversial issue [14, 8]. Both conclusions are based on a small number of data with large uncertainty. Future observations should be able to increase the number of sources and decrease the uncertainty in these measurements, and provide a conclusive answer to this issue [8].

If some kind of jets were powered by the black hole spin, the combination of spin measurements and estimates of the jet power could be used to test the Kerr metric [1, 2, 16].

D.2.2 Supermassive Black Holes

In the case of AGN, only a small fraction of them, around 1%, exhibit relativistic kpc-scale jets. One of the most spectacular example is Cygnus A, see Fig. D.1. Radio images of this object show two very collimated jets from the very center of the galaxy,

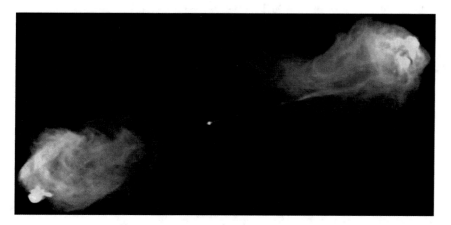

Fig. D.1 Radio image of Cygnus A. The *bright dot* at the *center* is the location of the supermassive *black hole*, where the two relativistic jets are generated. The jets are stopped by the intergalactic medium, forming two giant lobes. Image courtesy of NRAO/AUI

where it is supposed to be located its supermassive black hole. The two jets extend well outside the galaxy, for hundreds of kpc.

Jets dominate the spectrum of a source at radio frequencies. There are apparently two distinct populations of AGN: radio-loud AGN and radio-quiet AGN [17]. This is particularly evident on the plane optical luminosity vs radio luminosity. For the same optical luminosity, radio-loud AGN have a radio luminosity 3–4 orders of magnitude higher than that of radio-quiet AGN. These two populations seem really to follow two different tracks with a gap between them.

The origin of the radio-quiet/radio-loud dichotomy is not understood [17]. Some authors have also doubted about the actual existence of this dichotomy, suggesting it is a bias of observations.

In the case of AGN with an accretion luminosity above 1% of the Eddington limit, the most natural interpretation is that their jets are the counterpart of the transient jets in black hole binaries. This conclusion may be supported by the consideration that microquasars show intermittent jets for a few percent of the time, which is similar to the fraction of radio-loud AGN [13]. The time scale of these systems is proportional to their mass, so intermittent jets in black hole binaries look like persistent jets in AGN. However, in the case of AGN with a luminosity below 1% of the Eddington limit, this explanation cannot work: black hole binaries with a low accretion luminosity are all radio-loud.

Another popular interpretation is that the dichotomy is determined by the black hole spin. When the accretion luminosity is low, it turns out that radio-loud AGN are in elliptical galaxies, while radio-quiet AGN are mainly in spiral galaxies. Galaxies with different morphology have likely a different merger and accretion history. This, in turn, may have produced two populations of black holes, with high and low values of spin [20]. A difference in radio luminosity of 3–4 orders of magnitude between the two populations is impossible to explain if the jet power is proportional to Ω_H^2, but in the case of thick disks the jet power may scale as Ω_H^6 [18].

If jets are powered by the rotational energy of the accreting compact object, it is possible to extract energy and have an accretion efficiency $\eta > 1$, where $L_{acc} = \eta \dot{M}$. Some observations indicate that some AGN may have $\eta > 1$ [7, 9]. If these measurements are correct, the jet is extracting energy from the system, and it is likely that this is the rotational energy of the black hole; some version of the Blandford–Znajek mechanism is working. While in the past some GRMHD simulations have not been able to find high efficiency from the formation of jets, more recently Tchekhovskoy et al. [19] have presented simulations in which $\eta > 1$.

References

1. C. Bambi, Phys. Rev. D **85**, 043002 (2012), arXiv:1201.1638 [gr-qc].
2. C. Bambi, Phys. Rev. D **86**, 123013 (2012), arXiv:1204.6395 [gr-qc].
3. R.D. Blandford, D.G. Payne, Mon. Not. R. Astron. Soc. **199**, 883 (1982).
4. R.D. Blandford, R.L. Znajek, Mon. Not. R. Astron. Soc. **179**, 433 (1977).
5. R.P. Fender, T.M. Belloni, E. Gallo, Mon. Not. R. Astron. Soc. **355**, 1105 (2004), arXiv:astro-ph/0409360.

6. R. Fender, E. Gallo, D. Russell, Mon. Not. R. Astron. Soc. **406**, 1425 (2010), arXiv:1003.5516 [astro-ph.HE].

7. G. Ghisellini, F. Tavecchio, L. Foschini, G. Ghirlanda, L. Maraschi, A. Celotti, Mon. Not. R. Astron. Soc. **402**, 497 (2010), arXiv:0909.0932 [astro-ph.CO].

8. J.E. McClintock, R. Narayan, J.F. Steiner, Space Sci. Rev. **183**, 295 (2014), arXiv:1303.1583 [astro-ph.HE].

9. B.R. McNamara, M. Rohanizadegan, P.E.J. Nulsen, Astrophys. J. **727**, 39 (2011), arXiv:1007.1227 [astro-ph.CO].

10. D.L. Meier, Astrophys. J. **548**, L9 (2001), arXiv:astro-ph/0010231.

11. I.F. Mirabel, L.F. Rodriguez, Annu. Rev. Astron. Astrophys. **37**, 409 (1999), arXiv:astro-ph/9902062.

12. R. Narayan, J.E. McClintock, Mon. Not. R. Astron. Soc. **419**, L69 (2012), arXiv:1112.0569 [astro-ph.HE].

13. C. Nipoti, K.M. Blundell, J. Binney, Mon. Not. R. Astron. Soc. **361**, 633 (2005), arXiv:astro-ph/0505280.

14. D.M. Russell, E. Gallo, R.P. Fender, Mon. Not. R. Astron. Soc. **431**, 405 (2013), arXiv:1301.6771 [astro-ph.HE].

15. R. Penrose, Riv. Nuovo Cim. **1**, 252 (1969) [Gen. Rel. Grav. **34**, 1141 (2002)].

16. G. Pei, S. Nampalliwar, C. Bambi, M.J. Middleton, Eur. Phys. J. C **76**, 534 (2016), arXiv:1606.04643 [gr-qc].

17. M. Sikora, L. Stawarz, J.P. Lasota, Astrophys. J. **658**, 815 (2007), arXiv:astro-ph/0604095.

18. A. Tchekhovskoy, R. Narayan, J.C. McKinney, Astrophys. J. **711**, 50 (2010), arXiv:0911.2228 [astro-ph.HE].

19. A. Tchekhovskoy, R. Narayan, J.C. McKinney, Mon. Not. R. Astron. Soc. **418**, L79 (2011), arXiv:1108.0412 [astro-ph.HE].

20. M. Volonteri, M. Sikora, J.P. Lasota, Astrophys. J. **667**, 704 (2007), arXiv:0706.3900 [astro-ph].

21. J.A. Zensus, Annu. Rev. Astron. Astrophys. **35**, 607 (1997).

Appendix E
Thick Accretion Disks

The Novikov–Thorne model reviewed in Chap. 6 describes geometrically thin and optically thick accretion disks around black holes. It assumes that the self-gravitation of the disk and the gas pressure are negligible, so the fluid elements follow the geodesics of the background metric. If the gas pressure cannot be neglected, the disk can be geometrically thick and the fluid elements do not follow the geodesics of the background metric any longer. This appendix briefly reviews the so-called *Polish doughnut model* [3, 1]. It is a simple model to describes thick accretion disks in which the self-gravitation of the disk is still neglected but the gas pressure is taken into account.

E.1   Polish Doughnut Model

Like the Novikov–Thorne model for thin disks, even the Polish doughnut model can be formulated in a generic stationary, axisymmetric, and asymptotically flat spacetime. We can thus write the line element as

$$ds^2 = g_{tt}dt^2 + g_{rr}dr^2 + g_{\theta\theta}d\theta^2 + 2g_{t\phi}dtd\phi + g_{\phi\phi}d\phi^2 , \tag{E.1}$$

where the metric elements are independent of the t and ϕ coordinates. The disk is modeled as a perfect fluid with purely azimuthal flow. Its energy-momentum tensor is thus

$$T^{\mu\nu} = (\rho + P)u^\mu u^\nu + g^{\mu\nu}P , \quad u^\mu = \left(u^t, 0, 0, u^\phi\right) , \tag{E.2}$$

where ρ and P are, respectively, the energy density and the pressure of the fluid.

From the definitions of the specific energy of the fluid element $-u_t$, its angular velocity $\Omega = u^\phi/u^t$, and its angular momentum per unit energy $l = -u_\phi/u_t$, we have the following relations

© Springer Nature Singapore Pte Ltd. 2017
C. Bambi, *Black Holes: A Laboratory for Testing Strong Gravity*,
DOI 10.1007/978-981-10-4524-0

$$u_t = -\sqrt{\frac{g_{t\phi}^2 - g_{tt}g_{\phi\phi}}{g_{\phi\phi} + 2lg_{t\phi} + l^2 g_{tt}}} \,, \tag{E.3}$$

$$\Omega = -\frac{lg_{tt} + g_{t\phi}}{lg_{t\phi} + g_{\phi\phi}} \,, \tag{E.4}$$

$$l = -\frac{g_{t\phi} + \Omega g_{\phi\phi}}{g_{tt} + \Omega g_{t\phi}} \,. \tag{E.5}$$

It is worth noting that l is conserved for a stationary and axisymmetric flow in a stationary and axisymmetric spacetime [3].

The disk's structure can be inferred from the Euler equation, $\nabla_\nu T^{\mu\nu} = 0$. One finds

$$a^\mu = -\frac{g^{\mu\nu} + u^\mu u^\nu}{\rho + P} \partial_\nu P \,, \tag{E.6}$$

where $a^\mu = u^\nu \nabla_\nu u^\mu$ is the fluid 4-acceleration. If the pressure is independent of the t and ϕ coordinates (which follows from the assumption that the spacetime is stationary and axisymmetric) and if the equation of state is barotropic, i.e. $\rho = \rho(P)$, a^μ can be written as a gradient of a scalar potential $W(P)$

$$a_\mu = \partial_\mu W \,, \quad W(P) = -\int^P \frac{dP'}{\rho(P') + P'} \,. \tag{E.7}$$

After some algebra, it is possible to write Ω as a function of l, i.e. $\Omega = \Omega(l)$, and integrate the Euler equation to get W[2]

$$W = W_{\rm in} + \ln\frac{u_t}{u_t^{\rm in}} + \int_{l_{\rm in}}^l \frac{\Omega dl'}{1 - \Omega l'} \,, \tag{E.8}$$

where $W_{\rm in}$, $l_{\rm in}$, and $u_t^{\rm in}$ are, respectively, the potential, the angular momentum per unit energy, and (minus) the energy per unit mass at the inner edge of the fluid configuration. Actually, in Eq. (E.8) $W_{\rm in}$, $l_{\rm in}$, and $u_{\rm in}^t$ can be replaced by the value of W, l, and u^t at any other point of the fluid boundary. In the Newtonian limit, W reduces to the total potential, i.e. the sum of the gravitational potential and of the centrifugal one, and at infinity $W = 0$.

If the metric of the spacetime is known, there is only one unspecified function, $\Omega = \Omega(l)$, which characterizes the fluid rotation. In the zero-viscosity case, this function cannot be deduced from any equation, and it must be given as an assumption of the model. In this sense, the model is not very constrained as in the case of the Novikov–Thorne one, in which the radial structure of the disk follows from the equations of conservation of rest-mass, energy, and angular momentum.

---

[2]In the special case $l = $ const., Ω is not constant, but Eq. (E.8) is still correct and the integral vanishes.

Imposing a specific relation between Ω and l, we can find the equipotential surfaces $W = $ const. < 0, i.e. the surfaces of constant pressure describing the possible boundaries of the fluid configuration. One of these surfaces may have one (or more) sharp cusp(s), which may induce the accretion onto the compact object: like the cusp at the L_1 Lagrange point in a close binary system, the accreting gas can fill the Roche lobe and then be transferred to the compact object. The mechanism does not need the fluid viscosity to work, so the latter may be, at least in principle, very low.

A particularly simple case is the configuration with $l = $ const., which is marginally stable with respect to axisymmetric perturbations (the criterion for convective stability is simply that l does not have to decrease outward) [6, 5]. In this specific case, the integral in Eq. (E.8) vanishes and

$$W = \ln(-u_t) + \text{const.} .$$ (E.9)

In the Kerr spacetime, we find five qualitatively different configurations [1]. Figure E.1 shows these five possibilities in the case of a black hole with the spin parameter $a_* = 0.8$. We have:

1. $l < l_{\rm ms}$, where $l_{\rm ms}$ is the angular momentum per unit energy of the marginally stable equatorial circular orbit (or ISCO). In this case, there are no closed equipotential surfaces and therefore there is no accretion disk (the angular momentum of the fluid is too low).
2. $l = l_{\rm ms}$. This is the limiting case between $l < l_{\rm ms}$ and $l_{\rm ms} < l < l_{\rm mb}$. At the ISCO radius there is a flex (not a minimum) of W. The disk exists as an infinitesimally thin unstable ring at the ISCO.
3. $l_{\rm ms} < l < l_{\rm mb}$, where $l_{\rm mb}$ is the angular momentum per unit energy of the marginally bound equatorial circular orbit. There is only one equipotential surface $W = W_{\rm cusp}$ with a cusp on the equatorial plane. The cusp is located between the marginally bound and the marginally stable radius. Equipotential surfaces with $W > W_{\rm cusp}$ do not represent any disk configuration, while those with $W < W_{\rm cusp}$ describe the surface of non-accreting disks. Accretion starts when the gas fills the equipotential surface with the cusp.
4. $l = l_{\rm mb}$. This is the limiting case between $l_{\rm ms} < l < l_{\rm mb}$ and $l > l_{\rm mb}$. The cusp is located in the equatorial plane and belongs to the marginally closed equipotential surface $W = 0$. Accretion is possible in the limit of a disk of infinite size.
5. $l > l_{\rm mb}$. The angular momentum of the fluid is too high and no accretion is possible. There are no equipotential surfaces $W \leq 0$ with cusps.

In non-Kerr backgrounds, the picture may be slightly different [4]. For instance, there may not be marginally bound orbits and there may be disks with more than one cusp.

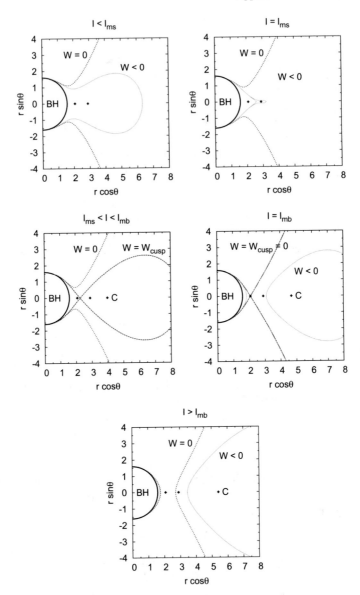

Fig. E.1 Equipotential surfaces around a Kerr black hole. The spin parameter is $a_* = 0.8$, which implies $l_{ms} = 2.712$ and $l_{mb} = 2.894$. *Top left panel* $l = 2.600$ and there is no disk. *Top right panel* $l = l_{ms}$ and it is possible an unstable ring at $r = r_{ISCO}$. *Central left panel* $l = 2.800$, there is an infinite number of disks without cusp, and there is a disk with a cusp. *Central right panel* $l = l_{mb}$, there is an infinite number of disks without cusp, and the disk with a cusp is that with $W = 0$. *Bottom panel* $l = 3.000$ and there is no disk with a cusp. See the text for more details

E.2   Evolution of the Spin Parameter

As done in Sect. 6.4 for the thin disk case, we can now consider the evolution of the
spin parameter of a black hole accreting from a thick disk. The argument is the same
and Eq. (6.39) becomes

$$\frac{da_*}{d \ln M} = \frac{1}{M} \frac{L_{in}}{E_{in}} - 2a_* , \qquad (E.10)$$

where E_{in} and L_{in} are, respectively, the specific energy and the specific angular
momentum of the gas at the inner edge of the disk.

Accretion from a thick disk is potentially more efficient to spin a black hole up [2].
If we consider Eq. (E.10) in the Kerr metric, we can see that the equilibrium spin
parameter is still 1 but it can be reached after accreting a smaller amount of matter
with respect to the thin disk case. If we take into account the radiation emitted by the
disk and captured by the black hole, the counterpart of Eq. (6.43), we can potentially
get closer to 1 than with a thin disk, because of the lower radiative efficiency. If we
consider the extreme case with $l = l_{mb}$, $E_{in} = 1$ and $\eta = 1 - E_{in} = 0$. The situation
in a realistic case is actually more complicated and it depends on the viscosity of the
accreting fluid, but numerical calculations suggest that it is indeed possible to get a
spin parameter very close to 1 [8].

E.3   Spectrum of an Ion Torus

The Polish doughnut model can be provided with a microscopic model for the accreting gas and be used to model the spectrum of black holes with a geometrically thick
disk. While it is a relatively simple model and more realistic scenarios would require
GRMHD simulations, it may capture some key-features of the accretion spectrum
of the source. For instance, the Polish doughnut model has been employed in [9, 10]
to try to describe the spectrum of the accretion structure around SgrA*.

Straub et al. [9] and Vincent et al. [10] use the Polish doughnut model to describe a
radiatively inefficient and advection dominated accretion flow like the one expected in
the case of SgrA*. They employ a gas pressure dominated optically thin 2-temperature
plasma model for the microscopic physics. The model has seven parameters: the
black hole spin of the background metric, the viewing angle of the observer, and five
parameters for the accretion flow and its microphysics:

background metric	spin parameter	a_*
observer	viewing angle	i
ion torus	dimensionless specific angular momentum	λ
	magnetic to total pressure ratio	β
	polytropic index	n
	energy density at the center	ρ_c
	electron temperature at the center	T_c

where

$$\lambda = \frac{l - l_{ms}}{l_{mb} - l_{ms}} \tag{E.11}$$

ranges from 0 to 1.

For a non-relativistic gas with negligible radiation pressure, the polytropic index is $n = 3/2$. The pressure P is the sum of the magnetic pressure P_m and the gas pressure P_{gas}, which, for simplicity, can be supposed to have a constant ratio

$$P = P_m + P_{gas}, \quad P_m = \frac{B^2}{24\pi} = \beta P. \tag{E.12}$$

B is the intensity of the magnetic field. The gas is described by a 2-temperature plasma, so the gas pressure is the sum of the ion and electron contributions. The pressure P, the energy density ρ, and the electron temperature T can be written as [9, 10]

$$P = K\rho^{5/3}, \tag{E.13}$$

$$\rho = \frac{1}{K^n} \left[\left(K\rho_c^{1/n} + 1 \right)^{\omega} - 1 \right]^n, \tag{E.14}$$

$$T = T_c \left(\frac{\rho}{\rho_c} \right)^{1/n}, \tag{E.15}$$

where

$$K = \frac{1}{\rho_c^{1/n}} \exp \left(\frac{W_c - W_s}{n+1} - 1 \right), \tag{E.16}$$

and W_c and W_s denote the scalar potential W evaluated, respectively, at the center and at the surface of the accretion disk.

With this set-up, we can compute the spectrum of an ion torus. The observed specific flux is given by

$$F_{obs}(\nu_{obs}) = \int I_{obs} d\tilde{\Omega}. \tag{E.17}$$

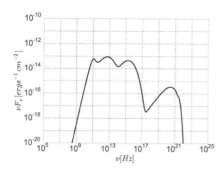 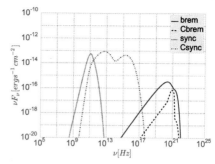

Fig. E.2 *Left panel* spectrum of an ion torus for the following set of parameters: $a/M = 0.5$, $i = 60°$, $\lambda = 0.3$, $\beta = 0.1$, $n = 3/2$, $\rho_c = 10^{-17}$ g/cm³, and $T_c/T_v = 0.02$. T_v is the viral temperature. *Right panel* contribution from every radiative mechanism to the spectrum in the *left panel* bremsstrahlung emission (*blue solid line*), Comptonization of bremsstrahlung emission (*violet dotted line*), synchrotron radiation (*red dashed line*), and Comptonization of synchrotron radiation (*green dashed-dotted line*). From [7], reproduced by permission of IOP Publishing. All rights reserved

The specific intensity $I_{\rm obs}$ can be evaluated by integrating along every photon path the emission and absorption contributions

$$I_{\rm obs}(\nu_e) = \int g^3 \Big[j_e(\nu_e) - \alpha(\nu_e) I_e(\nu_e) \Big] d\ell \,, \qquad (E.18)$$

where j_e and α are, respectively, the emission and absorption coefficient, $d\ell = u^\mu k_\mu d\tau$ is the infinitesimal proper length as measured in the rest frame of the emitter, and τ is the affine parameter of the photon trajectory.

Neglecting absorption ($\alpha = 0$) and assuming that the electromagnetic spectrum of the accretion structure is produced by bremsstrahlung, synchrotron processes, and inverse Compton scattering of both bremsstrahlung and synchrotron photons off free electrons in the medium, the emission coefficient can be written as

$$j_e(\nu_e) = j_e^{\rm brem}(\nu_e) + j_e^{\rm sync}(\nu_e) + j_e^{\rm Cbrem}(\nu_e) + j_e^{\rm Csyn}(\nu_e) \,. \qquad (E.19)$$

Figure E.2 shows an example of spectrum from this model (left panel) and the contributions from every emission mechanism (right panel).

References

1. M. Abramowicz, M. Jaroszynski, M. Sikora, Astron. Astrophys. **63**, 221 (1978).
2. M.A. Abramowicz, J.P. Lasota, Acta Astron. **30**, 35 (1980).
3. M. Kozlowski, M. Jaroszynski, M. Abramowicz, Astron. Astrophys. **63**, 209 (1978).
4. Z. Li, C. Bambi, JCAP **1303**, 031 (2013), arXiv:1212.5848 [gr-qc].
5. F. Daigne, J.A. Font, Mon. Not. R. Astron. Soc. **349**, 841 (2004), arXiv:astro-ph/0311618.

6. J.A. Font, F. Daigne, Mon. Not. R. Astron. Soc. **334**, 383 (2002), arXiv:astro-ph/0203403.
7. N. Lin, Z. Li, J. Arthur, R. Asquith, C. Bambi, JCAP **1509**, 09, 038 (2015), arXiv:1505.05329 [gr-qc].
8. A. Sadowski, M. Bursa, M. Abramowicz, W. Kluzniak, J.-P. Lasota, R. Moderski, M. Safarzadeh, Astron. Astrophys. **532**, A41 (2011), arXiv:1102.2456 [astro-ph.HE].
9. O. Straub, F.H. Vincent, M.A. Abramowicz, E. Gourgoulhon, T. Paumard, Astron. Astrophys. **543**, A83 (2012), arXiv:1203.2618 [astro-ph.GA].
10. F.H. Vincent, W. Yan, M.A. Abramowicz, A.A. Zdziarski, O. Straub, Astron. Astrophys. **574**, A48 (2015), arXiv:1406.0353 [astro-ph.GA].

Appendix F
Astrophysical Constants

Quantity	Symbol, Equation	Value
Speed of Light in Vacuum	c	$2.99792458 \cdot 10^8$ m s^{-1}
Newton's Constant	G_N	$6.6738(8) \cdot 10^{-11}$ m^3 kg^{-1} s^{-2}
Boltzmann Constant	k_B	$1.3806488(13) \cdot 10^{-23}$ J K^{-1}
		$8.6173324(78) \cdot 10^{-5}$ eV K^{-1}
Stefan–Boltzmann Constant	σ_{SB}	$5.670373(21) \cdot 10^{-8}$ W m^{-2} K^{-4}
Permeability of Free Space	μ_0	$4\pi \cdot 10^{-7}$ N A^{-2}
Permittivity of Free Space	$\varepsilon_0 = 1/(\mu_0 c^2)$	$8.854187817... \cdot 10^{-12}$ F m^{-1}
Electron Charge Magnitude	e	$1.602176565(35) \cdot 10^{-19}$ C
Electron Mass	m_e	$9.10938291(40) \cdot 10^{-31}$ kg
Proton Mass	m_p	$1.672621777(74) \cdot 10^{-27}$ kg
Classical Electron Radius	$r_e = e^2/(4\pi\epsilon_0 m_e c^2)$	$2.8179403267(27) \cdot 10^{-15}$ m
Thomson Cross Section	$\sigma_{Th} = 8\pi r_e^2/3$	$0.6652458734(13) \cdot 10^{-30}$ m^{-2}
Astronomic Unit	AU	$1.49597870700 \cdot 10^{11}$ m
Parsec (1 AU/1 arc sec)	pc	$3.08567758149 \cdot 10^{16}$ m
Solar Mass	$M_\odot$	$1.9885(2) \cdot 10^{30}$ kg
Solar Luminosity	$L_\odot$	$3.828 \cdot 10^{26}$ W
		$3.828 \cdot 10^{33}$ erg s^{-1}
Schwarschild Radius of the Sun	$r_{Sch} = 2G_N M_\odot c^{-2}$	$2.953250077(2) \cdot 10^3$ m
Gravitational Radius of the Sun	$r_g = G_N M_\odot c^{-2}$	$1.476625038(1) \cdot 10^3$ m
Characteristic Time $\tau = r_g/c$	$G_N M_\odot c^{-3}$	$4.922549095 \cdot 10^{-6}$ s
Characteristic Frequency $\nu = 1/\tau$	$1/(G_N M_\odot c^{-3})$	$2.030254466 \cdot 10^5$ Hz

© Springer Nature Singapore Pte Ltd. 2017
C. Bambi, *Black Holes: A Laboratory for Testing Strong Gravity*,
DOI 10.1007/978-981-10-4524-0

Appendix G
Glossary

Active galactic nucleus (AGN). Compact region at the center of a galaxy powered by the mass accretion onto a supermassive black hole. The term is also used to indicate the central supermassive black hole. See Appendix C about the AGN classification.

Black hole binary (BHB). Binary system in which one of the two objects is a stellar-mass black hole.

Binary black hole. Binary system in which both objects are black holes. The term can be used to indicate either a binary system of two stellar-mass black holes or a binary system of two supermassive black holes.

Blazar. A class of particularly bright quasars. According to the unified AGN model, blazars are those radio-loud AGN in which the relativistic jet points in the direction of Earth. See Appendix C for more details.

Bolometric luminosity. Total electromagnetic luminosity of an object, namely the electromagnetic luminosity integrated over all wavelengths.

Broad line region (BLR). It is the region "close" to a supermassive black hole, where cold material orbits with a velocity of 1,000–5,000 km/s. This material produces broad lines because of the Doppler boosting.

Corona. This term is used to indicate some kind of hot, usually optically-thin, electron cloud which enshrouds the central disk and acts as a source of X-rays. See Chap. 8 and Fig. 8.1 for more details.

Eddington luminosity. It is the maximum luminosity for an object. Such a luminosity is reached when the pressure of the radiation luminosity balances the gravitational force towards the object. The Eddington luminosity of an object of mass M is (assuming a ionized gas of protons and electrons)

$$L_{Edd} = \frac{4\pi G_N M m_p c}{\sigma_{Th}} = 1.26 \cdot 10^{38} \left(\frac{M}{M_\odot} \right) \, erg/s, \qquad (G.1)$$

© Springer Nature Singapore Pte Ltd. 2017
C. Bambi, *Black Holes: A Laboratory for Testing Strong Gravity*,
DOI 10.1007/978-981-10-4524-0

where G_N is Newton's gravitational constant, m_p is the proton mass, c is the speed of light, and σ_{Th} is the electron Thomson cross section.

In the special case of a black hole, the Eddington luminosity sets the maximum accretion luminosity. If we write $L_{Edd} = \eta_r \dot{M}_{Edd} c^2$, where η_r is the radiative efficiency, the Eddington accretion rate is

$$\dot{M}_{Edd} = 1.40 \cdot 10^{18} \left(\frac{0.1}{\eta_r}\right) \left(\frac{M}{M_\odot}\right) \text{ g/s}$$

$$= 2.22 \cdot 10^{-8} \left(\frac{0.1}{\eta_r}\right) \left(\frac{M}{M_\odot}\right) M_\odot/\text{yr} . \qquad (G.2)$$

Effective area. It is somehow a measurement of the efficiency of the optics and the detector of an instrument at detecting photons. It is usually measured in units of cm^2. See Sect. 5.1.2 for more details.

Hardness. It is the ratio between the luminosities in the hard and soft X-ray bands, but the exact definition may vary. For instance, it may be the ratio between the luminosity of the 6–10 keV band and of the 2–6 keV band, i.e. $H = L_{6-10\,keV}/L_{2-6\,keV}$.

Hard X-ray band. There is not a universal definition of "hard X-ray band", so the exact energy range must be specified. In some contexts, it may indicate, for instance, the 6–10 keV band. In other contexts, it may be used for the X-ray spectrum above 10 keV.

Hawking radiation. If quantum effects are taken into account, black holes radiate (almost) like a blackbody with the temperature $T_{BH} = \kappa/(2\pi)$, where κ is the surface gravity at the event horizon. Such a radiation is called the Hawking radiation. For a Schwarzschild black hole, the temperature is

$$T_{BH} = \frac{1}{8\pi M} = 5.32 \cdot 10^{-12} \left(\frac{M_\odot}{M}\right) \text{ eV} . \qquad (G.3)$$

The luminosity of a black hole due to Hawking radiation depends on the black hole mass and the particle content of the theory. The black hole luminosity can be written as

$$L_{BH} = \sigma 4\pi R^2 T_{BH}^4 \sim 10^{-21} \left(\frac{M_\odot}{M}\right)^2 \text{ erg/s} , \qquad (G.4)$$

where σ is some constant that takes into account the particle content and also depends on the black hole mass. The constant σ would be the Stefan–Boltzmann constant if the black hole emission were exactly that of a blackbody. A black hole emits any particle of the theory (not only electromagnetic radiation) and the spectrum of the emitted particles deviates from the blackbody one due to the mass and the spin of the particles, the finite size of the black hole, etc. The value of L_{BH} is extremely low for $M \geq M_\odot$, and the Hawking radiation can be unlikely detected from astrophysical black holes, even in the future. This radiation causes the "evaporation" of the black hole: the black hole emits particles and its mass decreases. The evaporation time can be obtained from

$$\frac{dM}{dt} = L_{BH} = 4\pi R^2 \sigma T_{BH}^4 = \frac{\sigma}{4^4\pi^3}\frac{1}{M^2}. \tag{G.5}$$

Integrating by parts, we find (assuming for simplicity $\sigma = $ const.)

$$\tau_{evap} = \int_0^{\tau_{evap}} dt \sim \frac{\sigma}{4^4\pi^3}\int_0^M \tilde{M}^2 d\tilde{M} \sim 10^{65}\left(\frac{M}{M_\odot}\right)^3 \text{ yrs}. \tag{G.6}$$

High-mass X-ray binary (HMXB). X-ray binary with a high-mass ($M > 10\ M_\odot$) stellar companion. HMXBs are often persistent sources, because mass accretion onto the compact object originates from the wind of the companion star, which is a relatively stable process. The compact object may be either a black hole or a neutron star.

High-soft state. It is one of the two "historical" states of a black hole binary. The spectrum is soft and dominated by the thermal component of the accretion disk. See Sect. 4.5 for more details.

Geometrically thin/thick disk. An accretion disk is geometrically thin (thick) if the disk opening angle is $h/r \ll 1$ ($h/r \sim 1$), where h is the semi-thickness of the disk at the radial coordinate r.

Innermost stable circular orbit (ISCO). It is the circular orbit separating unstable and stable circular orbits. See Sects. 3.1 and 3.2 for more details.

Low-hard state. It is one of the two "historical" states of a black hole binary. The spectrum is hard and dominated by the power-law component. See Sect. 4.5 for more details.

Low-mass X-ray binary (LMXB). X-ray binary with a low-mass ($M < 3\ M_\odot$) stellar companion. LMXBs are often transient sources, because mass accretion onto the compact object occurs via Roche lobe overflow from the companion star, which is not a stable process. The compact object may be either a black hole or a neutron star.

Marginally stable circular orbit. It is the circular orbit separating unstable and stable circular orbits. It is equivalent to innermost stable circular orbit (ISCO). See Sects. 3.1 and 3.2 for more details.

Narrow line region (NLR). It is the region "far" from a supermassive black hole, where cold material orbits with a velocity of 500–1,000 km/s. This material produces narrow lines because the effect of the Doppler boosting is weak.

Optical band. Spectrum of the electromagnetic radiation in the wavelength range 400–700 nm (energy range 1.7–3.3 eV). See Table 5.1 for more details.

Optically thin/thick. A medium is optically thin (thick) if the photon mean free path in the medium $l = 1/(\sigma n)$, where σ is the photon scattering cross-section in the medium and n is the number density of target particles in the medium, is $l \gg d$ ($l \ll d$), where d is the linear size of the medium under consideration.

Persistent X-ray source. It is a "persistent" X-ray source in the sky. It is the opposite of a transient X-ray source. In the context of X-ray binaries, it is often a high-mass X-ray binary (HMXB).

Photon orbit. It is a circular orbit for massless particles and it can be reached by massive particles in the limit of infinite energy. See Sects. 3.1 and 3.2 for more details.

Quasar. A class of bright AGN. According to the unified AGN model, quasars are those objects in which the line of sight is close to the axis of symmetry of the system. See Appendix C for more details.

Soft X-ray band. There is not a universal definition of "soft X-ray band", so the exact energy range must be specified. In some contexts, it may indicate, for instance, the 2–6 keV range. In other contexts, it may be used for the X-ray spectrum below 10 keV.

Synchrotron radiation. Electromagnetic radiation emitted by relativistic charged particles when their acceleration is perpendicular to their velocity. This is the typical emission mechanism of charged particles moving through magnetic fields. In the case of black holes, synchrotron radiation is often associated to the emission of jets.

Transient X-ray source. It is a "temporary" X-ray source in the sky. It is the opposite of a persistent X-ray source. In the context of X-ray binaries, it is often a low-mass X-ray binary (LMXB).

UV band. Spectrum of the electromagnetic radiation between the optical and X-ray bands. The photon wavelength is between 400 nm (3.3 eV) and 10 nm (124 eV). See Table 5.1 for more details.

Printed in the United States
By Bookmasters